Structural Biology

Structural Biology

Edited by **Gildroy Swan**

SYRAWOOD
PUBLISHING HOUSE

New York

Published by Syrawood Publishing House,
750 Third Avenue, 9th Floor,
New York, NY 10017, USA
www.syrawoodpublishinghouse.com

Structural Biology
Edited by Gildroy Swan

International Standard Book Number: 978-1-68286-025-0 (Hardback)

The publisher's policy is to use permanent paper from mills that operate a sustainable forestry policy. Furthermore, the publisher ensures that the text paper and cover boards used have met acceptable environmental accreditation standards.

Trademark Notice: Registered trademark of products or corporate names are used only for explanation and identification without intent to infringe.

Printed in the United States of America.

Contents

Preface   VII

Chapter 1   **Elements sequestered by arbuscular mycorrhizal spores in riverine soils: A preliminary assessment**   1
M. C. Pagano, A. I. C. Persiano, M. N. Cabello and M. R. Scotti

Chapter 2   **EGFP-FMRP forms proto-stress granules: A poor surrogate for endogenous FMRP**   7
Natalia Dolzhanskaya, Wen Xie, George Merz and Robert B. Denman

Chapter 3   **Crystal structure of the allosteric-defective chaperonin GroEL$_{E434K}$ mutant**   30
Aintzane Cabo-Bilbao, Ariel E. Mechaly, Jon Agirre, Silvia Spinelli, Begoña Sot, Arturo Muga
and Diego M.A. Guérin

Chapter 4   **The salicylic acid effect on the tomato (*Lycopersicum esculentum* Mill.) sugar, protein and proline contents under salinity stress (NaCl)**   36
Shahba, Zahra, Baghizadeh, Amin, Vakili Seid Mohamad Ali, Yazdanpanah Ali and
Yosefi Mehdi

Chapter 5   **Virulence prediction model (virprob) using amino acid and dipeptide composition for human pathogens**   43
S. B. Muley, V. Bastikar, S. Bothe, A. Meshram and N. Roy

Chapter 6   **Epigenetic regulation of PGC1 α in human type 2 diabetes**   49
Y. Dhanusha Yesudhas

Chapter 7   **Using electrical impedance tomography in following up skin conductivity change for different sonophoresis conditions**   58
Mamdouh M. Shawki and Abdel-Rahman M. Hereba

Chapter 8   **The use of a continuity equation of fluid mechanics to reduce the abnormality of the cardiovascular system: A control mechanics of the human heart**   66
L. S. Taura, I. B. Ishiyaku and A. H. Kawo

Chapter 9   **Effect of the tannoid enriched fraction of *Emblica officinalis* on α-crystallin chaperone activity under hyperglycemic conditions in lens organ culture**   78
P. Anil Kumar, P. Yadagiri Reddy, P. Suryanarayana and G. Bhanuprakash Reddy

Chapter 10 **Changes in airway resistance with cumulative numbers of cigarettes smoked**   86
Almaasfeh Sultan

Chapter 11 **Modeling and proposed mechanism of two radical scavengers through docking to curtail the action of ribonucleotide reductase**   95
Sampath Natarajan and Rita Mathews

Chapter 12   **Possible role of 2, 2'- (Diazinodimethylidyne) di - (o-phenylene) dibenzoate, a novel**
                hydrazine as an anti – HIV agent   **106**
                Rita Ghosh, Dipanjan Guha, Sudipta Bhowmik and Angshuman Bagchi

Chapter 13   **Determination of instantaneous arterial blood pressure from bio-impedance signal**   **111**
                Sofiene Mansouri, Halima Mahjoubi and Ridha Ben Salah

Chapter 14   **Ultra-low doses of melafen affect the energy of mitochondria**   **118**
                I. V. Zhigacheva, E. B. Burlakova, I. P. Generozova, A. G. Shugaev and S. G. Fattahov

Chapter 15   **Influence of gibberellic acid and arbuscular mycorrhizae inoculation on carbon**
                metabolism, growth, and diterpene accumulation in _Taxus wallichiana_ Zuccarini var. mairei   **126**
                A. Misra, N. K. Srivastava, A. K. Srivastava and S. K. Chattopadhyay

Chapter 16   **BPES analyses of a new diffusion-advection equation for fluid flow in blood vessels**
                under different bio-physico-geometrical conditions   **132**
                M. Dada, O. B. Awojoyogbe, K. Boubaker and O. S. Ojambati

Chapter 17   **Comparative study of inhibition of drug potencies of c- Abl human kinase**
                inhibitors: A computational and molecular docking study   **139**
                D. Kshatresh Dubey, K. Amit Chaubey, Azra Parveen and P. Rajendra Ojha

Chapter 18   **Chlorosoma: How can it contribute to photosynthesis of green bacteria?**   **147**
                A. Y. Borisov

Chapter 19   **Spectroscopic approach of the interaction study of ceftriaxone and human serum**
                albumin   **150**
                Abu Teir M. M., Ghithan J., Abu-Taha M. I., Darwish S. M. and Abu-hadid M. M.

Chapter 20   **A thermodynamic investigation of bovine carbonic anhydrase II interaction with**
                cobalt ion at 300 and 310K   **162**
                G. Rezaei Behbehani, A. Divsalar, A. A. Saboury and Z. Rezaei

Chapter 21   **pH uniquely modulates protein arginine methylation**   **168**
                Wen Xie, George Merz and Robert B. Denman

                Permissions

                List of Contributors

Preface

Structural biology is an interdisciplinary field which incorporates concepts of biochemistry and molecular biology to study and analyze biological structures such as nucleic acids and proteins. This book discusses the fundamental as well as modern approaches to understand structural biology with particular emphasis on macromolecules. It explains in detail some existent theories as well as innovative concepts revolving around structure of protein networks, application of atomic and molecular data and computational modelling. The aim of this book is to present researches that have transformed this discipline and aided its advancement. Scientists and students actively engaged in this field will find this book full of crucial and unexplored concepts.

This book is a comprehensive compilation of works of different researchers from varied parts of the world. It includes valuable experiences of the researchers with the sole objective of providing the readers (learners) with a proper knowledge of the concerned field. This book will be beneficial in evoking inspiration and enhancing the knowledge of the interested readers.

In the end, I would like to extend my heartiest thanks to the authors who worked with great determination on their chapters. I also appreciate the publisher's support in the course of the book. I would also like to deeply acknowledge my family who stood by me as a source of inspiration during the project.

Editor

Elements sequestered by arbuscular mycorrhizal spores in riverine soils: A preliminary assessment

M. C. Pagano[1]*, A. I. C. Persiano[1], M. N. Cabello[2] and M. R. Scotti[3]

[1]Electron Microscopy and Microanalysis Laboratory, Physics Department, Federal University of Minas Gerais, Brazil.
[2]National University of La Plata, Argentina.
[3]Institute of Biological Sciences, Federal University of Minas Gerais, Av. Antônio Carlos, 6627, Pampulha,
CEP: 31270-901, Belo Horizonte, MG, Brazil.

The elemental composition of spore wall of arbuscular mycorrhizal fungi (AMF) was analyzed by energy-dispersive X-ray spectrometry (EDS) in a preliminary assessment. Measurements of AMF spores were made for riparian soils in both urban (rivers subjected to different human activities) and farm sites in southeastern Brazil. Spore populations belonging to four genera: *Acaulospora*, *Gigaspora*, *Glomus* and *Scutellospora* were analyzed. The results suggest that Glomeraceae spores sequester more elements than Acaulosporaceae and Gigasporaceae; however, the presence of nickel was observed in one *Scutellospora* species. These data showed the elements sequestered by AMF spores in riparian sites presenting different conditions of disturbance.

Key words: Arbuscular mycorrhizal spores, urban sites, energy-dispersive X-ray spectrometry, riparian forest.

INTRODUCTION

Arbuscular mycorrhizal fungi (AMF) are obligatory dependent on host-plants for the resources necessary to produce reproductive propagules, the external spores (Read 2003). AMF are affected by disturbs in the ecosystems, like global change (Rillig et al., 2002), or heavy metal pollution (Meharg and Cairney, 2000); however, plant symbiotic mycorrhizal fungi can accumulate metals from soil components (Gadd, 2005). There are several reports of metal-tolerant AMF (Orlowska et al., 2002; Hildebrandt et al., 2007; Soares and Siqueira 2008); however, there is to the best of our knowledge no report of metal-tolerant AMF from riparian soils in Brazil.

Morphological studies of small structures are possible using energy-dispersive spectrometers and wavelength-dispersive spectrometers (WDS) coupled to a scanning electron microscope (SEM); however, these methods do not detect minor and trace elements (Przybylowicz et al., 2004). Not only is the energy-dispersive X-ray

spectrometry (EDS) technique limited to the detection of elements with an ordinal number between 10 and 25, such as aluminum (Al), calcium (Ca), potassium (K), magnesium (Mg), phosphorus (P) and sulfur (S) (Leapman and Hunt, 1991), but also peak values of micronutrients and lighter elements, such as nitrogen (N), cannot be clearly distinguished from the background (Bücking et al., 1998). Weiersbye et al. (1999) showed several elements (heavy metals) in AMF spores from uranium mine tailings in South Africa using Micro-Pixe mapping; however they did not identify isolated AMF species. Moreover, Cruz (2004) showed that quantitative light element microanalysis of AMF spores employing EDS is a technique still little explored and which may inform the chemical spectrum of AMF spores and point the differences among species. Within a research project to study the diversity, role and potential of AMF in riparian areas of Velhas river basin in Minas Gerais State, Brazil, different restored sites with native woody species and other presenting native vegetation were evaluated.

The aim of the present study was to investigate the elemental concentration of identified AMF genera or species isolated from riparian soils in southeastern Brazil.

*Corresponding author. E-mail: marpagano@gmail.com.

2

Figure 1. Location of sampling sites, Southeast Minas Gerais, Brazil. Site codes as in Table 1.

Table 1. Study sites, characteristics and land use in Minas Gerais, Brazil.

Site code	River	Coordinates	Locality	Elevation (m.a.s.l.)	Vegetation	River conditions	AMF spores[#]
P	Gaia stream	19º 52' S 43º 47'W	Sabará (Reserve forest)	735	Native herbaceous and woody species (Atlantic Forest and the Cerrado savannas)	Natural	*Scutellospora aurigloba* *Glomus* sp.
SR	Sabará river	19º53'32"S 43º48'31" W	Sabará (Urban site)	637	Restored with native woody species	Urban	*Acaulospora* sp. *S. gregaria*
SD	Sabará river	19º53'32"S 43º48'31" W	Sabará (Urban site)	637	Herbaceous cover	Urban	*S. fulgida*
VR	Velhas river	43º 51' 58" S 19º 50' 51" W	Sabará (Farm)	662	Restored with native woody species	Disturbed	*S. reticulata*

[#]Spores evaluated for microanalysis.

MATERIALS AND METHODS

Study sites

The study sites are located in the southeast of Brazil, in Mina Gerais State. The sites are: riparian forests at Sabará and Velhas rivers, belonging to the São Francisco basin, in the south of Minas Gerais State. Spores were isolated from restored sites at urban or farming areas, presenting sandy soils, and from a preserved forest upstream of the urban areas (Figure 1 and Table 1), presenting a vegetation dominated by trees, which is a transition between the Atlantic Forest and the Cerrado savannas (Rizzini, 1997). The climate is tropical (Aw) with temperatures between 22 and 23°C, and mean annual rainfall is 1.400 mm.

Sample collection

Soil samples (500 g) were collected to a depth of 25 cm from the soil surface of the four studied sites in 2007 and 2008 during visits at the riparian sites in Minas Gerais. Soils were air-dried and stored until processed. AMF spores were recovered from 100 g soil

Table 2. Semi quantitative analyses by EDS of some AMF species (spores) in riparian soils, Minas Gerais, Brazil.

AMF species	Elements												
	C	O	Na	Al	Si	P	S	Cl	Ca	Fe	K	Ni	Mg
Acaulosporaceae													
Acaulospora sp.	84.9b	4.8ab	0.04ab	0.27a	4.15a	ND	0.58c	1.32bc	1.35a	0.69ab	0.02ns	ND	1.6ns
Gigasporaceae													
Scutellospora aurigloba	93.1*a	3.2b	0.01b	<0.01b	0.3b	0.04	0.5c	2.5b	ND	ND	ND	ND	ND
S. fulgida	93.5a	5.15ab	0.01b	0.01b	0.56b	ND	0.6c	ND	0.1c	ND	<0.01	ND	ND
S. gregaria	89.75ab	5.6a	<0.01b	0.12ab	1.39ab	ND	1ab	0.15c	0.43bc	0.37ab	0.01	ND	0.85
S. reticulata	88.7ab	3.07b	ND	0.26ab	2.38ab	ND	1.15a	0.28c	1.72a	0.99a	0.02	1.28	ND
Glomeraceae													
Glomus sp.	82.6b	4.2b	0.08a	0.13ab	3.5a	<0.01	0.68bc	5.13a	1.27ab	0.25b	0.02	ND	2.04

*Spore wall atomic %. Data are means of five replicate entire spores. 10 kV potential, electron beam 20 nA, 200 s. ND = not detected. Different letters indicate significant differences as determined by Tukey's test ($p < 0.05$). Ns = non-significant

samples, separated by wet sieving (Gerdemann and Nicolson 1963), decanting, sucrose centrifugation and flotation (Walker et al., 1982) and were collected on 0.5, 0.062 and 0.037 mm sieves. Distilled water was used during the sieving process to avoid impurities on the spore surfaces. Samples were mounted in microscopic slides and were dried.

Microanalyses

AMF spores were analyzed in this work by EDS to examine the sequestration of elements by these structures. The chemical composition of the AMF spores (five replicates), for the assigned elements (Table 2) in the periodical table, was measured by electron probe microanalysis with energy-dispersive X-ray and the sample composition was determined by the ZAF method for semi quantitative analysis, as described by Goldstein et al. (1992). EDS measurements were made on a JXA-8900RL, JEOL WD/ED COMBINED MICROANALYZER, following Goldstein et al. (1992), at the Electron Microscopy and Microanalysis Laboratory, Physics Department, Federal University of Minas Gerais, Brazil.

Samples were mounted in microscopic slides and were dried on a support membrane prior to carbon coating. The entire slide was coated. Selected areas of spore wall (100 μm^2 in size) were used for microanalyses. Data were subjected to one-way ANOVA using Statistica 7.0, Statsoft, and means were compared by Tukey's test ($p < 0.05$). Average concentrations of elements in the spores were used in the analysis. A control spectrum taken from an area of adjacent support membrane close to the spore (carbon coated) was observed. The elemental mapping was done by data extracted from arbitrarily selected micro areas of the selected spore.

RESULTS

The results revealed differences in elemental composition between spores isolated from riparian areas (Table 2). Figure 2 shows micrographs and their EDS-measured composition spectra of the analyzed spores from the riparian site. The spore wall mainly contained Carbon (C), and varying proportions of oxygen (O), silicon (Si), Al and Ca. Spectra from spores showed a major C peak and smaller peaks of Si and Al. The control spectrum from the Adjacent support membrane did not have any significant significant characteristic peaks.

With respect to microanalyses of spores, relative high values for C (Table 2) were obtained in this work for three *Scutellospora* species. On the other hand, *Scutellospora fulgida* and *Scutellospora aurigloba* showed lower amounts of all the elements, except for C.

Significant differences in the elemental concentration of spore wall between *Glomus* and *Scutellospora* were found. In general, some species of *Scutellospora* were characterized by higher concentration of S than spores of *Glomus*. The distribution of Ca and Cl in spores was related to the AMF species, and the concentration of P was low in all samples. Nickel was detected in only one spore type (*Scutellospora reticulata*) isolated from restored riparian site of Velhas River (Table 1), which is presented in Figure 3, showing an elemental map distribution of some selected elements. The distribution of C was high in the interconnecting ridges that form a reticulum and in the spines. Al, S and Si showed lower intensity than C. Ni, P and Ca were relatively homogeneous distributed.

DISCUSSION

The presence of C observed in the spores in our study is in line with the report by Cruz (2004). As expected, C was detected in high proportions in six AMF species, since it is the main element in the organic compounds of the spore structures.

Moreover, in the present study, the microanalysis of the spores showed four elements (K, Ca, Ni and Fe) in common with spores analyzed for uranium mine tailings (Weiersbye et al., 1999). The presence of Ni, which at

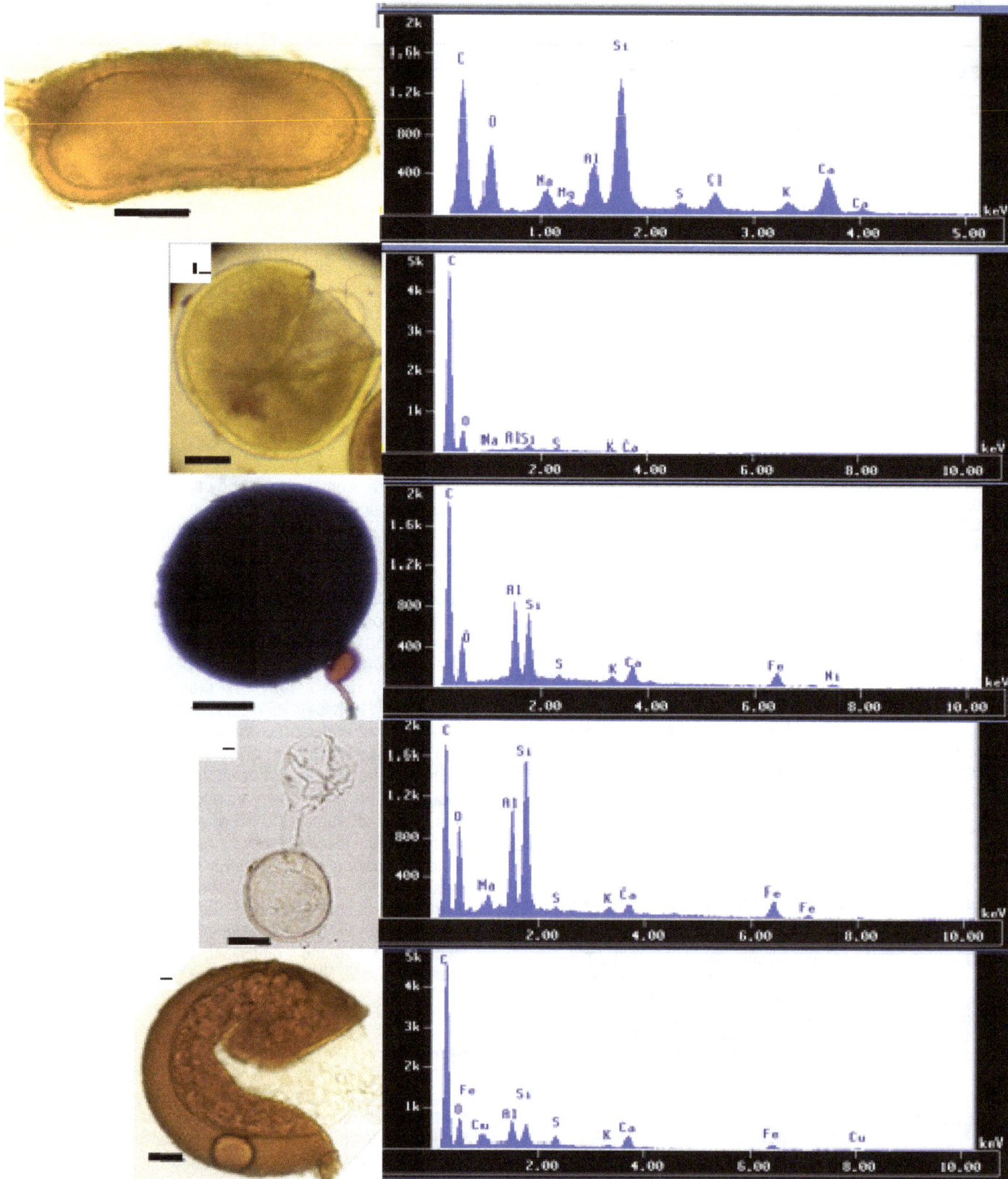

Figure 2. Photomicrographs of some AM spores found at riparian areas, Brazil and spectra of X-ray microanalysis (EDS) of spore wall. a = *Glomus* sp.; b = *S. aurigloba*; c = *S. reticulata*; d = *Acaulospora* sp. 1 and e = *S. gregaria*. The peaks correspond to carbon (C), oxygen (O); iron (Fe); sodium (Na); magnesium (Mg); copper (Cu), aluminum (Al); silicon (Si), phosphorus (P), sulfur (S); chloride (Cl); potassium (K); calcium (Ca) and nickel (Ni). (Bar, a = 25; b, e = 50 µm; d, c = 100 µm).

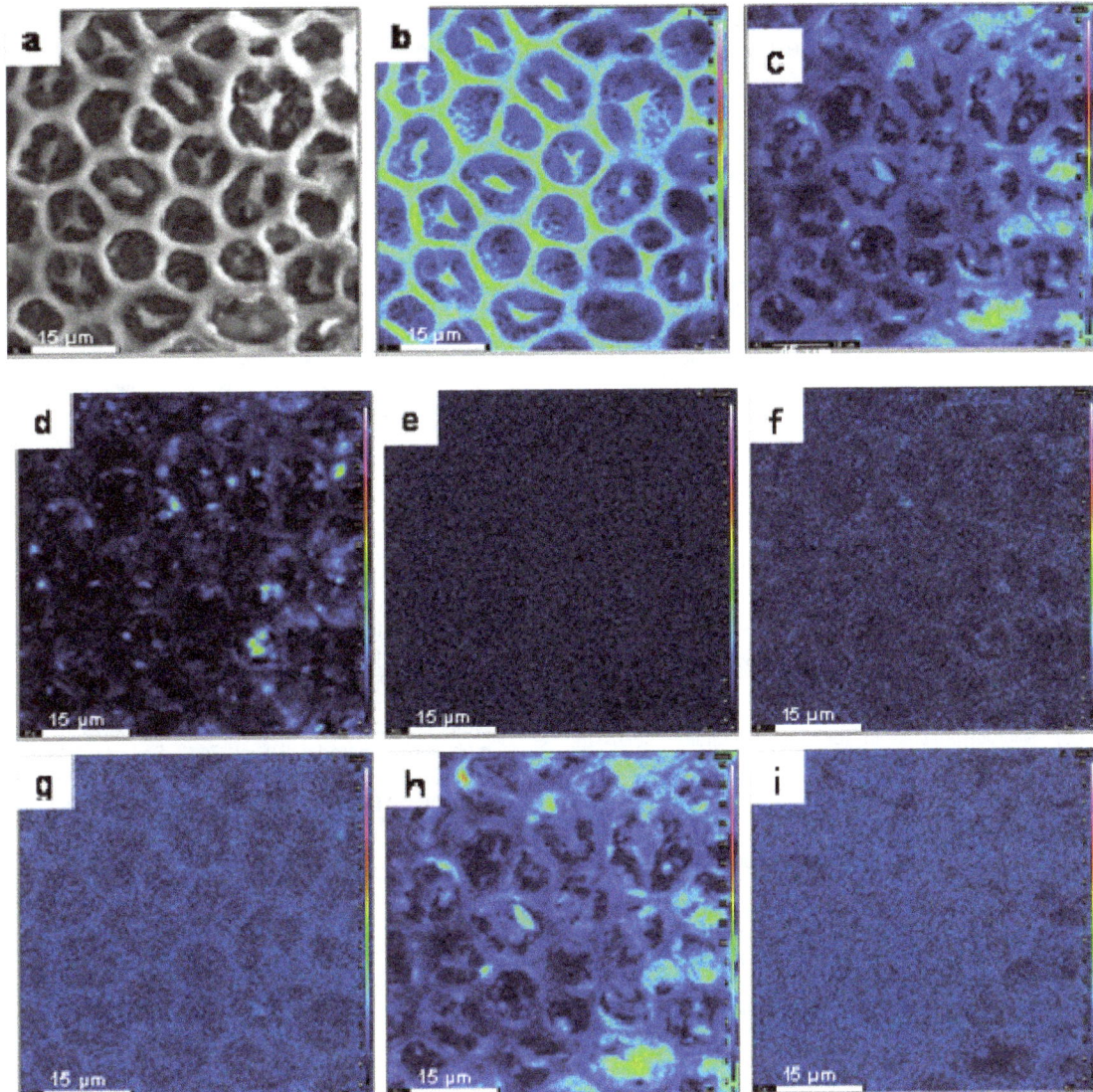

Figure 3. WDS elemental map showing the distribution of some elements in *S. reticulata*, isolated from soil of the riparian vegetation of Velhas River. Resolution: 900 × 900 pixels points. (a) Micrograph of the spore, showing the wall ornamentation; (b) map showing the distribution of C; (c) Al; (d) Fe; (e) Ni; (f) P; (g) S; (h) Si and (i) Ca. Bar, 10 μm.

higher concentrations can lead to poisoning (heavy metal), detected in only one AMF spore (*S. reticulata*) isolated from restored riparian forest at the farm site (Velhas River), may be related to riverine soil pollution. Orlowska et al., (2008) showed that external AM hyphae can bind Ni, influencing the uptake of heavy metals by plants. The presence of chemical elements in AMF spores is important to clarify their structure, since it allows characterizing these elements as a part of the components in the spores, and distinguishing them by their nutrient composition, which is useful for spore survival. Furthermore, the sequestered elements may reflect a polluted environmental condition.

The chemical composition of suggested indicator genus or species for disturbed (*Acaulospora*, *S. gregaria* and

S. reticulata) or for more pristine sites (*S. aurigloba*) in the riparian forest showed that spores from the disturbed site could present more nutrient richness in their wall than spores from natural sites. The spores of *S. aurigloba* isolated from the preserved riparian site contained less chemical elements. On the other hand, the results suggest that other AMF species present a rich element composition.

This study is, to the best of our knowledge, the first report of information on metals sequestered among the AMF species or genera from urban and farm sites. The results suggest that *Glomus*, a dominant genus in the sites (Pagano, MC unpublished), has high nutrient richness in the spore wall, and that the presence of the heavy metal Ni in spores can be investigated by EDS.

Finally, this technique also allows the seasonal variation of nutrient amounts in AMF spores to be investigated in the riparian vegetation of Minas Gerais's Rivers subject to different land uses.

ACKNOWLEDGEMENTS

Dr Marcela Pagano is grateful to FAPEMIG (Process 311/07) for Post-doctoral scholarship granted. The authors are grateful to FAPEMIG. Marta N. Cabello is a research scientist from the Comisión de Investigaciones Científicas (CIC) Bs. As. Province - Argentina. Ana M. Penna and Luis R. A. Garcia are gratefully acknowledged for all assistance.

REFERENCES

Bucking H, Beckmann S, Heyser W, Kottke I (1998). Elemental contents in vacuolar granules of ectomycorrhizal fungi measured by EELS and EDXS. A comparison of different methods and preparation techniques. Micron 29: 53-61.

Cruz AF (2004). Element storage in spores of *Gigaspora margarita* Becker and Hall measured by electron energy loss spectroscopy (EELS). Acta Bot. Bras. 18: 473-480

Gadd GM (2005). Microorganisms in toxic metal-polluted soils. In: Buscot, F., Varma, A. (Eds.). Microorganisms in soils: roles in genetics and functions, Springer-Verlag, Berlin, Germany pp. 325-356.

Gerdemann JW, Nicolson TH (1963). Spores of mycorrhizal Endogone species extracted from soil by wet sieving and decanting. Trans. Br. Mycol. Soc. 46: 235-244.

Goldstein J, Newbury DE, Echlin P, Joy DC, Romig AD, Lyman CE, Fiori C, Lifshin E (1992). Scanning Electron Microscopy and X-Ray Microanalysis: A Text for Biologists, Materials Scientists and Geologists. Plenum Press, New York pp. 805.

Hildebrandt U, Regvar M, Bothe H (2007). Arbuscular mycorrhiza and heavy metal tolerance Phytochemistry 68: 139-146.

Leapman RD, Hunt JA (1991). Comparison of detection limits for EELS and EDXS. Microsc. Microanal. Microstruct. 2: 231-244.

Meharg AA, Cairney JWG (2000). Co-evolution of mycorrhizal symbionts and their hosts to metal-contaminated environments. Adv. Ecol. Res. 30: 69-112.

Orlowska E, Zubek S, Jurkiewicz A, Szarek-Łukaszewska G, Turnau K (2002). Influence of restoration on arbuscular mycorrhiza of *Biscutella Laevigata* L. (Brassicaceae) and *Plantago lanceolata* L. (Plantaginaceae) from calamine spoil mounds Mycorrhiza 12: 153-160.

Orlowska E, Mesjasz-Przybyłowicz J, Przybyłowicz W, Turnau K (2008). Nuclear microprobe studies of elemental distribution in mycorrhizal and non-mycorrhizal roots of Ni-hyper accumulator *Berkheya coddii*. X-Ray Spectrom. 37: 129-132.

Przybylowicz WJ, Mesjasz-Przybylwics J, Migula P, Turnau K, Nakoniecny M, Augustyniak M, Glowacka E (2004). Elemental microanalysis in ecophysiology using ion microbeam. Nuclear Instruments and Methods in Physics Research 219-220: 57-66.

Read DJ (2003). Towards ecological relevance-Progress and pitfalls in the path towards an understanding of mycorrhizal functions in nature. In: van der Heijden MGA, Sanders IR (eds). Mycorrhizal ecology. Springer, Berlin pp. 3-24.

Rillig MC, Treseder KK, Allen MF (2002). Mycorrhizal fungi and global change. In: Ecological Studies Series 157: Mycorrhizal Ecology, Springer Verlag pp.135-160.

Rizzini CT (1997). Tratado de fitogeografia do Brasil: aspectos ecológicos, sociológicos e florísticos. Âmbito Cultural Edições Ltda., São Paulo pp. 747.

Soares CRFS, Siqueira JO (2008). Mycorrhiza and phosphate protection of tropical grass species against heavy metal toxicity in multi-contaminated soil. Biol. Fertil. Soils 44: 833-841.

Walker C, Mize W, McNabb HS (1982). Populations of endogonaceous fungi at two populations in central Iowa. Can. J. Bot. 60: 2518-2529.

Weiersbye IM, Straker CJ, Przybylowicz WJ (1999). Micro-PIXE mapping of elemental distribution in arbuscular mycorrhizal roots of the grass, *Cynodon dactylon*, from gold and uranium mine tailings. Nuclear Instruments and Methods in Physics Research- NIMB 158: 335-343.

EGFP-FMRP forms proto-stress granules: A poor surrogate for endogenous FMRP

Natalia Dolzhanskaya[1], Wen Xie[1], George Merz[2] and Robert B. Denman[1]*

[1]Department of Molecular Biology and New York State Institute for Basic Research in Developmental Disabilities, 1050 Forest Hill Road Staten Island, New York 10314, U.S.A.
[2]Department of Developmental Neurobiology, New York State Institute for Basic Research in Developmental Disabilities, 1050 Forest Hill Road, Staten Island, New York 10314, U.S.A.

Overexpressed autofluorescent-tagged versions of the Fragile mental retardation protein (FMRP) such as EGFP-FMRP have been used in protein-protein interaction studies and in studies of the composition, the formation and the localization of neuronal granules. However, the question of whether these molecules truly recapitulate the properties of the endogenous protein has not been addressed. Here we demonstrate that overexpressed EGFP-FMRP forms three distinct granule types based on colocalization with various marker proteins. The majority of EGFP-FMRP-containing granules are larger and more amorphous than known granule types. Consistent with this, there is only partial colocalization with stress granule or P-body markers. Nevertheless, agents such as sodium arsenite, which create endogenous stress granules and P-bodies and hippuristanol, which induces stress granule formation, drive EGFP-FMRP exclusively into stress granules. Additionally, whereas inhibiting methyl-protein formation alters the composition of endogenous FMRP-containing stress granules, we found that such treatment had little effect on the formation of EGFP-FMRP granules, or their composition. Altogether these data suggest that many overexpressed EGFP-FMRP granules represent proto-stress granules requiring external stimuli for their conversion. More importantly, the inherent heterogeneity of these granules suggests that caution should be used in extrapolating results obtained with autofluorescent-tagged surrogates of FMRP to endogenous FMRP granules.

Key words: EGFP, FMRP, stress granules, P-bodies, protein arginine methylation.

INTRODUCTION

Autofluorescent protein tags (AFPs) have been widely used as tools to study a variety of biological processes

*Corresponding author. E-mail: rbdenman@yahoo.com.

Abbreviations: aDMA, Asymmetric dimethylarginine; AdOx, adenosine 2', 3'-dialdehyde; AFPs, autofluorescent proteins; DRG, dorsal root ganglion; EGFP, enhanced green fluorescence protein; FMRP, fragile X mental retardation protein FRAP, fluorescence recovery after photobleaching; FXR1P, fragile X related 1 mental retardation protein; GFP, green fluorescent protein; Dcp1a, decapping protein factor 1a; eIF2, eucaryotic initiation factor 2; PRMT, protein arginine methyltransferase; RBP, RNA binding protein; RNAi, RNA interference; sDMA, symmetric dimethylarginine; TIA1, T-cell internal antigen 1; tM1, tM2, thresholded Manders coefficients.

including, nervous system development (Brand, 1999), developmental abnormalities (Detrich III, 2008), neural stem cell development (Encinas and Enikolpov, 2008), localization of proteins in specific organelles (Di Giorgi et al., 1999), protein-protein interactions (Kedersha et al., 2005), protein-RNA interactions (Rackham and Brown, 2004), protein (Pierce and Vale, 1999) and mRNA trafficking (Querido and Chartrand, 2008) and membrane dynamics (Lippincott-Schwartz et al., 1999). Green fluorescent protein (GFP) and the myriad of spectral variants comprising the AFPs are relatively small proteins (Mr 27), but are much larger than many other widely used tags. As such, it is imperative to demonstrate that the AFP tag does not alter the transport, the localization or the functional properties of the protein being tagged (Lippincott-Schwartz et al., 1999). For example, Özlu et al. (2005) used RNA interference (RNAi) in combination with a TXL-1-GFP fusion to demonstrate that the transgene could functionally replace endogenous TXL-1 at the centrosome (Özlu et al., 2005).On the other hand, there appears to be a significant discrepancy between the localization of endogenous P58TFL with endogenous P-body markers and that of GFP-P58TFL and transfected P-body markers, suggesting that the fusion protein may not adequately mimic the endogenous form (Bloch and Nobre, 2010; Minagawa et al., 2009; Minagawa and Matsui, 2010). The fragile X mental retardation protein, FMRP, is a RNA binding protein that plays an important role in controlling translation in neurons (Bassell and Warren, 2008). Studies from a number of laboratories have demonstrated that endogenous FMRP associates with macromolecular granules in the brain, in the neurites of cultured neuronal cells and in the cell bodies of non-neuronal cells (Aschrafi et al., 2005; Didiot et al., 2008; Dolzhanskaya et al., 2006a; Dolzhanskaya et al., 2006b; Kanai et al., 2004; Xie et al., 2009). Immunofluorescence studies have shown that these granules tend to be small and not that prevalent. In contrast, AFP-tagged FMRPs have been used by a number of laboratories to assess various questions concerning neuronal granule make up and dynamics (Antar et al., 2005; Barbee et al., 2006; Castren et al., 2001; Cziko et al., 2009; Darnell et al., 2005; De Diego Otero et al., 2002; Dictenberg et al., 2008; Levenga et al., 2009; Ling et al., 2004; Pfeiffer and Huber, 2007). These granules tend to be much larger than endogenous FMRP granules and also much more prevalent in dendrites. However, the question of what type(s) of granules AFP-FMRP represents has not been adequately addressed.

Here we have examined EGFP-FMRP expression in HeLa cells using a variety of known granule marker proteins and under various conditions known to induce the formation of specific granule types or modulate their composition. Our studies demonstrate that EGFP-FMRP granules fall into at least three different functional classes, the majority of which appears to be a proto-stress granule. The implications of these results are discussed.

MATERIALS AND METHODS

Reagents

Sodium arsenite, adenosine 2', 3'-dialdehyde (AdOx) and emetine were purchased from Sigma. Hippuristanol was a kind gift from Jerry Pelletier (McGill Cancer Center, McGill University, Montreal, Quebec, Canada).

Plasmids and transfection

A mammalian expression plasmid containing an enhanced green fluorescent-FMRP fusion protein under the control of a cytomegalovirus promoter pEGFP-FMRP (Antar et al., 2004) was a kind gift of Dr. Gary Bassell (Emory University, Atlanta, GA). The parent vector, pEGFP-C2, was purchased from Clontech. Cells (3 × 10^5/35 mm dish) were transfected with 1 μg of plasmid DNA using Lipofectamine Plus (Invitrogen).

Proteins and antibodies

FMRP mAb-2160, which recognizes an epitope in the N-terminus of human FMRP, and normal mouse serum were purchased from Chemicon. Phospho-eIF2α pAb (Ser52) (KAP-CP131) and Hsp70c mAb (HSP-820) were obtained from StressGen. TIA1 pAb (sc-1751) and FXR1P pAb (sc-10552) were purchased from Santa Cruz. Asymmetric dimethylarginine pAb (ASYM24) and symmetric dimethylarginine pAb (SYM10) were purchased from Millipore. Dimethylarginine pAb (mRG) was obtained from CH3 Biosystems. Protein arginine methyltransferase antibody, PRMT1 was obtained from Abcam (ab70724), while PRMT3 pAb (07-256), PRMT4 pAb (AB3345), PRMT5 pAb (07-405) and PRMT7 pAb (07-639) were obtained from Millipore. hDcp1a pAb was a kind gift of J. Lykke-Andersen (University of Colorado, Boulder, CO).

Cell culture

HeLa cells were grown at 37°C in 5% CO_2 and maintained in DMEM supplemented with 10% FBS, 100 U/ml penicillin and 100 μg /ml streptomycin. In some cases, the cells were treated with 20 μM AdOx, 0.5 mM sodium arsenite, 10 μM hippuristanol, or with 10 μg/ml emetine and processed as indicated (Dolzhanskaya et al., 2006a).

Gene expression in cultured cells

Western blotting was carried out according to protocols set forth by the manufacturer for each antibody or as previously described (Dolzhanskaya et al., 2006a). For immunostaining, cells were grown on poly-L-lysine coated coverslips in the presence or absence of 20 μM AdOx and in the presence or absence of 0.5 mM sodium arsenite as indicated. The cells were fixed in 2% paraformaldehyde for 10 min and washed with PBS and then blocked in (RPMI1640 base medium, 0.05 % saponin, 0.1% sodium azide, 2% goat serum) for 30 min at room temperature. Subsequently, the cells were stained with antibodies to FMRP (1:500), TIA-1 (1:100), Hsp70 (1:300), FXR1P (1:50), hDcp1 (1:200), ASYM24 (1:100), mRG (1:150), or SYM10 (1:100) for 1 h. This was followed by incubation with Alexa Fluor secondary antibodies (1:500 dilutions) for 30 min at room temperature. Finally, the coverslips were washed in RPMI1640 base medium, 1% goat serum and mounted in buffered glycerol. Fluorescence was detected with an Eclipse 90i dual laser-scanning confocal microscope (NIKON). Images were acquired at 20 - 100× magnification. For presentation purposes the images were sometimes cropped to highlight certain points. All cropping was uniform within the particular figure.

Image analyses

Relative EGFP-FMRP granule size (pixels^2) was determined from select confocal images. Granules in each image file were defined using the threshold function in Image J (http://rsb.info.nih.gov.80/ij/). Granule areas were then calculated using the analyze particles feature. For calculations all images were taken at 40× magnification and the entire image was quantified; thereby ensuring a uniform pixel size. The circularity of the granules, 4π (area/perimeter^2), was calculated similarly. The data were then imported into Excel for subsequent analysis. For presentation, granule size distributions were all normalized to 500 particles and the normalized distributions were compared using the F-test for two sample variance. Protein co-localization measurements were performed by counting single- and double-stained granules of acquired images (Thomas et al., 2004). Co-localization was determined using the JaCoP Plugin (Bolte and Cordelieres, 2006; Didiot et al., 2008).

Line scans of the cellular fluorescence intensity (Biron et al., 2004) were also acquired using Image J software.

FRAP

Fluorescence recovery after photobleaching (FRAP) was measured with modification (Wang et al., 2008). HeLa cells were cultured on 40 mm coverslips and transfected with pEGFP–FMRP. Twenty four hours the cells were mounted in Sykes-Moore chambers (Bellco Glass Inc., Vineland, NJ) on a Nikon 90i microscope coupled to a Nikon C1 three-laser scanning confocal system (NIKON Instruments Inc., Melville, NY). The cells were maintained at 37°C with an air curtain incubator (NevTek, Williamsville, Va). Baseline images of EGFP-FMRP-expressing cells were acquired and select regions of interest (ROI) within the transfected cells were bleached to 10% or less of their initial intensity with the argon laser. Time-lapse images of the recovery were obtained at 30 s intervals.

Image data before and after bleaching was first concatenated and then converted into animated GIF files and processed using Image J software. The mean fluorescence within select bleached and control ROI was obtained using the analyze particles function, while fluorescence of particular EGFP–FMRP granules was measured using the linescan feature. Data were exported into Microsoft Excel for subsequent analysis. In some cases, cells were treated with arsenite 0.5 mM for 2 min prior to bleaching to induce stress granules. Arsenite was maintained throughout the 20 min FRAP time-lapse. A total of five arsenite-treated cells and five untreated cells were examined.

RESULTS

EGFP-FMRP expression results in a heterogeneous array of granules

Fluorescence microscopy of primary hippocampal neurons and Neuro 2A cells transfected with a plasmid encoding an EGFP-FMRP fusion protein has shown that its expression was in the form of granules that were confined largely to the cytoplasm (Antar et al., 2004; Darnell et al., 2005; Pfeiffer and Huber, 2007). Here, we show that HeLa cells transfected with the same plasmid express fluorescent granules with an assortment of sizes. In contrast, transfection of the parent vector shows that EGFP exhibits diffuse, non-granular cytoplasmic expression, Figure 1a. Analysis of the distribution of EGFP-FMRP granules reveals that they fall into three different size categories (small, intermediate and large), Figure 1b. In contrast, endogenous FMRP granules were much more homogeneous and mainly correspond in size to the smaller size EGFP-FMRP granules, Figure 1c. These data are consistent with two competing hypotheses. One possibility is that individual EGFP-FMRP granules have disparate compositions and possibly differing functions. The other likelihood is that the population of EGFP-FMRP granules represents nascent core particles of the same composition, but with differing aggregation states.

EGFP-FMRP expression and stress granule markers

Previous studies have shown that endogenous FMRP forms a variety of granule types including stress granules. Stress granules can be formed in response to heat shock (Mazroui et al., 2002), oxidative stress (Thomas et al., 2004), RNA binding protein over expression (Solomon et al., 2007), expression of a phosphomimetic mutant form of eukaryotic initiation factor 2 (eIF-2) (McEwen et al., 2005) and interfering with eIF4A activity (Mazroui et al., 2006). To determine whether the granules formed by expression of EGFP-FMRP were stress granules, HeLa cells were transiently transfected with an EGFP-FMRP expression vector. Twenty four hours later the cells were immunostained with anti-TIA1, an RNA binding protein which detects core stress granules (Kedersha et al., 2005) and then visualized by confocal microscopy. As shown in Figure 2a (panel a), punctate EGFP-FMRP expression was observed in the HeLa cell cytoplasm. In contrast, but as expected, TIA1 was found in both the cytoplasm and the nucleus, Figure 2a (panels b and e). In the cytoplasm TIA1 and EGFP-FMRP co-localized in large perinuclear granules; however, there were a host of EGFP-FMRP granules that did not contain TIA1. Nevertheless, it did not appear that granule size was a determining factor in the co-localization of TIA1 and EGFP-FMRP, rather there appeared to be cells with extensive co-localization, and cells with poor co-localization (Figure 2a; panels c and f) and Figure 2b. However, overall cells exhibiting extensive colocalization were in the minority. In these latter cells, TIA1 was largely found in the nucleus. Interestingly, treating cells with either hippuristanol, or arsenite, which induces stress granule formation resulted in a marked increase in the colocalization of EGFP-FMRP and TIA1, Figure 2a (panel i).

To quantify the overall extent of colocalization between EGFP-FMRP and TIA1 in the transfected HeLa cells we performed colocalization analyses. By manually masking nuclear staining and setting threshold limits on each fluorescence channel we were able to differentiate granular versus non-granular staining. Table 1 shows the thresholded Manders coefficients tM1 (fraction of red in the green channel) and tM2 (fraction of green in the red channel) for the EGFP-FMRP granules under different conditions. In the absence of treatment the majority of EGFP-FMRP granules were devoid of TIA1; however, following treatment with either sodium arsenite or hippuristanol there was a significant shift of TIA1 into these granules.

We next examined colocalization between EGFP-FMRP granules and another stress granule marker, FXR1P. Endogenous FXR1P extensively colocalizes with endogenous FMRP in arsenite-induced stress granules (Dolzhanskaya et al., 2006a, 2006b) these data are recapitulated in Figure 3a. However, as shown in Figure 3b (panels d-f), colocalization between EGFP-FMRP granules and FXR1P was also incomplete. Again however, in the presence of arsenite EGFP-FMRP granules almost completely colocalized with FXR1P, Figure 3c (panel i).

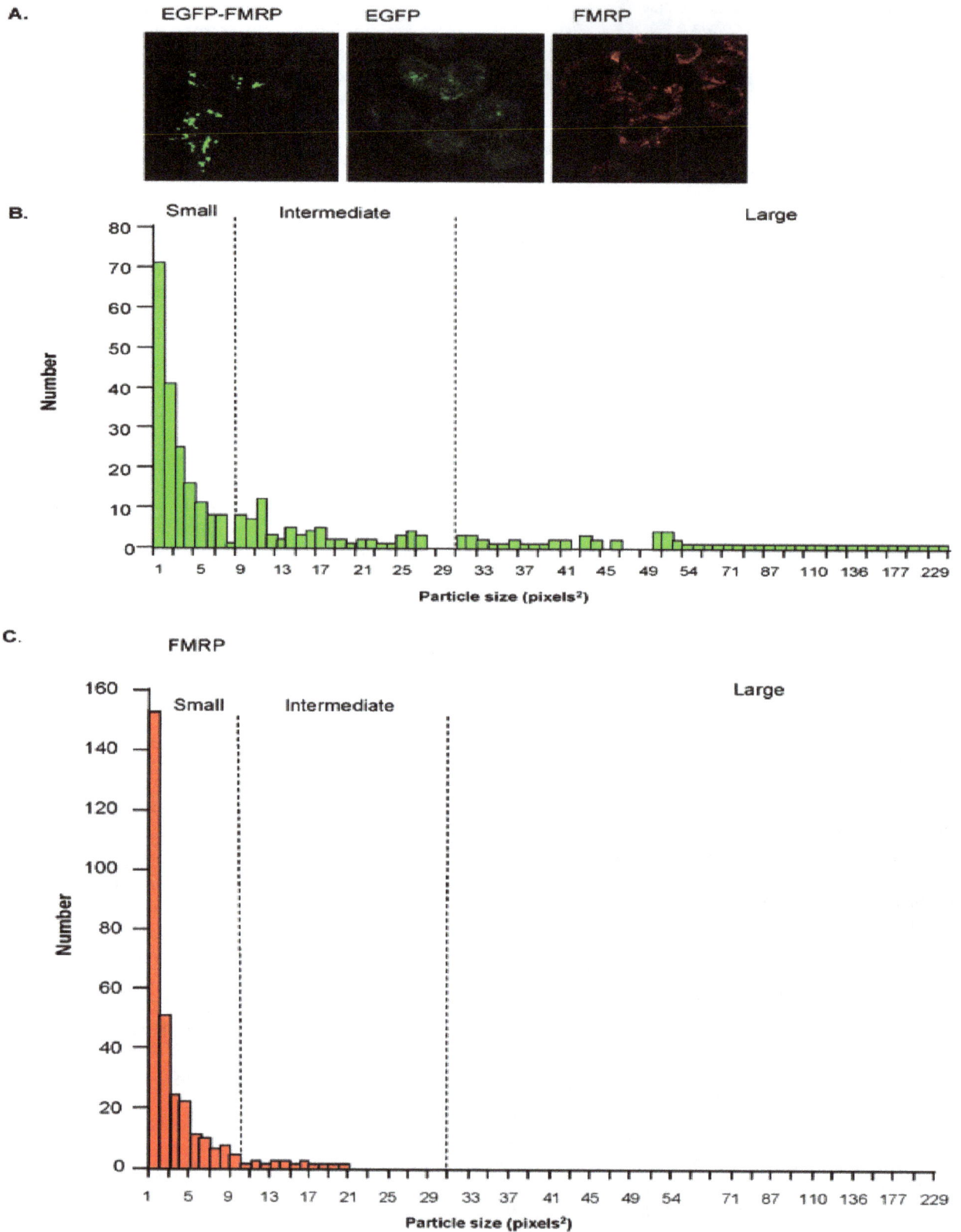

Figure 1. EGFP-FMRP granules are heterogeneous. (A) Confocal images of HeLa cells expressing EGFP-FMRP, EGFP, or endogenous FMRP. (B) Size distribution of EGFP-FMRP granules in HeLa cells. Analysis was based on 9 transfected cells. Note the three classes of granules (small, intermediate, large) based on size. (C) Size distribution of endogenous FMRP granules. Analysis was based on 27 cells. Comparison of the two distributions shows that endogenous FMRP granules are significantly smaller (P < 0.00084).

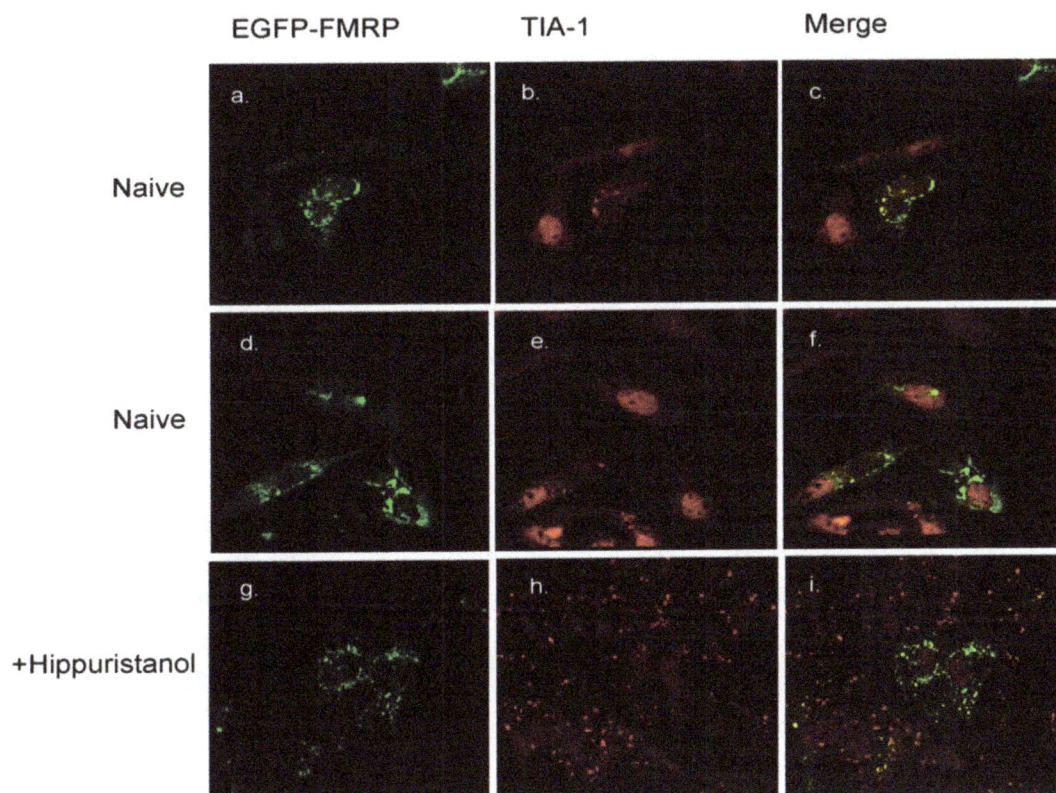

Figure 2A. EGFP-FMRP granules partially colocalize with the stress granule marker TIA1. (A) Panels a-f show EGFP-FMRP (green), TIA1 (red) and merged images of transfected HeLa cells. Panels g-i show cells treated with 10 μM hippuristanol for 30 min. Notice that all of the cells contain TIA1 stress granules (red).

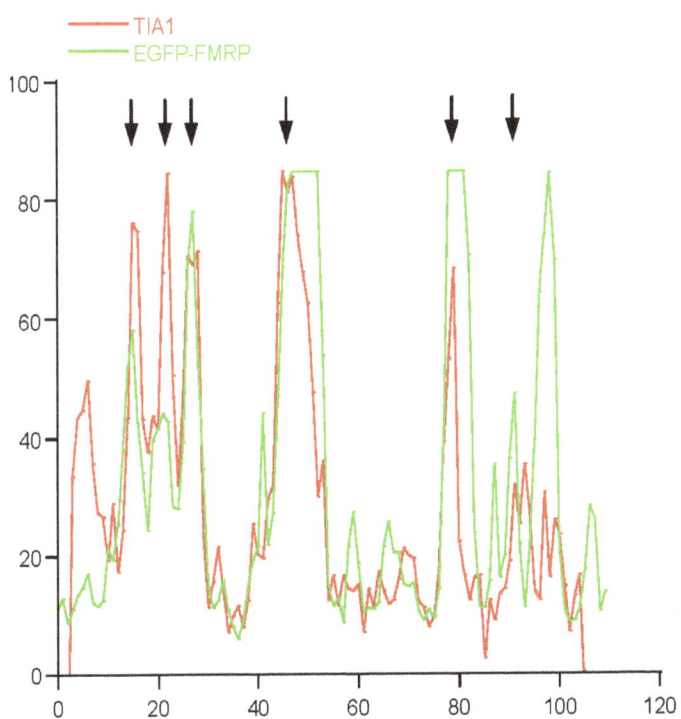

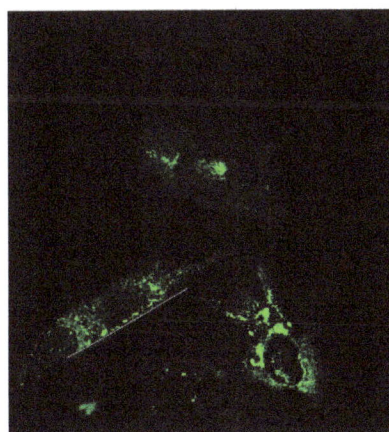

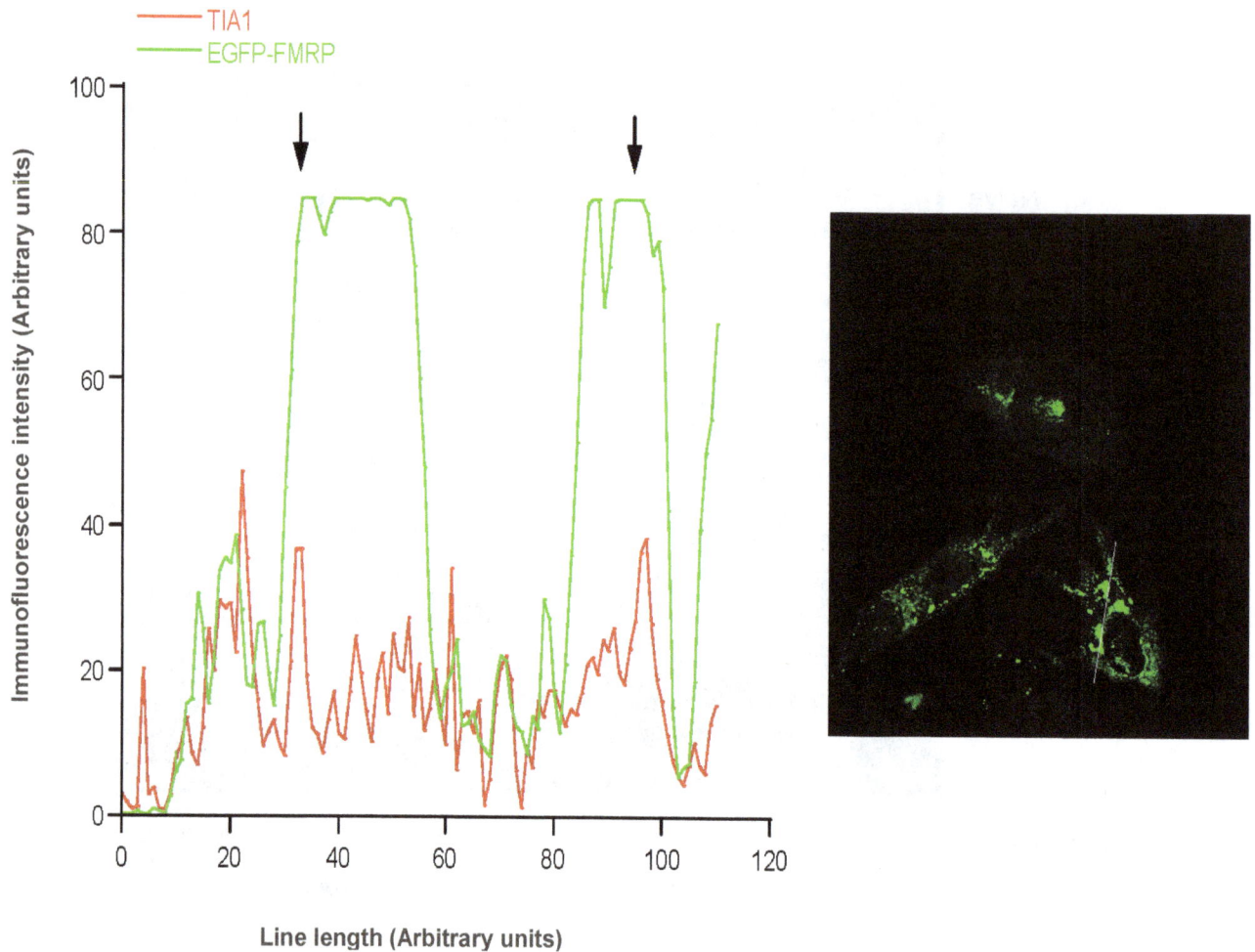

Figure 2. (B) Linescans of the red and green fluorescent channels were acquired from untreated HeLa cells expressing EGFP-FMRP and immunostained with TIA1 (Figure 1A) in which there was good colocalization of the two proteins (upper panel) or poor colocalization of the two proteins (lower panel). The arrows mark the individual EGFP-FMRP granules. The granules that were scanned are marked by a white line on the images shown to the right of the graphs.

Table 1. EGFP-FMRP colocalization analysis.

Pair	Treatment	tM1[d]	tM2[d]
EGFP-FMRP/TIA1	None	0.069 (0.015)	0.057 (0.021)
	AdOx[a]	0.068 (0.023)	0.061 (0.007)
	Arsenite[b]	0.521 (0.014) [†]	0.449 (0.027) [†]
	Hippuristanol[c]	0.228 (0.028) [†]	0.373 (0.003) [†]
EGFP-FMRP/Dcp1a	None	0.024 (0.014)	0.007 (0.003)
	AdOx[a]	0.029 (0.011)	0.006 (0.003)
	Arsenite[b]	0.028 (0.003)	0.010 (0.005)
	Hippuristanol[c]	0.111 .045) [†]	0.126 (0.047) [†]

a. HeLa cells treated with 10 μM AdOx for 24 h prior to analysis. b. HeLa cells treated with 0.5 mM sodium arsenite for 20 min prior to analysis. c. HeLa cells treated with 10 μM hippuristanol for 30 min. prior to analysis. d. Thresholded Manders coefficients (Bolte and Cordelieres, 2006) for EGFP-FMRP granules, tM1 and tM2 were calculated using the JaCoP module in Image J. The means and standard deviations, shown in parentheses, for at least 25 transfected cells per treatment are presented; fluorescence from non-transfected cells in the field was manually masked in order to examine only the colocalization of EGFP-FMRP transfected cells. †Treatment is significantly different from non-treated ($P > 0.005$, ANOVA).

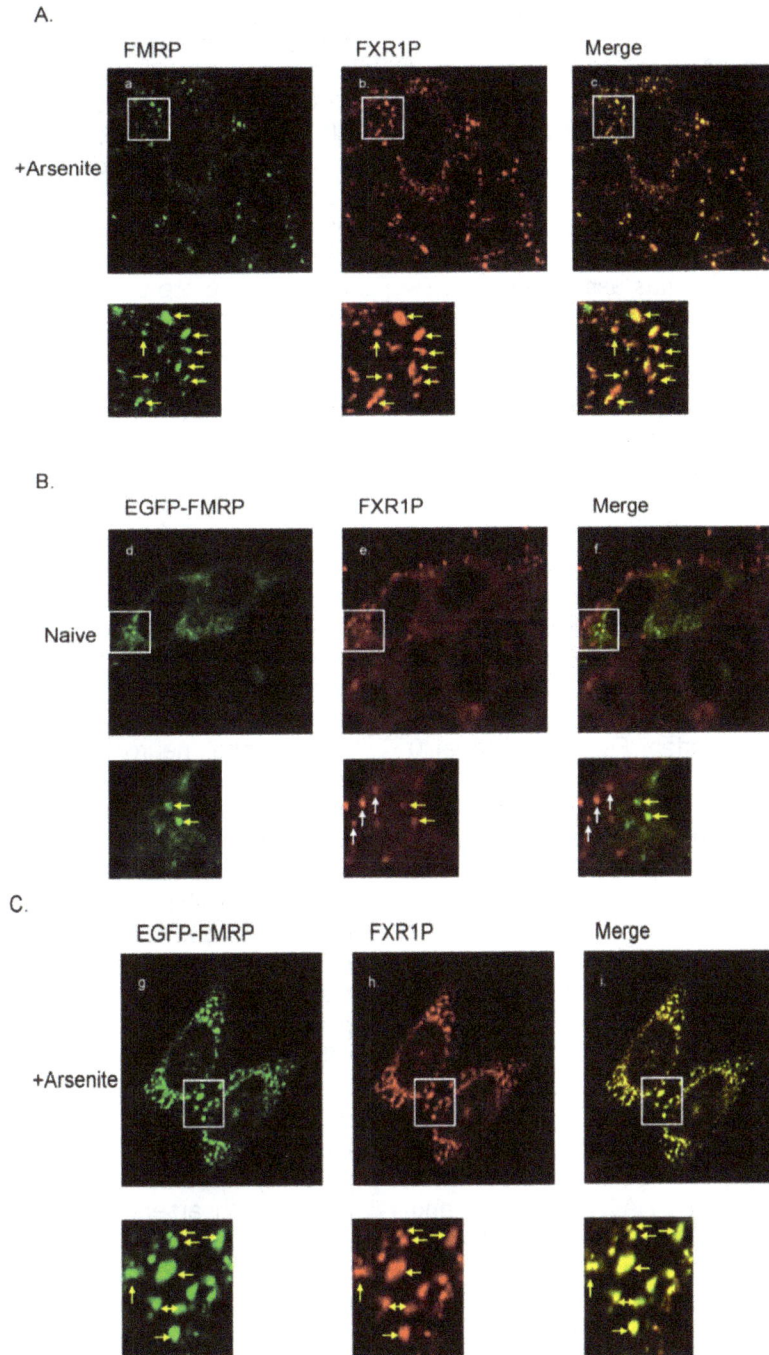

Figure 3. EGFP-FMRP granules partially colocalize with the stress granule marker FXR1P. (A) Panels a-c show endogenous FRMP (green), endogenous FXR1P (red) and merged images of HeLa cells treated with 0.5 mM arsenite for 20 minutes. Boxed areas show magnified views of select FMRP granules reveal the near complete colocalization of the granules. (B) Panels d-f show EGFP-FMRP (green), endogenous FXR1P (red) and merged images of transfected HeLa cells. Boxed areas show magnified views of select EGFP-FMRP granules. White arrows mark FXR1P-only granules; yellow arrows mark colocalized granules and green arrows mark EGFP-FMRP-only granules. (C) Panels g-i show EGFP-FMRP (green), endogenous FXR1P (red) and merged images of transfected HeLa cells treated with 0.5 mM arsenite for 20 min, 24 h post-transfection. Boxed areas show magnified views of select EGFP-FMRP granules. White arrows mark FXR1P-only granules; yellow arrows mark colocalized granules and green arrows mark EGFP-FMRP-only granules.

EGFP-FMRP granules and the P-body marker Dcp1a

P-bodies are another type of granule, distinct from, but associated with stress granules (Anderson and Kedersha, 2006). Studies have shown that certain proteins shuttle between stress granules and P-bodies, while others do not, marking them as core constituents (Kedersha et al., 2005). The decapping factor, Dcp1a, is a core constituent of P-bodies (Cougot et al., 2004). To determine whether EGFP-FMRP was associated with P-bodies HeLa cells were transiently transfected with an EGFP-FMRP expression vector. Twenty four hours later the cells were immunostained with anti-Dcp1a and visualized by confocal microscopy. Figure 4a (panel c) shows that EGFP-FMRP granules occasionally completely overlap with Dcp1a-containing granules, indicating that a portion of EGFP-FMRP was present in P-bodies. Quantitative colocalization analyses revealed however, that like the stress granule markers, the extent of colocalization was very weak, Table 1. Treating cells with sodium arsenite for 20 min following transfection produces stress granules and P-bodies (Kedersha et al., 2007). Under these conditions Dcp1a granules associate more closely with EGFP-FMRP granules, although they generally do not completely overlap, Figure 4b (panel f) and Table 1.

Thus, exogenously expressed EGFP-FMRP sorts into at least three granule types: stress granules, P-bodies and a unique granule, which represents the largest fraction of this set.

The physical characteristics of EGFP-FMRP granules differ from stress granules and P-bodies

To further characterize EGFP-FMRP granules we undertook morphometric analyses of two readily quantifiable features, granule shape and granule size. The results were compared to arsenite-induced TIA1 granules, hippuristanol-induced TIA1 granules and endogenous Dcp1a-containing P-bodies. As expected, we found that the Dcp1a-containing P-bodies were significantly smaller than arsenite-induced stress granules, hippuristanol-induced stress granules, and EGFP-FMRP granules. On the other hand, EGFP-FMRP granules were both substantially larger and more amorphous than the other granules, Table 2. These data again highlight differences between EGFP-FMRP granules and other known granule types.

The dynamic properties of EGFP-FMRP granules mimic stress granules

Emetine inhibits protein synthesis elongation (Gay et al., 1989; Grollman and Huang, 1976; Yamasaki et al., 2007) altering the dynamic equilibrium between polyribosomes and stress granules. It was found that in DU-145 cells treated with arsenite the addition of emetine at concentrations that inhibit protein synthesis effectively "dissolved" TIA1-containing stress granules (Kedersha et al., 2000). Therefore, we used emetine to assess the dynamic interrelationship between EGFP-FMRP-containing granules and polyribosomes. We first demonstrated that endogenous TIA1 stress granules and endogenous FMRP stress granules in HeLa cells treated with arsenite are substantially altered by emetine, Figure 5a. Specifically, the sizes of the granules in the emetine-treated cells were much smaller than the arsenite-induced stress granules (Figure 5b and c). We then applied this paradigm to assess the effect of emetine on EGFP-FMRP granules. Here we found that both untreated EGFP-FMRP granules and arsenite-treated EGFP-FMRP granules responded to emetine in much the same way as TIA1 and FMRP stress granules (Figure 6a, b and c).

Fluorescence recovery after photobleaching is a method that is often used to characterize the dynamic properties of granules (Kedersha et al., 2005, 2008). For example, Antar et al. (2005) examined the effect nocodazole had on EGFP-FMRP granules in cultured hippocampal neurons and found their transport within dendrites was microtubule-dependent. Here, a similar analysis was performed on EGFP-FMRP granules comparing them to ones formed following arsenite treatment. In the absence of treatment EGFP-FMRP granules exhibited a range of motilities. Some granules were quite stationary, while others were highly mobile. FRAP analysis of EGFP-FMRP transfected HeLa cells in the absence of treatment revealed that the EGFP-FMRP granules recovered 50% of their unbleached intensity within 10 min, implying that there is a dynamic exchange between non-granular EGFP-FMRP and the granules, Figure 7a, b and c. To determine whether conditions that induce the formation of stress granules alters the recovery of EGFP-FMRP granules, HeLa cells were treated with arsenite and FRAP was performed on the resulting perinuclear granules. The results showed that the treatment increased the rate of recovery slightly, but not significantly, Figure 7d.

EGFP-FMRP granules contain methylated proteins, but do not require methylated proteins for their formation

Previously we showed that endogenous FMRP-containing stress granules contain asymmetric dimethylarginine-modified proteins (aDMA-proteins) and that the cell's methylation state affected their composition (Dolzhanskaya et al., 2006a). To examine whether EGFP-FMRP granules contained asymmetrically dimethylated (aDMA) proteins HeLa cells, transiently expressing EGFP-FMRP, were incubated in the absence

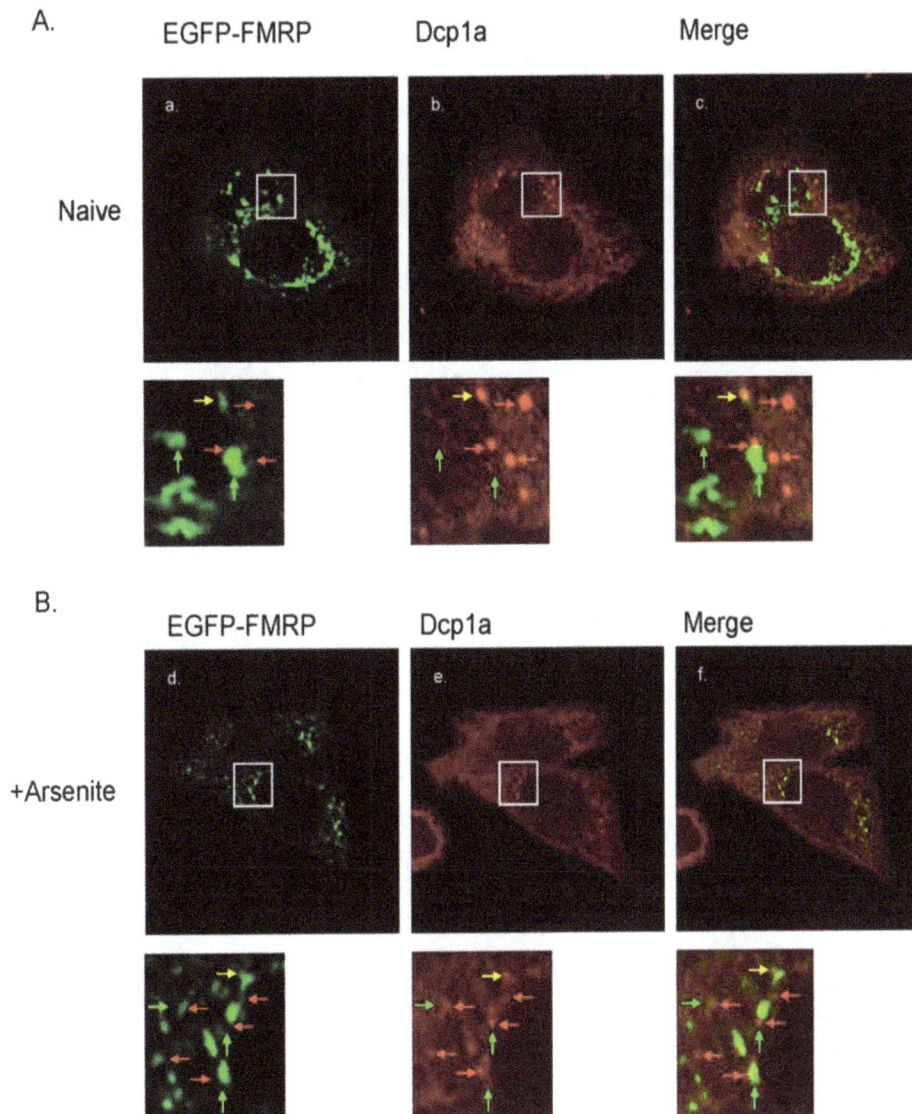

Figure 4. EGFP-FMRP granules weakly colocalize with the P-body marker protein Dcp1a. (A) Panels a-c show EGFP-FMRP (green), Dcp1a (red) and merged images of transfected untreated HeLa cells. Boxed areas presented below show magnified views of select EGFP-FMRP granules. Red arrows mark Dcp1a-only granules; yellow arrows mark colocalized granules and green arrows mark EGFP-FMRP-only granules. (B) Panels d-f show comparable images of transfected HeLa cells treated with 0.5 mM arsenite for 20 min, 24 h post-transfection. Boxed areas presented below show magnified views of select EGFP-FMRP granules. Red arrows mark Dcp1a-only granules; yellow arrows mark colocalized granules and green arrows mark EGFP-FMRP-only granules.

Table 2. Physical characteristics of EGFP-FMRP granules.

Measure/granule	EGFP-FMRP	Stress granules[c]	Stress granules[d]	P-bodies[e]
Circularity[a]	0.87 (0.2)[†]	0.91 (0.15)	0.91 (0.14)	0.95 (0.13)
Area[b]	16.2 (47)[‡]	13.8 (29)	14.4 (19)	4.4 (8)

a. Circularity measures the deviation of the granule from a perfect circle, scored as 1; the mean and (standard deviation) are shown. b. Area is given in squared pixels; the mean and (standard deviation) are shown. c. HeLa cell arsenite-induced TIA1 granules. d. HeLa cell hippuristanol-induced TIA1 granules.e. HeLa cell endogenous Dcp1a granules.† Circularity of 1000 EGFP-FMRP granules was significantly different from that of 1000 arsenite-induced granules, hippurstanol-induced granules and 1200 endogenous P-bodies ($P < 1 \times 10^{-32}$, ANOVA). ‡ Area of 1000 EGFP-FMRP granules was significantly different from that of 1000 arsenite-induced granules, 1000 hippuristanol-induced granules and 1200 endogenous P-bodies ($P < 6 \times 10^{-19}$, ANOVA).

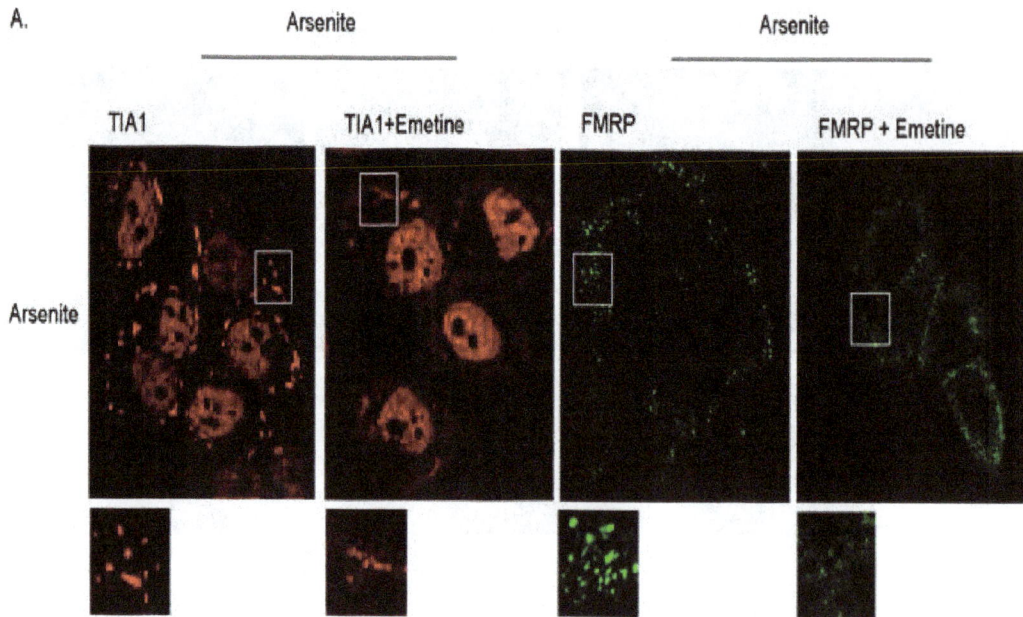

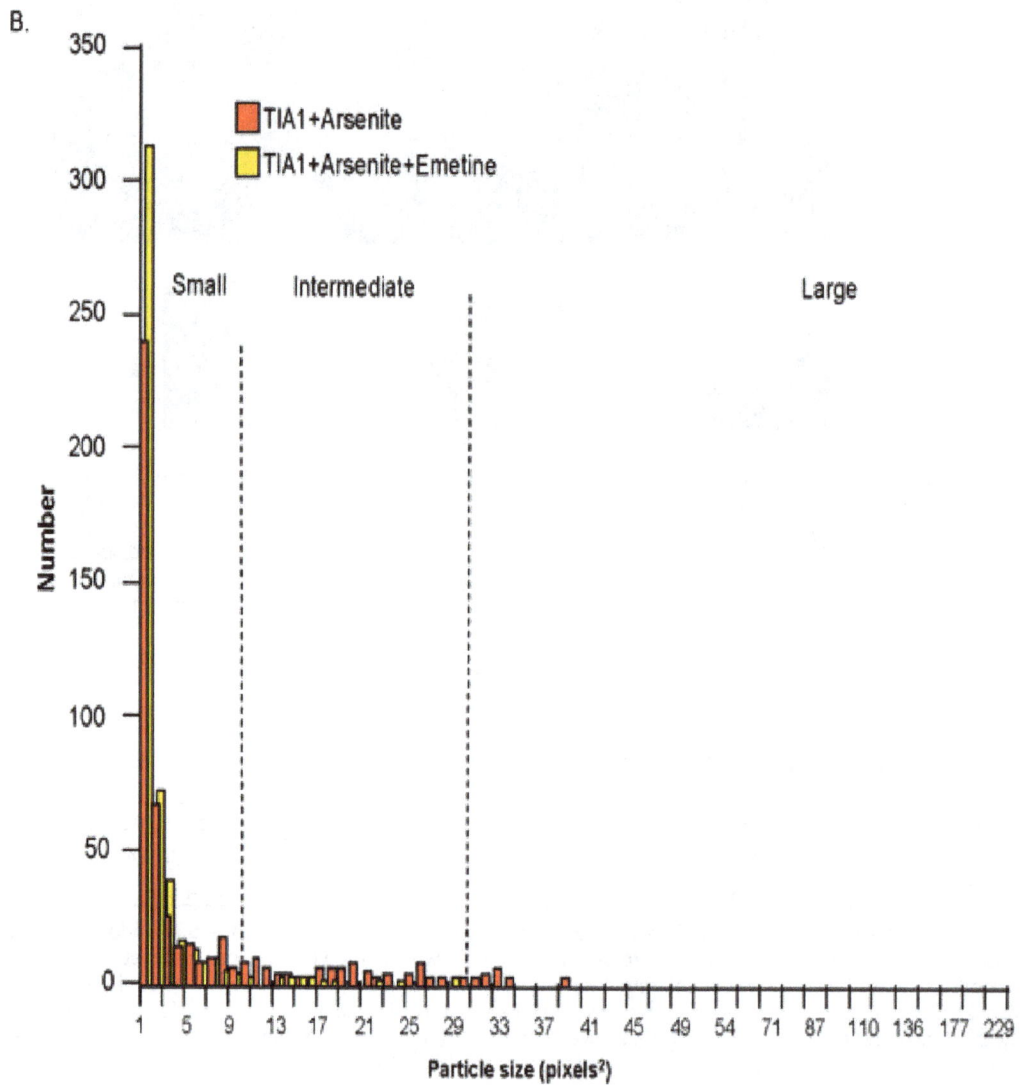

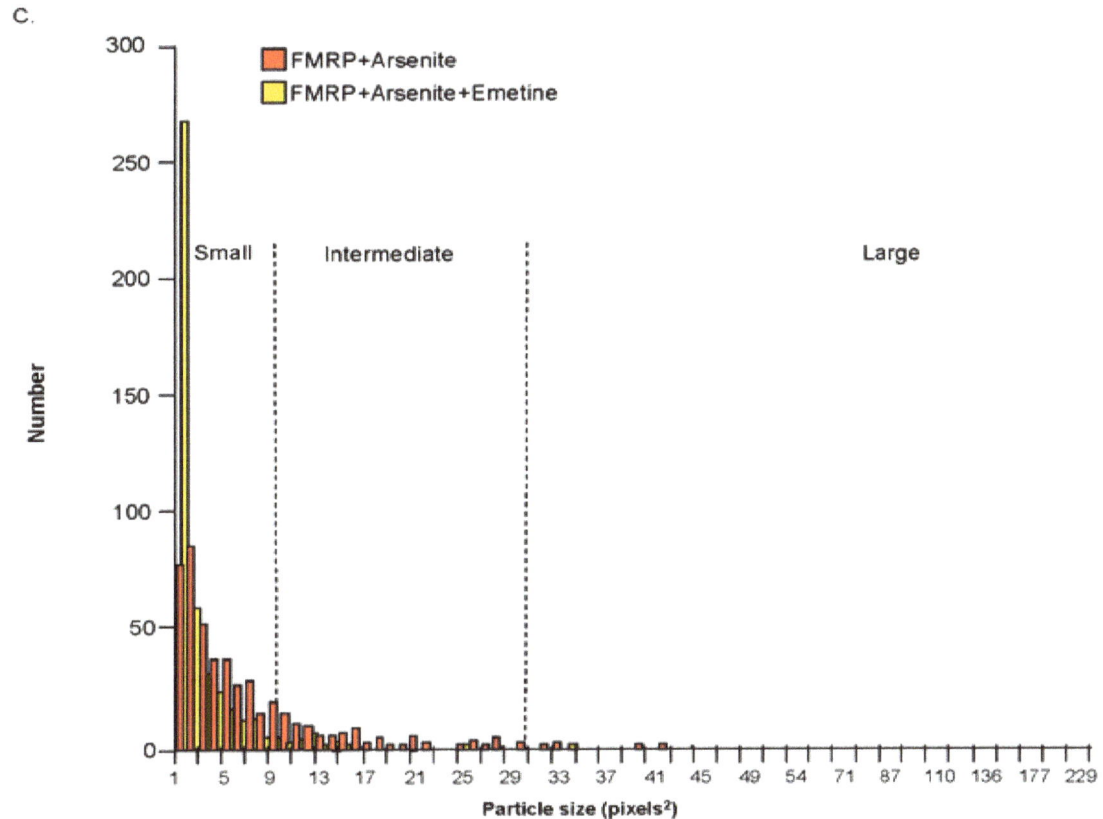

Figure 5. Endogenous TIA1 and endogenous FMRP stress granules respond to emetine. (A) Panels a-d show endogenous TIA1 (red), endogenous FMRP (green) HeLa cells treated with 0.5 mM arsenite for 20 min or treated with 0.5 mM arsenite and then 10 μg/ml of emetine for 1 h. Boxed areas presented below show magnified views of select stress granules. Size distribution of HeLa cell endogenous (B) TIA1 (C) endogenous FMRP stress granules in the presence and absence of emetine. Analysis was based on 10 cells for each protein. Comparison of the two distributions shows that emetine significantly reduces the size of both classes of stress granules ($P < 0.049$ and $P < 4.5 \times 10^{-6}$, respectively, F-test for two sample covariance).

A.

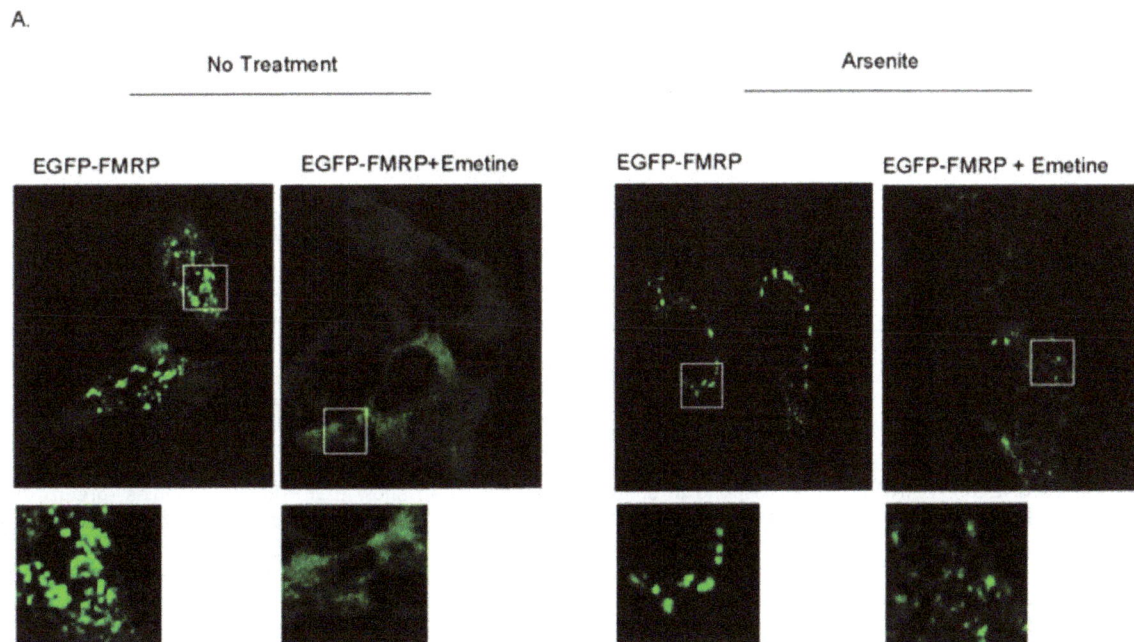

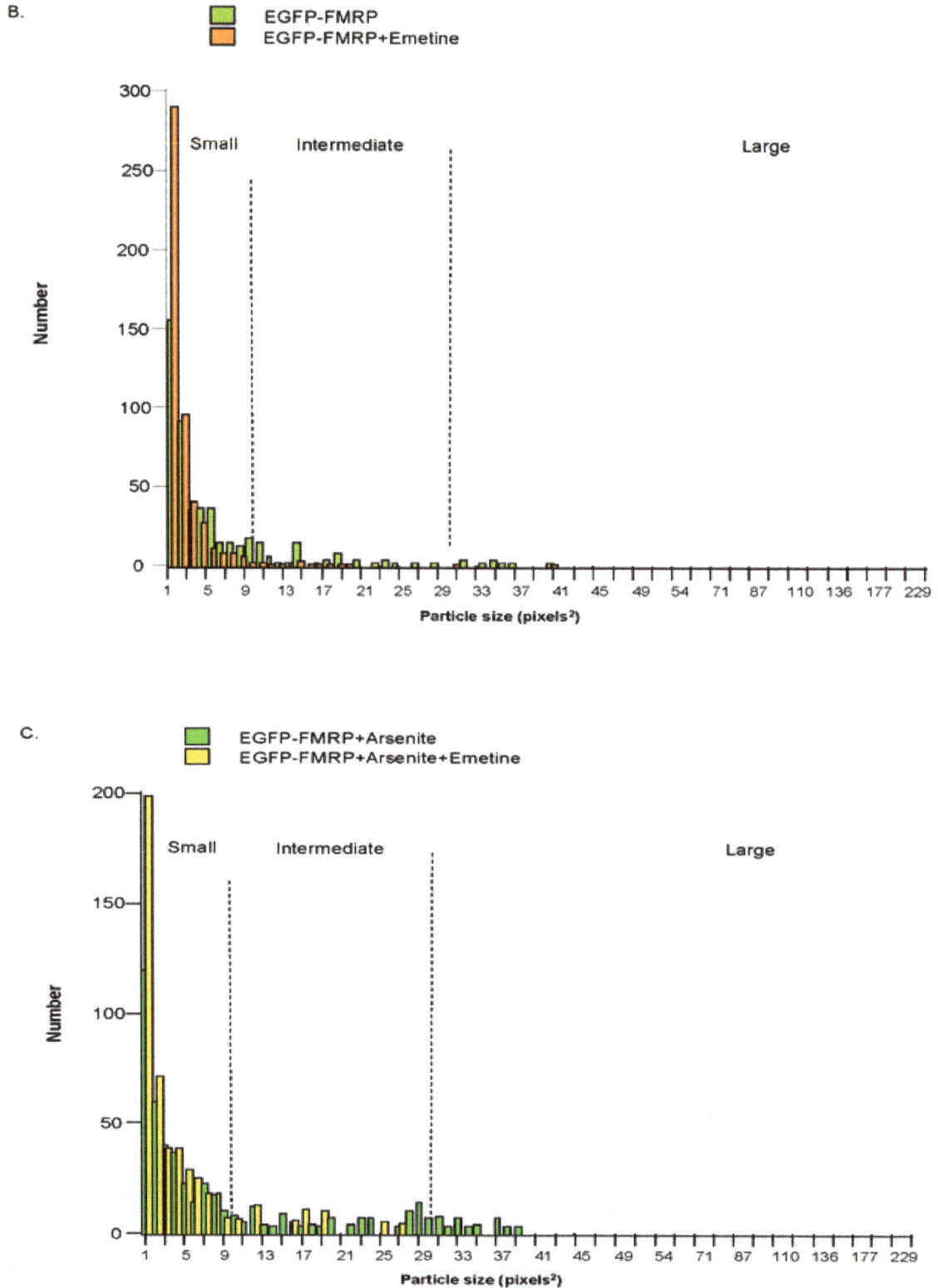

Figure 6. EGFP-FMRP granules respond to emetine. (A) Panels a-d show EGFP-FMRP granules (green) in untreated HeLa cells (a, b) or HeLa cells treated with 0.5 mM arsenite for 20 min or treated with 0.5 mM arsenite and then 10 µg/ml of emetine for 1 h (c, d). All treatments were performed 24 h post-transfection. Boxed areas presented below show magnified views of select stress granules. Size distribution of HeLa cell EGFP-FMRP granules in the absence or presence of emetine and in the (B) absence or (C) presence of arsenite. Analysis was based on 10 transfected cells for each protein. Comparison of the two distributions shows that emetine significantly reduces the size of both classes of stress granules ($P < 0.00084$ and $P < 0.0023$, respectively).

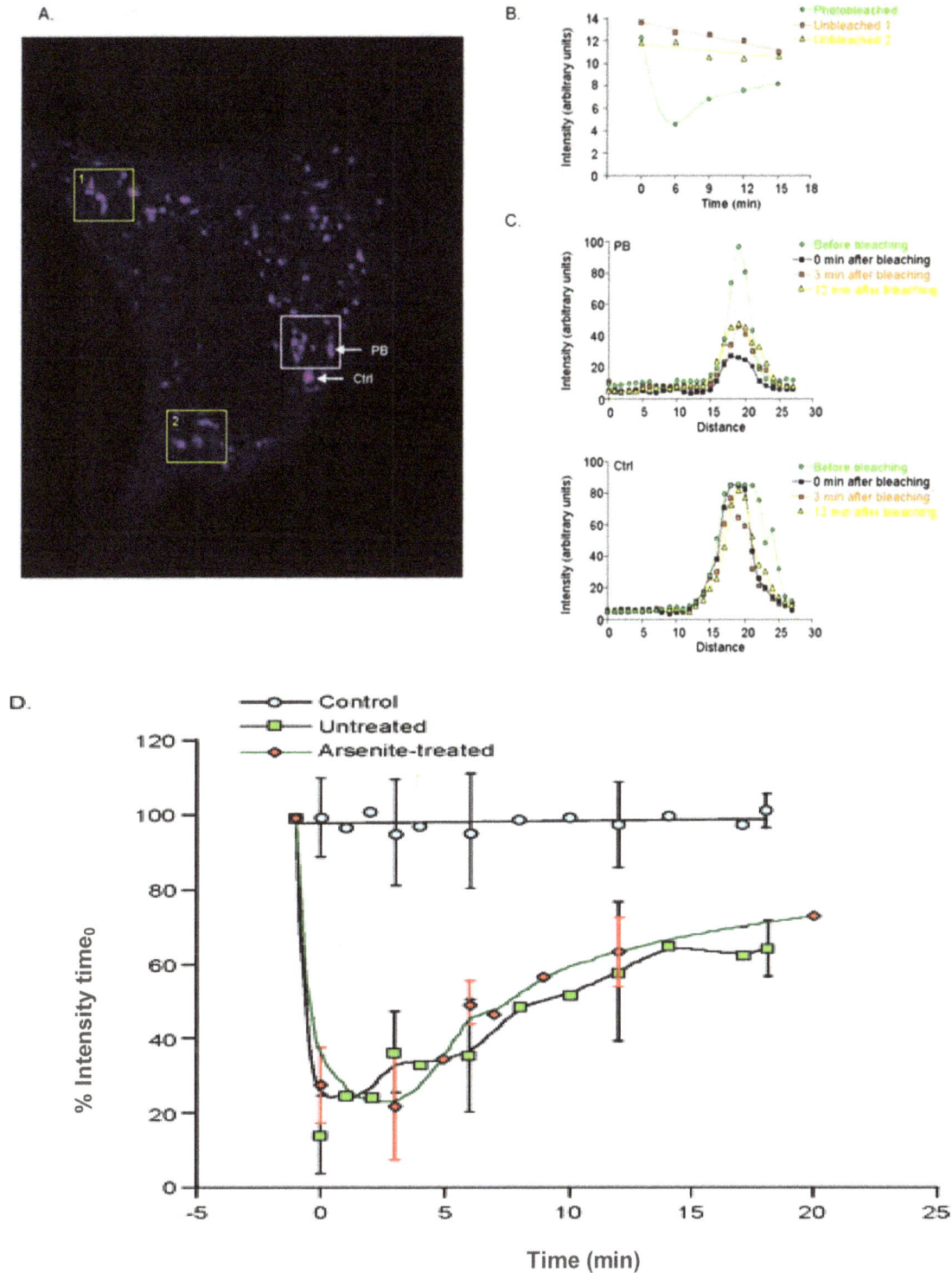

Figure 7. EGFP-FMRP granules are dynamic. (A) Pseudo colored image of EGFP-FMRP expressing HeLa cell. White box shows a region of interest (ROI) prior to photobleaching, while the yellow boxes show comparable ROIs of unbleached EGFP-FMRP granules. Arrows point to a granule that was photobleached (PB) and a nearby control granule (Ctrl) that was not. (B) Time-dependent changes in the intensity of the ROIs shown in (A). (C) Time-dependent changes in PB (upper) and Ctrl (lower) granules shown in (A). (D) Effect of arsenite treatment on EGFP-FMRP photobleaching recovery rates. The intensity at time 0 for 25 non-bleached EGFP-FMRP granules, 23 arsenite-treated photobleached EGFP-FMRP granules and 39 photobleached untreated EGFP-FMRP granules from five different cells was plotted. The difference between the photobleached arsenite-treated granules was not significantly different from the untreated granules (P > 0.2, ANOVA). For presentation clarity error bars are shown on every third time point.

or presence of AdOx for twenty-four hours. Subsequently, the cells were immunostained with two antibodies that detect different sets of aDMA proteins. For the untreated cells, cytoplasmic EGFP-FMRP granules exhibited extensive colocalization with the proteins detected by these two antibodies; Figure 8a (panels a-c and f-h). Importantly, AdOx treatment resulted a marked reduction in immunoreactivity, both overall and in the granules, Figure 8a (panels c-e and l-k), indicating that the staining was indeed due to aDMA proteins. Quantitative analyses showed that under the conditions used the extent of the reduction was ca. 50%, Figure 8b.

The aforementioned data suggested that reduced methylation of endogenous proteins does not drastically alter the composition of EGFP-FMRP granules; however, it does not inform one as to whether asymmetric dimethylation affects their formation. To address this question, HeLa cells were pre-treated with AdOx for twenty-four hours prior to transfection. Subsequently, the cells were transfected with the EGFP-FMRP expression vector in AdOx-containing media. As shown in Figure 9a, EGFP-FMRP granules were readily detected even in the continued presence of AdOx implying that full protein methylation is not required for the formation of these granules. In concert with these data treatment with AdOx had no effect on the size distribution of EGFP-FMRP granules (Figure 9b).

We next assessed whether EGFP-FMRP granules contained symmetrically dimethylated (sDMA) proteins. In the absence of treatment we found that the symmetrically dimethylated protein immunoreactivity localized primarily to the nucleus, Figure 10a (panel b). Notably, the cytoplasmic staining partially associated with EGFP-FMRP granules, indicating that these granules are heterogeneous with respect to sDMA-containing proteins, Figure 10b (panel c). However, treating the cells with arsenite following transfection resulted in the recruitment of a significant amount of sDMA proteins into cytoplasmic granules, and in EGFP-FMRP expressing cells; under these conditions almost all of the EGFP-FMRP granules contained sDMA-modified proteins, Figure 10 a and b (panels g-i). Treating the cells with AdOx prior to arsenite treatment significantly decreased the sDMA protein immunoreactivity Figure 10 a and b (panel k), although the extent of colocalization with EGFP-FMRP granules was unchanged, Figure 10b (panels h and k). Consistent with the immunostaining results, Western blot analyses showed that AdOx treatment decreased the amount of aDMA and sDMA in various proteins, albeit to different extents. In contrast, the expression of five different protein arginine methyltransferases (PRMTs) was unaffected by treatment with AdOx (Figure 11).

As these data unequivocally demonstrate sDMA is harbored in arsenite-treated EGFP-FMRP granules we next endeavored to ascertain whether endogenous stress granules also contained sDMA-modified proteins. To this end we treated HeLa cells with arsenite and then immunostained them with antibodies to TIA1 and sDMA. We observed significant colocalization between sDMA-modified proteins and TIA1-containing stress granules (Figure 12).

DISCUSSION

Heterologously expressed EGFP-FMRP has been used as a surrogate marker for endogenous FMRP by several laboratories in a wide variety of studies (Antar et al., 2005; Castren et al., 2001; Cougot et al., 2008; Darnell et al., 2005; Davidovic et al., 2004; De Diego Otero et al., 2002; Dictenberg et al., 2008; Levenga et al., 2009; Pfeiffer and Huber, 2007). Like endogenous FMRP, EGFP-FMRP was found to co-sediment with polyribosomes in cultured neurons (Darnell et al., 2005). Additionally, Antar et al. (2004) showed that EGFP-FMRP was transported to dendrites in the form of granules (Antar et al., 2004). Subsequent work showed that these granules associated with mRNA and kinesin transport motors (Wang et al., 2008) and were responsive to stimulation by the mGluR agonist DHPG. These too are properties of endogenous FMRP (Hou et al., 2006; Kanai et al., 2004).

However, important questions regarding these granules remained to be addressed. Were they uniformly composed? Did they correspond to one or more types of known granules? Could they be modified by various treatments that are known to affect the formation and/or composition of known granule types? Answering these questions in light of the well-established properties of endogenous FMRP granules (Didiot et al., 2008; Dolzhanskaya et al., 2006a, 2006b) was one of the objectives of this study.

In contradistinction to endogenous FMRP granules, which are small and while pervasive in cultured cells not particularly prevalent, (Cougot et al., 2008; Dolzhanskaya et al., 2006a, 2006b) EGFP-FMRP granules are large and amorphous. In fact, they tend to be larger and more non-uniformly round than stress granules (Table 2). Underlying this size/shape heterogeneity we found that EGFP-FMRP granules also are non-uniformly composed. Endogenous FMRP absent environmental stress does not colocalize with stress granule markers; however, in the presence of a stressor such as heat or arsenite FMRP is recruited into stress granules (Didiot et al., 2008; Dolzhanskaya et al., 2006a; Mazroui et al., 2002). Using well-known stress granule markers (TIA1 and FXR1P) we found that EGFP-FMRP granules weakly colocalize with these proteins in the absence of a bonafide stressor; however, stressing EGFP-FMRP-expressing cells with either arsenite or hippuristanol resulted in a marked increase in colocalization, (Figures 2 and 3; Table 1). While the overall colocalization of TIA1 or FXR1P in the absence of stress was low we noticed that in some cells nearly all of the EGFP-FMRP granules

a

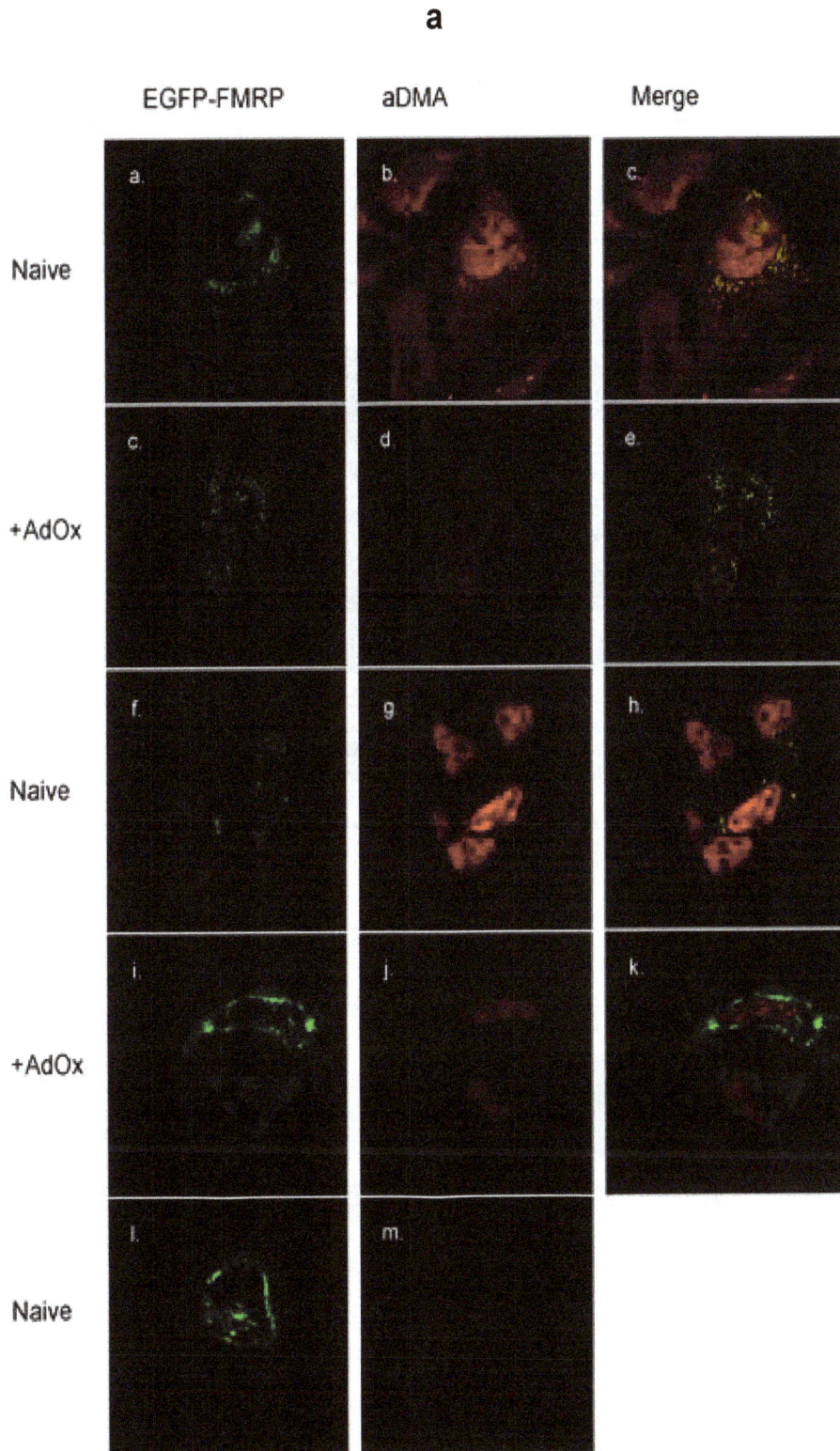

Figure 8A. EGFP-FMRP granules contain asymmetrically dimethylated proteins. (A) HeLa cells expressing EGFP-FMRP were immunostained with antibodies directed to asymmetrically dimethylated proteins. Panels a-e show images of EGFP-FMRP (green), ASYM24 (red), and their merged counterpart whereas panels f-k show images of EGFP-FMRP (green), mRG (red), and their merged counterpart. Panels l-m demonstrate lack of immunostaining in the absence of primary antibody. As indicated, the cells were either treated or not treated with 20 μM AdOx for 24 h following transfection.

b

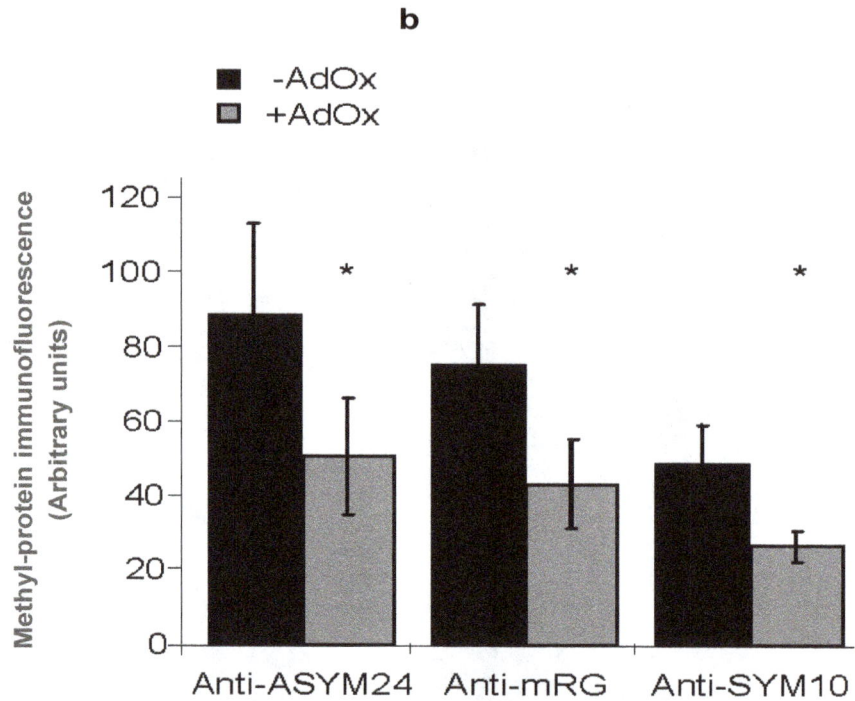

Figure 8B. Quantification of anti-ASYM24 pAb, anti-mRG pAb and anti-SYM10 immunostaining in HeLa cells grown in the absence (-) or presence (+) of 20 μM AdOx. The fluorescence intensity of the individual cells was determined using the histogram function in Adobe Photoshop as previously described (Dolzhanskaya et al., 2006a). The values for 50 - 75 cells/antibody for each treatment are plotted. The reduction in fluorescence intensity in the presence of AdOx was significantly less than in its absence (ASYM24 P < 5 × 10-15, mRG P < 4 × 10^{-15}, SYM10 P < 5 × 10^{-17} ANOVA).

a

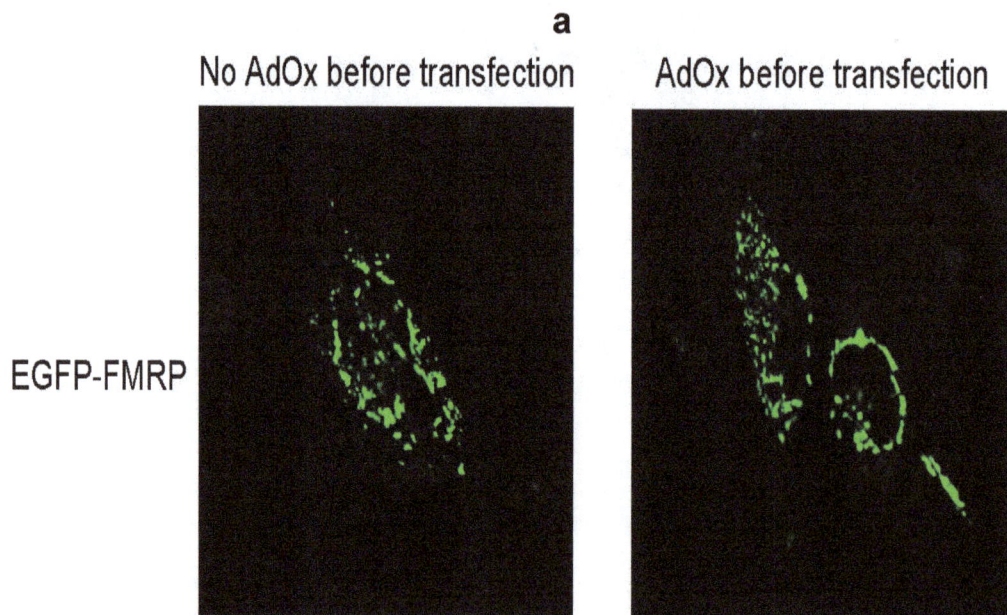

Figure 9A. EGFP-FMRP granule expression does not require full protein methylation. (A) HeLa cells were either treated or not treated with 20 μM AdOx for 24 h prior to transfection. Subsequently, the cells were transfected with pEGFP-FMRP and imaged 24 h later. Note: AdOx was re-applied to the set of cells that were pre-treated with AdOx.

b

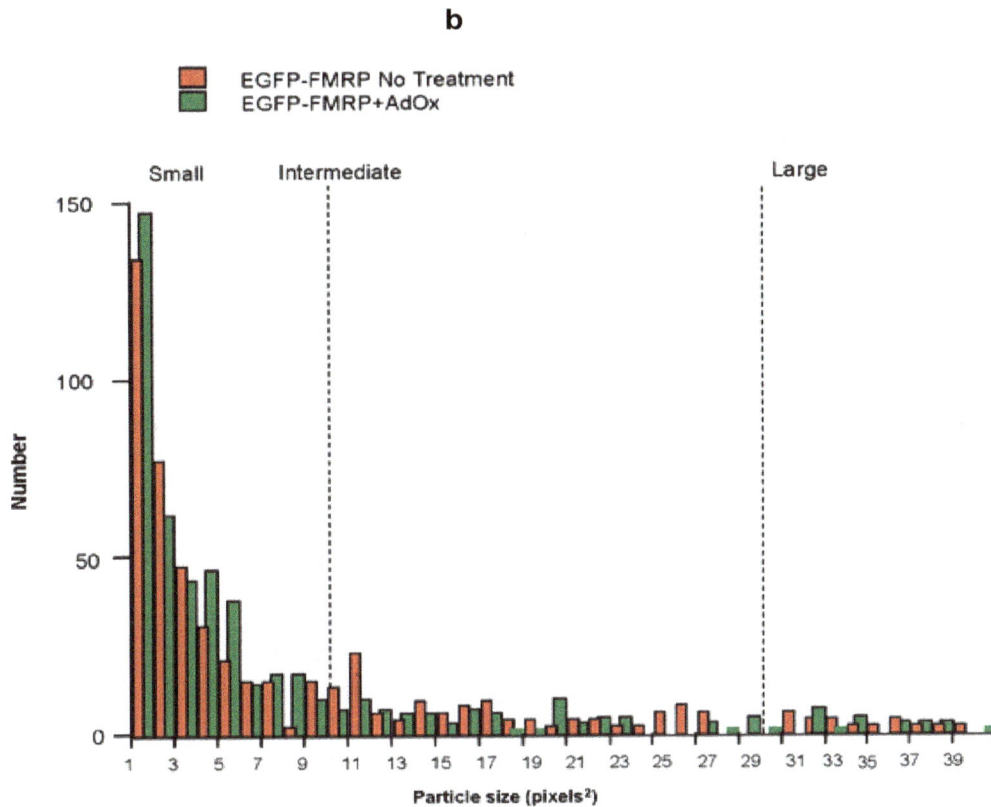

Figure 9B. Size distribution of HeLa cell arsenite-treated EGFP-FMRP granules in the presence and absence of AdOx. Analysis was based on 10 transfected cells/treatment. Comparison of the two distributions shows that AdOx does not significantly reduce the size of EGFP-FMRP granules (P > 0.4).

colocalized with these markers. Currently, the molecular and cellular basis of this phenomenon is unknown. It may be related in some way to the cell cycle or it may simply be a stochastic phenomenon and we are investigating these possibilities. More importantly however, the data demonstrate that EGFP-FMRP granules are biochemically heterogeneous and this non-uniformity must be considered when interpreting studies that use these granules as markers.

Contradicting data exists regarding the association of FMRP with P-bodies. Several lines of evidence have shown that FMRP, or its *Drosophila* counterpart dFMR1, interact specifically with components that are associated with P-bodies (Anderson and Kedersha, 2006). Ishizuka et al. (2002) first demonstrated that dFMR1 interacted with Ago2 and the RNA helicase Dmp68 (Ishizuka et al., 2002); more recently Cheever et al showed that FMRP interacts with Dicer in a phosphorylation-dependent fashion (Cheever and Ceman, 2009). Moreover, immunofluorescence studies showed that endogenous FMRP partially overlapped the staining of both Ago3 and Ago4 in rat dorsal root ganglion (DRG) processes (Hengst et al., 2006), while EGFP-FMRP and GFP-dFMR1 colocalized with Dcp1 (Barbee et al., 2006;

Cougot et al., 2008). Nevertheless, Didiot et al. (2008) failed to observe much localization of endogenous FMRP with Dcp1a in HeLa cells (Didiot et al., 2008). Here we found a weak colocalization between EGFP-FMRP granules and the P-body marker Dcp1a, Figure 6 and Table 1. We hypothesize that the differing results may be related to differences in the species and tissue expression of P-body components (González-González et al., 2008) and in the heterogeneity of Ago-containing granules (González-González et al., 2008; Iwasaki et al., 2009).

We also investigated whether EGFP-FMRP granules were in dynamic equilibrium with polyribosomes and/or the pool of non-granular cytoplasmic EGFP-FMRP by emetine treatment (Kedersha et al., 2000) and FRAP (Lippincott-Schwartz et al., 1999), respectively. Here we found that EGFP-FMRP granules mimic the properties of endogenous FMRP, implying that there are certain "core characteristics" of granules that are invariant (Buchan and Parker, 2009).

Large ribonucleoprotein granules are often regulated by one or more post-translational modifications (Qi et al., 2008; Yoon et al., 2010). Stress granules, in particular, undergo a myriad of such reactions (Carpio et al., 2010;

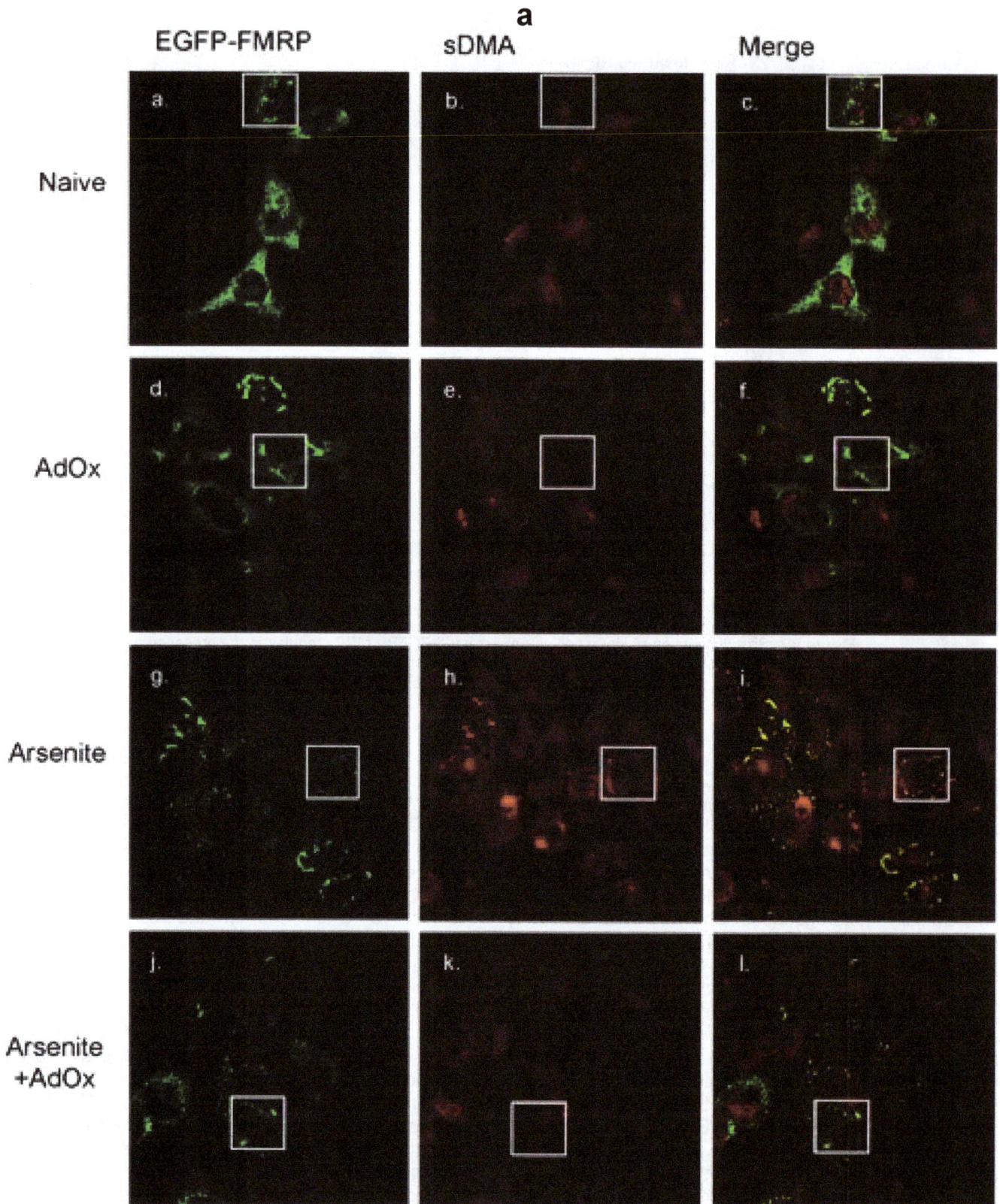

Figure 10A. EGFP-FMRP granules contain symmetrically dimethylated proteins. (A) HeLa cells expressing EGFP-FMRP were immunostained with an antibody directed to symmetrically dimethylated proteins (sDMA), SYM10. Panels a-f show images of EGFP-FMRP (green), sDMA (red), and their merged counterparts in untreated cells and cells treated with 20 μM AdOx for 24 h following transfection. Panels g-l show cells comparably treated cells that were either treated or not treated with 0.5 mM arsenite for 20 min, 24 h post-transfection.

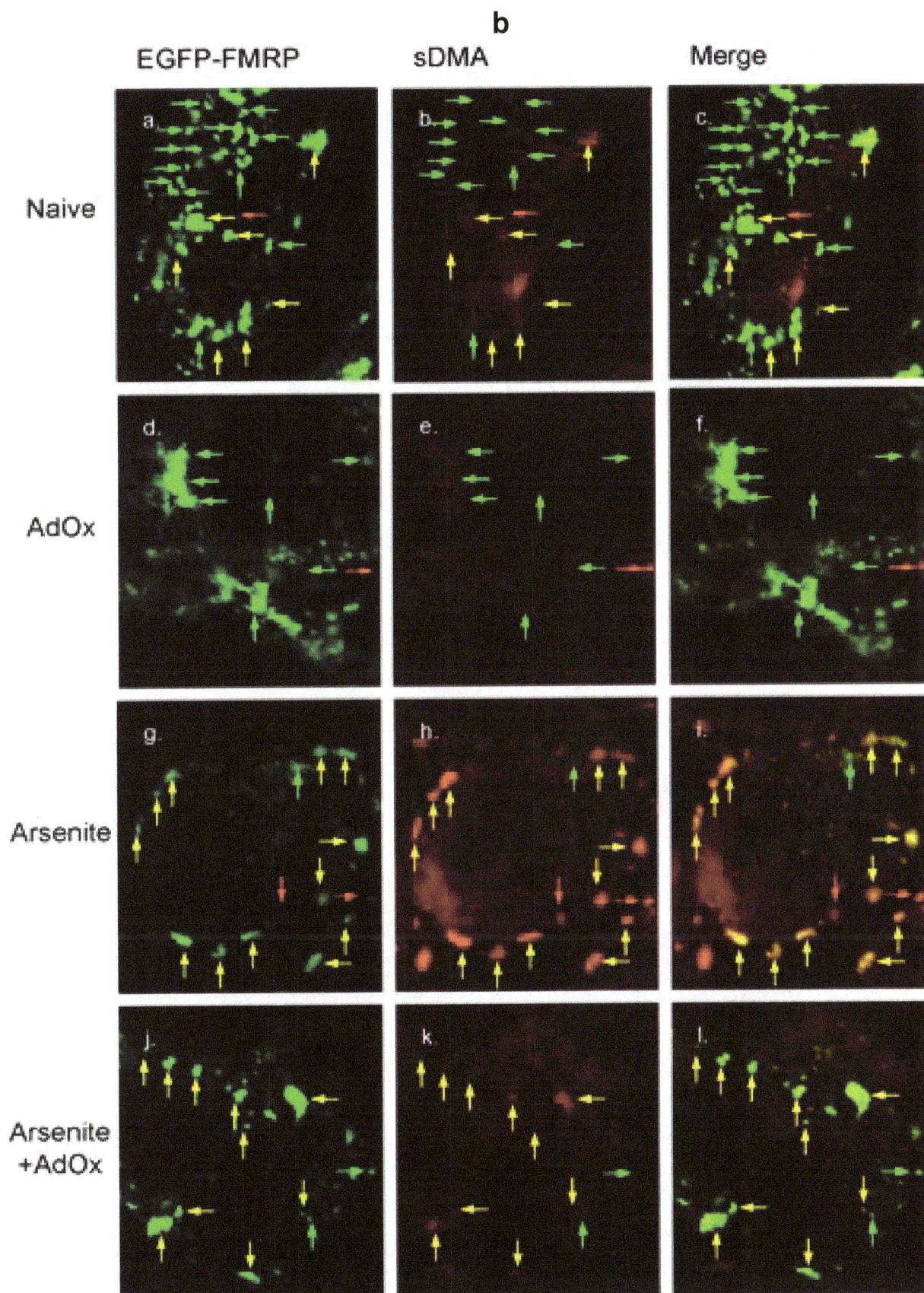

Figure 10B. Boxed areas from (A) show magnified views of select cells; yellow arrows mark colocalizing granules, green arrows mark EGFP-FMRP only granules and red arrows mark sDMA only granules.

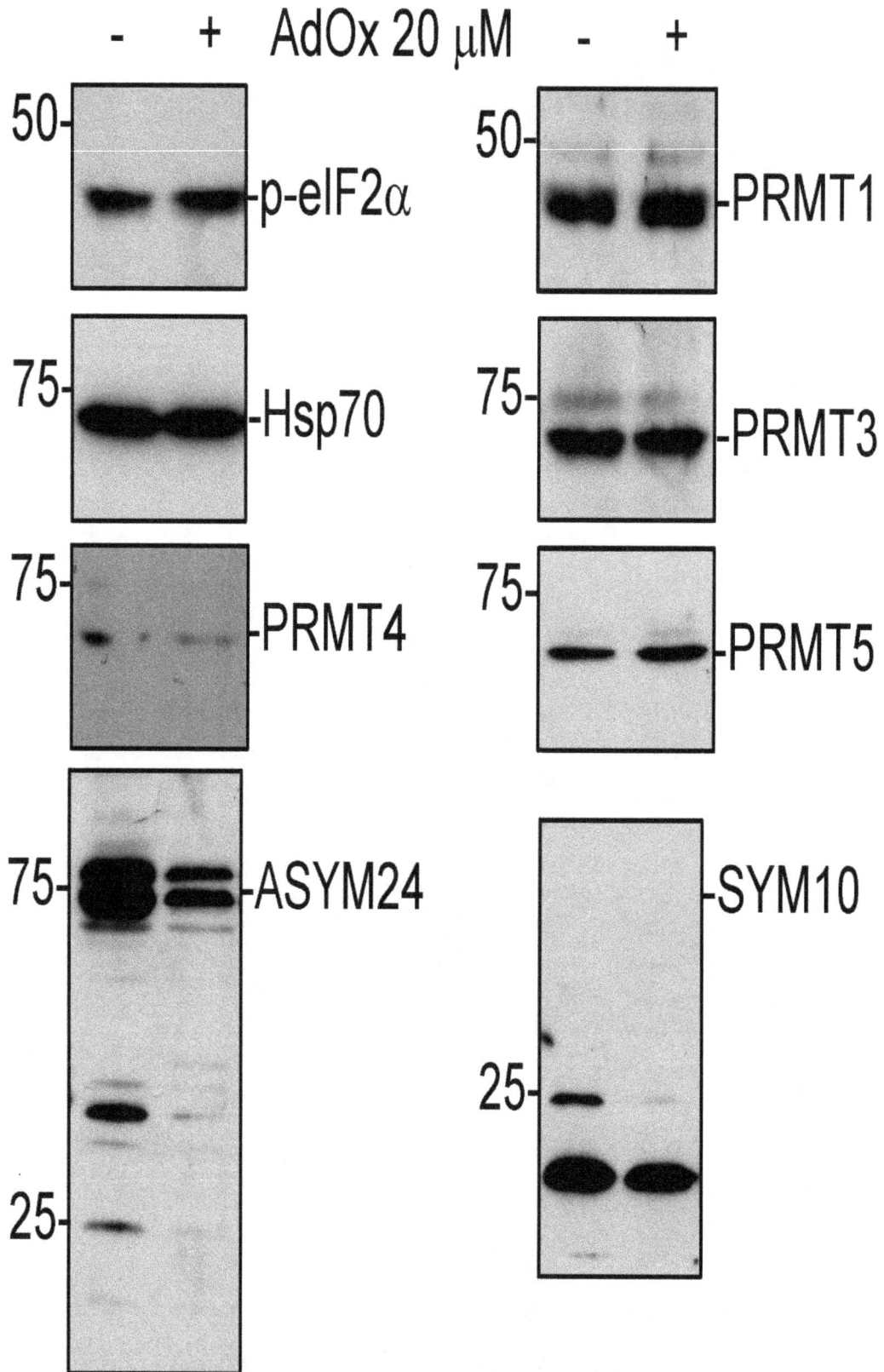

Figure 11. AdOx inhibits cellular methylation of proteins that colocalize in EGFP-FMRP granules. Representative Western blots of HeLa cell proteins (30 μg) probed with antibodies directed to methylated proteins, ASYM24 pAb (left) and SYM10 pAb (right), protein arginine methyltransferases (PRMTs) 1, 3, 4, and 5 and various load controls.

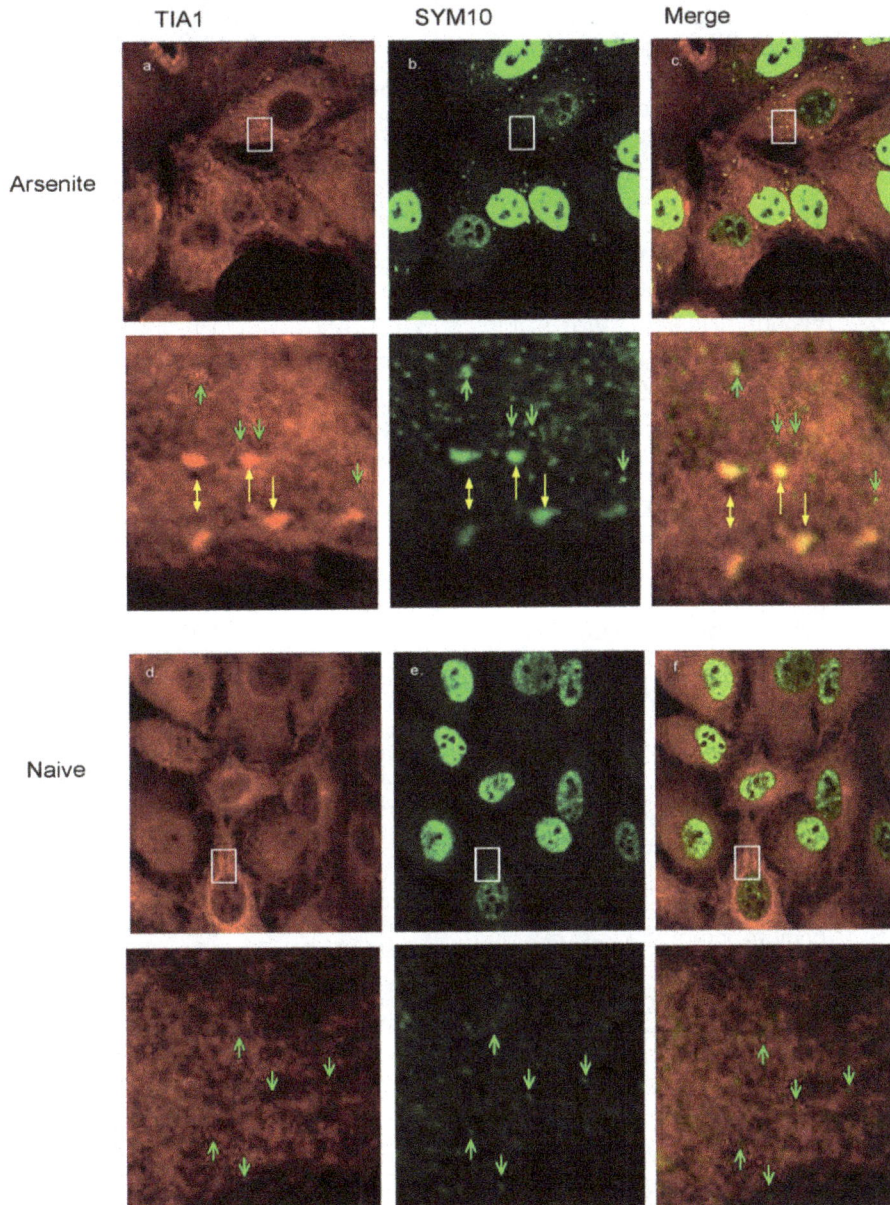

Figure 12. Endogenous TIA1 stress granules contain sDMA proteins. HeLa cells were treated with 0.5 mM sodium arsenite (panels a-c) or not treated panels (d-f). Subsequently, the cells were immunostained with antibodies that detect the stress granule marker TIA1 (red) and SYM10 (green) and subject to confocal microscopy. Magnified views of select regions of the upper panels are presented below to clearly show the granules. Large TIA1-containing stress granules that colocalize with symmetrically dimethylated proteins are marked with yellow arrows. These are contrasted with smaller granules harboring symmetrically dimethylated proteins, but which do not colocalize with TIA1 (green arrows). Note that these smaller granules are constitutively present in untreated cells. We speculate that these may be related to the 7S PRMT5 complex, the 20S methylosome or the SMN complex.

Kwon et al., 2007; Mazroui et al., 2007; Wasserman et al., 2009), among which is protein arginine methylation. We have previously demonstrated that native stress granules contain asymmetrically dimethylated (aDMA) proteins and that pharmacological inhibition of protein methylation altered their FXR1P content; in contrast, endogenous FMRP granules were devoid of aDMA-containing proteins (Dolzhanskaya et al., 2006a). Thus, it was of interest to determine whether EGFP-FMRP granules associated with methylated proteins and if so what role they played in granule formation. We found that EGFP-FMRP granules, like stress granules, always contained aDMA- modified proteins and sometimes contained sDMA-modified proteins, Figure 8 and Figure 10a, b (panel c). Interestingly, we also determined that endogenous stress granules harbored sDMA-containing proteins and because the sDMA antibody used overwhelmingly recognizes splicing-related proteins (Boisvert et al., 2003) these data provide a potential link between mRNA splicing and stress granules. Future studies must address the identities of the sDMA-containing proteins to confirm this connection. Finally, using AdOx to significantly reduce endogenous protein methylation we discovered that protein methylation is not required for the formation of EGFP-FMRP granules, Figure 9. Thus, although methylation clearly can affect protein dimerization (Dolzhanskaya, 2006a) it appears less intimately involved in the assembly/disassembly of granules than phosphorylation, deacetylation and ubiquitination.

Overall, our work has shown both similarities and marked differences between EGFP-FMRP granules and three defined granule types: endogenous FMRP granules, stress granules and P-bodies. The fact that EGFP-FMRP granules uniformly contain aDMA-modified proteins distinguishes them from endogenous FMRP granules. Thus, EGFP-FMRP is not an appropriate surrogate for FMRP, and studies that have used it as a marker may need to be reinterpreted in light of this new information. Additionally, because of the weak overlap with stress granule and P-body markers the vast majority of EGFP-FMRP is not normally present in stress granules or P-bodies. However, triggering stress granule and P-body formation with arsenite (Kedersha et al., 2007), or stress granule formation with hippuristanol (Mazroui et al., 2006) resulted in the near-complete recruitment of EGFP-FMRP into stress granules, Figures 2 and 3. Thus, these data are consistent with the view that EGFP-FMRP granules are uniquely poised to become stress granules; hence, we have called them proto-stress granules.

ACKNOWLEDGEMENTS

We thank Drs. Ying-Ju Sung and W. Ted Brown for helpful discussions concerning this manuscript. These studies were generously supported by the New York State Research Foundation for Mental Hygiene and the FRAXA Research Foundation.

REFERENCES

Anderson P, Kedersha N (2006). RNA granules. J. Cell Biol., 172: 803-808.

Antar LN,Afroz R,Dictenberg J, BCarroll RC, Bassell GJ (2004). Metabotropic glutamate receptor activation regulates fragile X mental retardation protein and Fmr1 mRNA localization differentially in dendrites at synapses. J. Neurosci., 24: 2648.

Antar LN, Dictenberg JB, Plociniak M, Afroz R, Bassell GJ (2005). Localization of FMRP-associated mRNA granules requirement of microtubules for activity-dependent trafficking in hippocampal neurons. Genes Brain Behav., 4: 350.

Aschrafi A ,Cunningham BA , Edelman GM , Vanderklish PW (2005). The fragile X mental retardation protein and group I metabotropic glutamate receptors regulate levels of mRNA granules in brain. Proc. Nat'l Acad. Sci., 102: 2180.

Barbee SA, Estes PS, Cziko AM, Hillebrand J, Luedeman RA, Coller JM, Johnson N, Howlett IC, Geng C, Ueda R (2006). Staufen FMRP-containing neuronal RNPs are structurally and functionally related to somatic P bodies. Neuron, 52: 997.

Bassell GJ , Warren ST (2008). Fragile X syndrome: Loss of local mRNA regulation alters synaptic development function, 60: 201.

Biron VL, McManus KJ, Hu N, Hendzel MJ, Underhill DA(2004). Distinct dynamics distribution of histone methyl-lysine derivatives in mouse development. Dev. Biol., 276: 337

Bloch DB, Nobre R (2010). p58TFL does not localize to messenger RNA processing bodies Letter. Mol. Cancer Res., 8(131).

Boisvert FM, Cote J, Boulanger MC, Richard S(2003). A proteomic analysis of arginine methylated protein complexes. Mol. Cell Proteomics, 2: 1319.

Bolte S, Cordelieres FP. (2006). A guided tour into subcellular colocalization analysis in light microscopy J. Microsc., 224: 213.

Brand A (1999). GFP as a cell developmental marker in the Drosophila nervous system In Green Fluorescent Proteins vol. 58 eds KF, Sullivan S A, Kay San Diego, Academic Press, pp. 165-180.

Buchan JR, Parker R (2009). Eukaryotic stress granules: The ins and outs of translation. Mol Cell, 36: 932-941.

Carpio MA, LÃ³pezÃ Sambrooks C,Durand ES, Hallak ME (2010). The arginylation-dependent association of calreticulin with stress granules is regulated by calciu. Biochem. J., 429: 63-72.

Castren M, Haapasalo A, Oostra B,Castren E (2001). Subcellular localization of fragile X mental retardation protein with the I304N mutation in the RNA-binding domain in cultured hippocampal neurons .Cell. Mol. Neurobiol., 21: 29

Cheever A, Ceman S (2009). Phosphorylation of FMRP inhibits association with Dicer. RNA, 15: 362

Cougot N, Babajko S, Seraphin B (2004). Cytoplasmic foci are sites of mRNA decay in human cells. Cell. Mol. Neurobiol., 165: 31

Cougot N, Bhattacharyya SN,Tapia-Arancibia L,Bordonne R, Filipowicz W, Bertrand E, Rage F (2008). Dendrites of mammalian neurons contain specialized P-body-like structures that respond to neuronal activation. J. Neurosci., 28: 13793-13804.

Cziko AMJ, McCann CT, Howlett IC, Barbee SA, Duncan RP, Luedemann R, Zarnescu D, Zinsmaier KE, Parker RR, Ramaswami M (2009). Genetic modifiers of dFMR1 encode RNA-granule components in Drosophila. Genetics, 182:1051-1060.

Darnell, JC, Mostovetsky O, Darnell RB (2005). FMRP RNA targets: identification and validation. Genes Brain Behav., 4: 341-349.

Davidovic L, Huot MC, Khandjian E. (2004). Lost once, the fragile X mental retardation protein is now back on brainpolyribosomes. RNA Biol., 1: 125-127.

B,Willemsen R (2002). Transport of fragile X mental retardation proteinvia granules in neurites of PC12 Cells. Mol. Cell. Biol., 22: 8332-8341.

Detrich III HW (2008). Fluorescent proteins in zebrafish cell and developmental biology. In Fluorescent Proteins (ed. KF. Sullivan), San Diego: Academic Press, 85: 220-237.

Di Giorgi F, Ahmed Z, Bastianutto C, Brini M, Jouaville S, Marsault R,Murgia M,Pinton P,Pozzan T,Rizzuto R (1999). Targeting GFP to organelles. In Green Fluorescent Proteins (eds K F Sullivan ,SA Kay), San Diego: Academic Press, 58: 75-85.

Dictenberg JB, Swanger SA, Antar LN, Singer RH, Bassell GJ (2008). A direct role for FMRP in activity-dependent dendritic mRNA transport links filopodial-spine morphogenesis to fragile X syndrome. Dev. Cell, 14: 926-939.

Didiot MC, Subramanian M, Flatter E Mandel JL, Moine H (2008). Cells lacking the fragile X mental retardation protein (FMRP) have normal RISC activity but exhibit altered stress granule assembly. Mol. Biol. Cell, 20: 428-437.

Dolzhanskaya N, Merz G, Aletta JM, Denman RB (2006a). Methylation regulates FMRP's intracellular protein-protein and protein-RNA interactions. J. Cell. Sci., 119: 1933-1946.

Dolzhanskaya N, Merz G, Denman RB (2006b). Oxidative stress reveals heterogeneity of FMRP granules in PC12 cell neurites. Brain Res., 1112: 56-64.

Encinas, JM, Enikolpov G (2008). Identifying and quantifying neural stem cell and progenitor cells in the adult brain. In Fluorescent Proteins (ed. KF Sullivan), San Diego: Academic Press, 85: 244-270.

Gay DA, Sisodia SS, Cleveland DW (1989). Autoregulatory control of beta-tubulin mRNA stability is linked to translation elongation. Proceedings of the National Academy of Sciences of the United States of America, 86: 5763-5767.

González-González E, López-Casas PP, del Mazo J (2008). The expression patterns of genes involved in the RNAi pathways are tissue-dependent and differ in the germ and somatic cells of mouse testis. Biochim. Biophys. Acta (BBA) - Gene Regulatory Mechanisms, 1779: 306-311.

Grollman AP, Huang MT (1976). Protein Synthesis. New York: Marcel Dekker.

Hengst U, Cox LJ, Macosko EZ, Jaffrey SR (2006). Functional and selective RNA interference in developing axons and growth cones. J. Neurosci., 26: 5727-5732.

Hou L, Antion MD, Hu D, Spencer CM, Paylor R, Klann E (2006). Dynamic translational and proteasomal regulation of fragile X mental retardation protein controls mGluR-dependent Long-Term Depression. Neuron, 51: 441-454.

Ishizuka A, Siomi MC, Siomi H (2002). A Drosophila fragile X protein interacts with components of RNAi and ribosomal proteins. Genes Dev., 16: 2497-508.

Iwasaki S, Kawamata T, Tomari Y (2009). Drosophila Argonaute1 and Argonaute2 employ distinct mechanisms for translational repression. Mol. Cell, 34: 58-67.

Kanai Y, Dohmae N, Hirokawa N (2004). Kinesin transports RNA: isolation and characterization of an RNA-transporting granule. Neuron, 43: 513-525.

Kedersha N, Anderson P, Jon L (2007). Mammalian stress granules and processing bodies. In Methods in Enzymology, Academic Press, 431: 61-81.

Kedersha N, Cho MR, Li W, Yacono PW, Chen S, Gilks N, Golan DE Anderson P (2000). Dynamic shuttling of TIA-1 accompanies the recruitment of mRNA to mammalian stress granules. J. Cell Biol., 151: 1257-1268.

Kedersha N, Stoecklin G, Ayodele M, Yacono P, Lykke-Andersen J Fitzler MJ, Scheuner D, Kaufman RJ, Golan DE Anderson P (2005). Stress granules and processing bodies are dynamically linked sites of mRNP remodeling. J Cell Biol., 169: 871-884.

Kedersha N, Tisdale S, Hickman T, Anderson P, Maquat LE, Kiledjian M (2008). Real-time and quantitative imaging of mammalian stress granules and processing bodies. In Methods in Enzymology. Academic Press, 448: 521-552.

Kwon S,Zhang Y, Matthias P (2007). The deacetylase HDAC6 is anovel critical component of stress granules involved in the stress response. Genes Dev., 21: 3381-3394.

Levenga J, Buijsen RAM, Rifé M, Moine H, Nelson DL, Oostra BA, Willemsen R, de Vrij FMS (2009). Ultrastructural analysis of the functional domains in FMRP using primary hippocampal mouse neurons. Neurobiol. Dis., 35:241-250.

Ling SC, Fahrner PS, Greenough WT, Gelfand VI (2004). Transport of Drosophila fragile X mental retardation protein-containing ribonucleoprotein granules by kinesin-1 and cytoplasmic dynein. Proc Nat'l Acad. Sci., 101: 17428.

Lippincott-Schwartz J, Presley F, Zaal KJM, Hirschberg K, Miller CD Ellenberg J(1999). Monitoring the dynamics of membrane proteins tagged with green fluorescent protein. In Green Fluorescent Proteins vol. 58 eds KF, Sullivan SA, Kay San Diego: Academic Press, pp. 261-280.

Mazroui R, Di Marco S, Kaufman RJ, Gallouzi IE (2007). Inhibition of the ubiquitin-proteasome system induces stress granule formation. Mol. Biol. Cell, 18: 2603-2618.

Mazroui R, Huot ME, Tremblay S, Filion C, Labelle Y, Khandjian EW. (2002). Trapping of messenger RNA by fragile X mental retardation protein into cytoplasmic granules induces translation repression. Hum. Mol. Genet., 11: 3007-3017.

Mazroui R, Sukarieh R, Bordeleau ME, Kaufman RJ, Northcote P, Tanaka J, Gallouzi I, Pelletier J (2006). Inhibition of ribosome recruitment induces stress granule formation independent of eIF2{alpha} phosphorylation. Mol. Biol. Cell, 17: 4212-4219.

McEwen E, Kedersha N, Song B, Scheuner D, Gilks N, Han A, Chen JJ, Anderson P, Kaufman RJ (2005). Heme-regulated inhibitor kinase-mediated phosphorylation of eukaryotic translation initiation factor 2 inhibits translation, induces stress granule formation, and mediates survival upon arsenite exposure. J. Biol. Chem., 280: 16925-16933.

Minagawa K, Katayama Y, Nishikawa S, Yamamoto K, Sada A, Okamura A, Shimoyama M, Matsui T (2009). Inhibition of G1 to S phase progression by a novel zinc finger protein p58TFL at P-bodies. Mol. Cancer Res., 7: 880-889.

Minagawa K, Matsui T (2010). p58TFL does not localize to messenger RNA processing bodies - Response. Mol. Cancer Res., 8: 132-133.

Özlü N, Srayko M, Kinoshita K, Habermann BO ,Toole ET, Müller-Reichert T, Schmalz N, Desai A, Hyman AA (2005). An essential function of the C. elegans ortholog of TPX2 is to localize activated Aurora A kinase to mitotic spindles. Dev. Cell, 9: 237-248.

Pfeiffer BE, Huber KM (2007). Fragile X mental retardation protein induces synapse loss through acute postsynaptic translational regulation. J. Neurosci., 27: 3120-3130.

Pierce D, Vale RD (1999). Single-molecule fluorescence detection of green fluorescent protein and application to single protein dynamics. In Green Fluorescent Proteins (eds KF,Sullivan SA, Kay, San Diego: Academic Press, 58: 49-72.

Qi HH, Ongusaha PP, Myllyharju J, Cheng D, Pakkanen O, Shi Y, Lee SW, Peng J, Shi Y (2008). Prolyl 4-hydroxylation regulates Argonaute 2 stability. Nature, 455: 421-424.

Querido E, Chartrand P (2008). Using fluorescent proteins to study mRNA trafficking in living cells. In Fluorescent Proteins (ed. KF Sullivan), San Diego: Academic Press, 85: 274-291.

Rackham O, Brown CE (2004). Visualization of RNA-protein interactions in living cells: FMRP and IMP1 interact on mRNAs. EMBO J., 23: 3346-3355.

Solomon S, Xu Y, Wang B, David MD, Schubert P, Kennedy D, Schrader JW (2007). Distinct structural features of caprin-1 mediate its interaction with G3BP-1 and its induction of phosphorylation of eukaryotic translation initiation factor 2{alpha}, entry to cytoplasmic stress granules, and selective interaction with a subset of mRNAs. Mol. Cell Biol., 27: 2324-2342.

Thomas MG, Martinez Tosar LJ, Loschi M, Pasquini JM, Correale J, Kindler S, Boccaccio GL (2004). Staufen recruitment into stress granules does not affect early mRNA transport in oligodendrocytes. Mol. Biol. Cell, 16: 405-420.

Wang H, Dictenberg JB, Ku L, Li W, Bassell GJ, Feng Y (2008). Dynamic association of the fragile X mental retardation protein as a messenger ribonucleoprotein between microtubules and polyribosomes. Mol. Biol. Cell, 19: 105-114.

Wasserman T, Katsenelson K, Daniliuc S, Hasin T, Choder M, Aronheim A(2009). A novel JNK binding protein WDR62 is recruited to stress granules and mediates a non-classical JNK activation. Mol. Biol. Cell, 21: 117-130.

Xie W, Dolzhanskaya N, LaFauci G, Dobkin C, Denman RB (2009). Tissue and developmental regulation of fragile X mental retardation protein exon 15 isoforms. Neurobiol. Dis., 35: 52-62.

Yamasaki S, Stoecklin G, Kedersha N, Simarro M, Anderson P (2007). T-cell intracellular antigen-1 (TIA-1)-induced translational silencing promotes the decay of selected mRNAs. J. Biol. Chem., 282: 30070-30077.

Yoon JH, Choi EJ, Parker R (2010). Dcp2 phosphorylation by Ste20 modulates stress granule assembly and mRNA decay in Saccharomyces cerevisiae. J. Cell Biol., 189: 813-827.

3

Crystal structure of the allosteric-defective chaperonin GroEL$_{E434K}$ mutant

Aintzane Cabo-Bilbao[1], Ariel E. Mechaly[2], Jon Agirre[1], Silvia Spinelli[3], Begoña Sot[4], Arturo Muga[1] and Diego M.A. Guérin[1]*

[1]Unidad de Biofísica (UBF, CSIC-UPV/EHU) and Departamento de Bioquímica y Biología Molecular, UPV/EHU. Barrio Sarriena s/n, E-48940, Leioa, Spain.
[2]Unité de Biochimie Structurale, Institut Pasteur, 25 rue du Dr. Roux, F-75724 Paris, France.
[3]AFMB-CNRS, UMR 6098, 163, Av. de Luminy, 13288 Marseille Cedex 09, France.
[4]Centro Nacional de Biotecnología, CSIC, Campus de la Universidad Autónoma de Madrid, Darwin, 3, 28049 Madrid, Spain.

The chaperonin GroEL adopts a double-ring structure with various modes of allosteric communication. The simultaneous positive intra-ring and negative inter-ring cooperativities allow alternating functionality of the folding cavities in both protein rings. Mutation of glutamic acid 434 (located at the ring interface), to lysine alters the negative inter-ring cooperativity. The crystal structure of the mutant chaperonin GroEL$_{E434K}$ has been determined at low-resolution (4.5 Å) and has been compared to the wild-type GroEL and the allosteric-defective GroEL$_{E461K}$ mutant structures. Despite the allosteric-defective behavior of the GroEL$_{E434K}$ mutant, its structure remains strikingly similar to that of the wild-type GroEL.

Key words: Chaperonin, GroEL, cooperativity, twinning, low-resolution refinement.

INTRODUCTION

The bacterial GroEL-GroES complex assists unfolded proteins in achieving their native conformation through an ATP-dependent cyclic reaction (Houry et al., 1999). The GroEL monomer is organized into three domains (Figure 1A): an apical domain (186 residues) that interacts with substrate proteins and its co-chaperone GroES, a flexible intermediate domain (89 residues), and an equatorial domain (243 residues) that contains the ATP binding site and is responsible for the inter-ring communication. The biological GroEL molecule is made of fourteen identical 57 kDa protomers that assemble into a stacked, double-ring structure (Figures 1B and 1C), whereas GroES is a heptamer of seven identical 10 kDa subunits arranged in a lid-like single ring (Hunt et al., 1996). The GroEL-GroES interactions and stoichiometry are dependent on

the presence of Mg^{2+}, KCl, and a nonhydrolyzable ATP analogue (Gorovits et al., 1997; Llorca et al., 1997). The co-chaperone GroES alternately binds to the GroEL ring that interacts with non-native protein substrates. This process leads to isolation of the substrate in a folding chamber in which it eventually refolds into its correct conformation (Shtilerman et al., 1999). GroEL function is allosterically regulated, and both positive intra-ring and negative inter-ring cooperativities have been observed (Horovitz et al., 2001). Negative inter-ring cooperativity is thought to be mediated by, among other factors, ionic interactions at the so called 'right site' (RS) and the 'left-site' (LS) of the inter-ring interface. These interactions involve residues Glu461 and Arg452 (Bartolucci et al., 2005; PDB 1XCK) at the RS and the putative salt bridge between Glu434 and Lys105 (Llorca et al., 1997; Sot et al., 2003) at the LS. Mutation of residues Glu461 and Glu434 to lysine affects the ability of GroEL to distinguish between physiological and stress temperatures (Sot et

*Corresponding author. E-mail: diego.guerin@ehu.es

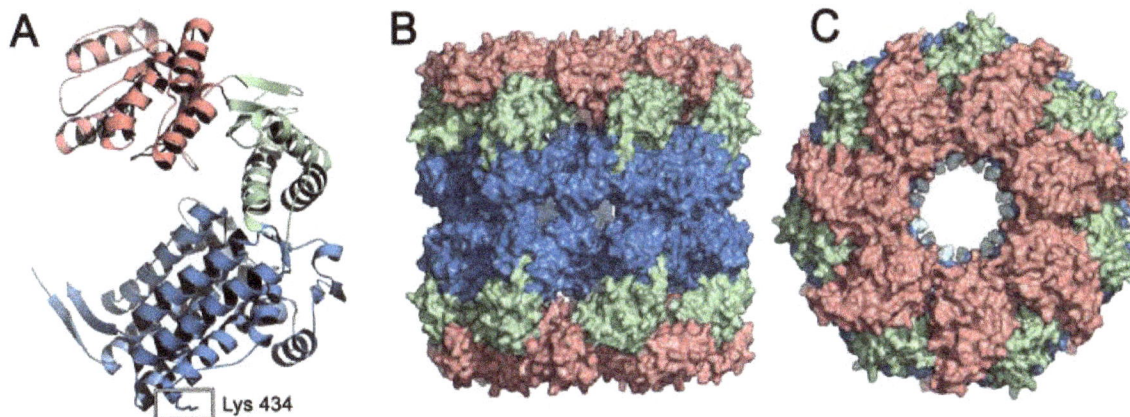

Figure 1. Overall structure. (A) Ribbon diagram of a GroEL$_{E434K}$ protomer, which consists of three functionally distinct domains: an apical domain (residues 189–377), an intermediate domain (residues 137–188 and 378–409) and an equatorial domain (residues 2–136 and 410–425). (B) Side view of the GroEL complex, consisting of two stacked homo-heptameric rings. (C) Each ring is formed by seven identical protomers. View along the 7-fold symmetry axis

al., 2002; 2003), and also decrease the chaperoning ATPase activity at 25°C (Sot et al., 2002). Up to date there is no structural information available on either of these two GroEL single mutants in complex with nucleotides. To better understand the role of the inter-ring interactions in the chaperonin inter-ring negative

MATERIALS AND METHODS

Protein expression and purification

Purification of GroEL$_{E434K}$ was carried out following a previously described protocol (Weissman et al., 1995). The GroEL mutant protein was produced in *E. coli* cells transformed with the plasmid pOF39, which causes overexpression of both GroEL and GroES in the bacteria (Fayet et al., 1986; 1989). Cells were grown in LB medium supplemented with chloramphenicol (25 µg/ml) at 37°C until A_{550} = 0.5. Overexpression was induced with 0.5 mg/ml arabinose for 4 h.

Cells were harvested by centrifugation and resuspended in 5 mM EDTA, 1 mM DTT, 1 mM PMSF, 50 mM Tris-HCl at pH 7.5, and 200 µg/ml lysozyme. After disrupting the cells by sonication, the lysate was centrifuged at 10,000 g for 1 h. The supernatant was precipitated with 70% (w/v) ammonium sulfate for 4 h at 4°C and centrifuged for 20 min at 15,000 g. The pellet was resuspended in 50 mM Tris-HCl at pH 7.5, 1 mM EDTA, 1 mM DTT, and 1 mM PMSF and then dialyzed against the same buffer. The sample was loaded onto an ion exchange FPLC column (Q-Sepharose High Performance 26/20) previously equilibrated with 1 mM EDTA, 1 mM DTT and 50 mM Tris-HCl at pH 7.5 and then washed at a rate of 3 ml/min with two column volumes of the equilibration buffer. The protein was eluted with five column volumes using a NaCl gradient from 0 to 1 M at a flow rate of 2 ml/min. Fractions containing GroEL$_{E434K}$ were concentrated by ultrafiltration in Centricon YM-100 filters (Millipore, Bedford M.A., U.S.A) up to 50 mg/ml and washed three times with 10 mM MgCl$_2$ and 50 mM Tris-acetate at pH 8.0 to eliminate most of the lower molecular weight contaminants. Other contaminants that co-purified bound to GroEL$_{E434K}$ were eliminated in a final step. The protein (2-10 mg/ml) was injected into a

cooperativity, we have studied two allosteric-defective mutants, GroEL$_{E461K}$ and GroEL$_{E434K}$. We have previously determined the crystal structure of GroEL$_{E461K}$ (Cabo-Bilbao et al., 2006) (PDB 2EU1). Here we report a 4.5 Å resolution structure of GroEL$_{E434K}$ from perfect hemihedral twinned crystals.

Reactive Red 120-Agarose (Sigma, St. Louis, MO, U.S.A.) column equilibrated with 5 mM MgCl$_2$ and 20 mM Tris-HCl at pH 7.5 and then eluted with 15 column volumes of 1.5 M NaCl, 0.02% NaN$_3$, and 20 mM Tris-HCl at pH 7.5 (Clark et al., 1998).

Crystallization, X-ray data collection and processing

Screening of crystallization conditions was carried out using a Mosquito nano-drop dispensing robot (TTP Labtech, UK). Micro-crystals grew at 4°C for two days in a drop equilibrated in a 50 µL reservoir containing the following precipitant solution: 1.2 M NaH$_2$PO$_4$, 0.8 M K$_2$HPO$_4$, 200 mM LiSO$_4$, and 100 mM CAPS at pH 10.5 (Appendix).

Larger crystals were grown by the hanging-drop vapor diffusion method at 4°C in the above-mentioned conditions, and slight modifications were made to either the pH, precipitant or protein concentration for improved growth in Linbro plates (Appendix). These crystals were dehydrated by incubating the drops containing the crystals in vapor diffusion equilibrium with increasing concentrations of Li$_2$SO$_4$. Finally, 4 M NaH$_2$PO$_4$ was employed as a cryoprotectant.

Diffraction data were collected at beamline ID14-1 at the European Synchrotron Radiation Facility (ESRF), France. Data were indexed and integrated using the Mosflm program (Leslie, 1992). Scaling and merging of the data were done with SCALA from the CCP4 suite (Collaborative Computational Project, 1994).

Structure determination and refinement

The GroEL$_{E434K}$ atomic structure was determined by molecular replacement using Phaser (McCoy et al., 2007), employing one ring (one heptamer) of the wild-type apo-chaperonin (PDB code: 1XCK)

Table 1. Data collection and refinement statistics. Values between parentheses are for the highest resolution shell. [1] $R_{merge} = \sum_{hkl} \sum_i | I_i(hkl) - <I_i(hkl)> | / \sum_{hkl} \sum_i I_i(hkl)$, being $I_i(hkl)$ the intensity of the hkl reflection and i the number of measurements of that reflection. [2] $R_{work} = \sum_{hkl} || F_{obs} - F_{calc} || / \sum_{hkl} | F_{obs} |$, being F_{obs} and F_{calc} the observed and calculated structure factors respectively. [3] R_{free} = R-value calculated for the 5% of reflections not used in the refinement process.

Data collection	
Unit-cell parameters (Å)	a=b= 172.0 c= 454.6
Space group	P3$_2$
Resolution range (Å)	20.0-4.5 (4.74-4.50)
No. of observations	252,352 (37,848)
No. of number unique observations	87,597 (13,001)
Mean I/σ(I)	6.0 (2.3)
Completeness	98.2 (99.7)
Multiplicity	2.9 (2.9)
R_{merge}[1]	0.150 (0.472)
Refinement	
R_{work}[2]/R_{free}[3]	0.17/0.24
No. of protein atoms	53,984
Average B factor	174.112
R.m.s.d. for bond lengths (Å)	0.03
R.m.s.d. for bond angles (°)	2.3
Ramachandran favored, residues in (%)	95.9
Ramachandran outliers, residues in (%)	1.2
Rotamers outliers, residues in (%)	1.9

as a search model. A clear solution containing two heptameric rings per asymmetric unit (log-likelihood gain = 7923.440) was found in space group P3$_2$. However, initial refinement attempts with the program phenix. refine (Afonine et al., 2005) failed, suggesting the presence of twinned data. Twinning was confirmed by analyzing the diffraction data with phenix. xtriage (Adams et al., 2010), as L and N(z) tests showed that the intensities from our dataset clearly deviated from the expected values for untwinned data. Application of the twinning law "h,-h-k,-l" allowed the refinement to proceed to completion. Refinement included TLS groups and restraints for NCS (14-fold) secondary structure, Ramachandran angles and the reference model (PDB code: 1XCK). Minor manual modifications to the model between refinement rounds were done using the program COOT (Emsley and Cowtan, 2004; Emsley et al., 2010). The final model of the GroEL$_{E434K}$ mutant was refined until a crystallographic R_{work} and R_{free} of 0.17 and 0.24 were reached, respectively (Table 1). Figures were prepared using the program PyMOL (DeLano, 2002). Atomic coordinates and structure factors of GroEL$_{E434K}$ have been deposited in the RCSB Protein Data Bank with the accession code 2YEY.

RESULTS AND DISCUSSION

Overall structure

Here we report the X-ray structure of the GroEL$_{E434K}$ allosteric-defective mutant (Sot et al., 2003) determined at 4.5 Å resolution from hemihedral twinned crystals. Despite the limited resolution of the data, modern refinement software (Brunger et al., 2009; Schroder et al.,

2010; Read, 2010) allows to determine the atomic structure accurately. In order to obtain a reliable model in terms of both secondary structure and crystallographic R factors, we used several restraints in the phenix. refine program. These restraints were as follows:

(i) 14-fold NCS,
(ii) The wild-type GroEL structure (PDB code 1XCK) as a reference model,
(iii) Ramachandran restraints.

The GroEL$_{E434K}$ crystal structure does not show any major structural differences when compared to the wild-type apo structure (Bartolucci et al., 2005) (PDB code 1XCK). When these structures are superimposed by their Cα, backbones, the r.m.s.d is approximately 0.5 Å. It is noteworthy that the GroEL$_{E434K}$ structure was determined in an unrelated crystalline environment as compared with the wild-type structure. Consequently, we conclude that crystal packing does not force the wild-type-like relative orientation of the rings.

Inter-ring interface

In our previously reported crystal structure of the allosteric-defective variant GroEL$_{E461K}$ (PDB code 2EU1), we observed a rotation of 22° between rings relative to

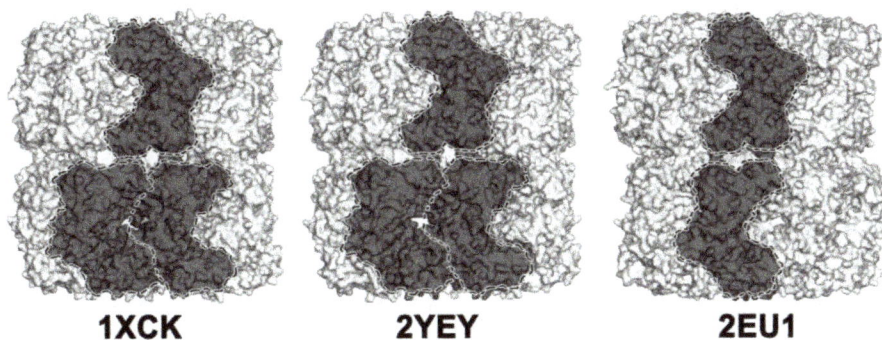

Figure 2. Comparison between wild-type GroEL, GroEL$_{E434K}$ and GroEL$_{E461K}$ mutants. In both wild-type GroEL (PDB 1XCK) and GroEL$_{E434K}$ (PDB 2YEY), each protomer contacts two protomers at the opposing ring (1:2 contact). In contrast, in the GroEL$_{E461K}$ mutant (PDB 2EU1), the upper ring is rotated 22° about the 7-fold axis, and consequently each protomer contacts only one protomer at the opposite ring (1:1 contact).

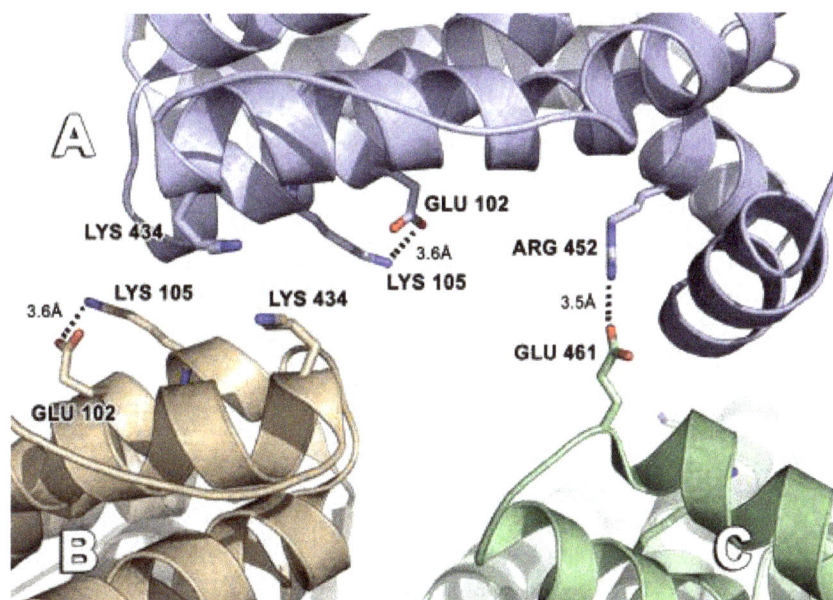

Figure 3. Ring interface of the GroEL$_{E434K}$ mutant. The top monomer is labeled 'A', while left and right contact sites are labeled 'B' and 'C,' respectively. There is a salt bridge between arginine 452 of the top monomer (A) and glutamic acid 461 of the right bottom monomer (C).

the wild-type GroEL structure (Bartolucci et al., 2005). In the structure of that mutant, we observed that the interface was not stabilized by any inter-ring salt bridges and that the inter-ring distance was slightly larger than that of the wild-type GroEL (Cabo-Bilbao et al., 2006). Despite also having its inter-ring communication altered, the GroEL$_{E434K}$ mutant maintains the same interface (1:2 contact) observed in the wild-type chaperonin (Figure 2), in contrast to the GroEL$_{E461K}$ mutant.

The replacement of glutamic acid 434 with lysine modifies the surface charge distribution of the inter-ring interface. A strong bend of lysines from both the top and bottom monomers is observed (Figure 3), which most likely occurs as a consequence of electrostatic repulsions between the positively charged side-chains on residues 434 on each monomer. Nevertheless, this repulsive effect is not strong enough to disrupt the salt bridge formed between Arg452 and Glu461 at the RS (Figure 3), which is the major inter-ring interaction in the wild-type GroEL structure (Bartolucci et al., 2005). Consequently, this salt bridge might be essential for the conservation of the 1:2 contact between opposing rings in the GroEL$_{E434K}$ variant.

Our results are consistent with the previously determined structure of the double mutant GroEL$_{D398A-E434A}$ (Wang and Boisvert, 2003) (PDB codes 1KP8 and 1J4Z), in which no major changes in the inter-ring interface were observed. It is interesting to note that while the functional effects of the E461K and E434K mutations are similar, i.e., they switch from a foldase to a holdase activity at 32 and 29°C, respectively (Sot et al., 2003), their respective inter-ring interfaces differ significantly. Thus, different molecular mechanisms might disrupt inter-ring allosteric signaling in these GroEL variants to render them inactive as foldases at physiological temperatures.

Conclusion

Although the low resolution of the data prevents a deeper structural analysis, we can conclude that, in agreement with previously determined GroEL structures containing mutations in residue E434, the E434K mutation does not seem to cause any obvious structural change with respect to the wild-type GroEL structure. From this study, we can conclude that in contrast to what we observed from the E461K mutant (Cabo-Bilbao et al., 2006), the communication pathway through the LS can be interrupted without inducing major changes at the inter-ring interface. Consequently, these results suggest that the communication pathway through this contact site most likely occurs by small structural modifications rather than large conformational changes.

ACKNOWLEDGMENTS

We acknowledge the European Synchrotron Radiation Facility (ESRF) for providing access to X-ray sources and the ID14-1 staff for support and assistance in data collection. A.C-B. and J.A. received fellowships from the Basque Government, and A.E.M. received a fellowship from the MEyC, Spain. D.M.A.G. is researcher from the Fundación Biofísica Bizkaia who received partial support from Bizkaia::Xede, Bizkaia, Basque Country, Spain. Part of this project was funded by ETORTEK, FEDER, MEC (BFU2010-15443), and the Basque Government (IT-358-07).

REFERENCES

Adams PD, Afonine PV, Bunkoczi G, Chen VB, Davis IW, Echols N, Headd JJ, Hung LW, Kapral GJ, Grosse-Kunstleve RW, McCoy AJ, Moriarty NW, Oeffner R, Read RJ, Richardson DC, Richardson JS, Terwilliger TC, Zwart PH (2010). PHENIX: a comprehensive Python-based system for macromolecular structure solution. Acta Crystallogr. D. Biol. Crystallogr., 66:213-221

Afonine PV, Grosse-Kunstleve RW, Adams PD (2005). A robust bulk-solvent correction and anisotropic scaling procedure. Acta.Crystallogr. D. Biol. Crystallogr., 61:850-855

Bartolucci C, Lamba D, Grazulis S, Manakova E, Heumann H (2005). Crystal structure of wild-type chaperonin GroEL. J. Mol. Biol., 354:940-951

Brunger AT, DeLaBarre B, Davies JM, Weis WI (2009). X-ray structure determination at low resolution. Acta. Crystallogr. D. Biol. Crystallogr., 65:128-133

Cabo-Bilbao A, Spinelli S, Sot B, Agirre J, Mechaly AE, Muga A, Guerin DM (2006). Crystal structure of the temperature-sensitive and allosteric-defective chaperonin GroELE461K. J. Struct. Biol., 155:482-492

Clark AC, Ramanathan R, Frieden C (1998). Purification of GroEL with low fluorescence background. Methods Enzymol., 290:100-118

Collaborative Computational Project N (1994). The CCP4 suite: programs for protein crystallography. Acta. Crystallogr. D. Biol. Crystallogr., 50:760-763

DeLano WL (2002). The PYMOL Molecular Graphics System.

Emsley P, Cowtan K (2004). Coot: model-building tools for molecular graphics. Acta. Crystallogr. D. Biol. Crystallogr., 60:2126-2132

Emsley P, Lohkamp B, Scott WG, Cowtan K (2010). Features and development of Coot. Acta. Crystallogr. D.Biol. Crystallogr., 66:486-501

Fayet O, Louarn JM, Georgopoulos C (1986). Suppression of the Escherichia coli dnaA46 mutation by amplification of the groES and groEL genes. Mol. Gen. Genet., 202:435-445

Fayet O, Ziegelhoffer T, Georgopoulos C (1989). The groES and groEL heat shock gene products of Escherichia coli are essential for bacterial growth at all temperatures. J. Bacteriol., 171:1379-1385

Gorovits BM, Ybarra J, Seale JW, Horowitz PM (1997). Conditions for nucleotide-dependent GroES-GroEL interactions. GroEL14(groES7)2 is favored by an asymmetric distribution of nucleotides. J. Biol. Chem., 272:26999-27004

Horovitz A, Fridmann Y, Kafri G, Yifrach O (2001). Review: allostery in chaperonins. J. Struct. Biol., 135:104-114

Houry WA, Frishman D, Eckerskorn C, Lottspeich F, Hartl FU (1999). Identification of in vivo substrates of the chaperonin GroEL. Nature 402:147-154

Hunt JF, Weaver AJ, Landry SJ, Gierasch L, Deisenhofer J (1996). The crystal structure of the GroES co-chaperonin at 2.8 A resolution. Nature, 379:37-45

Leslie AGW (1992). Recent changes to the MOSFLM package for processing film and image plate data. Joint CCP4 + ESF-EAMCB Newsletter on Protein Crystallography, 26:7

Llorca O, Perez-Perez J, Carrascosa JL, Galan A, Muga A, Valpuesta JM (1997). Effects of the inter-ring communication in GroEL structural and functional asymmetry. J. Biol. Chem., 272:32925-32932

McCoy AJ, Grosse-Kunstleve RW, Adams PD, Winn MD, Storoni LC, Read RJ (2007). Phaser crystallographic software. J. Appl. Crystallogr., 40:658-674

Read RJ (2010). From poor resolution to rich insight. Structure, 18:664-665

Schroder GF, Levitt M, Brunger AT (2010). Super-resolution biomolecular crystallography with low-resolution data. Nature, 464:1218-1222

Shtilerman M, Lorimer GH, Englander SW (1999). Chaperonin function: folding by forced unfolding. Science, 284:822-825

Sot B, Galan A, Valpuesta JM, Bertrand S, Muga A (2002). Salt bridges at the inter-ring interface regulate the thermostat of GroEL. J. Biol. Chem., 277:34024-34029

Sot B, Banuelos S, Valpuesta JM, Muga A (2003). GroEL stability and function. Contribution of the ionic interactions at the inter-ring contact sites. J. Biol. Chem., 278:32083-32090

Wang J, Boisvert DC (2003). Structural basis for GroEL-assisted protein folding from the crystal structure of (GroEL-KMgATP)14 at 2.0A resolution. J. Mol. Biol., 327:843-855

Weissman JS, Hohl CM, Kovalenko O, Kashi Y, Chen S, Braig K, Saibil HR, Fenton WA, Horwich AL (1995). Mechanism of GroEL action: productive release of polypeptide from a sequestered position under GroES. Cell, 83:577-587.

Appendix

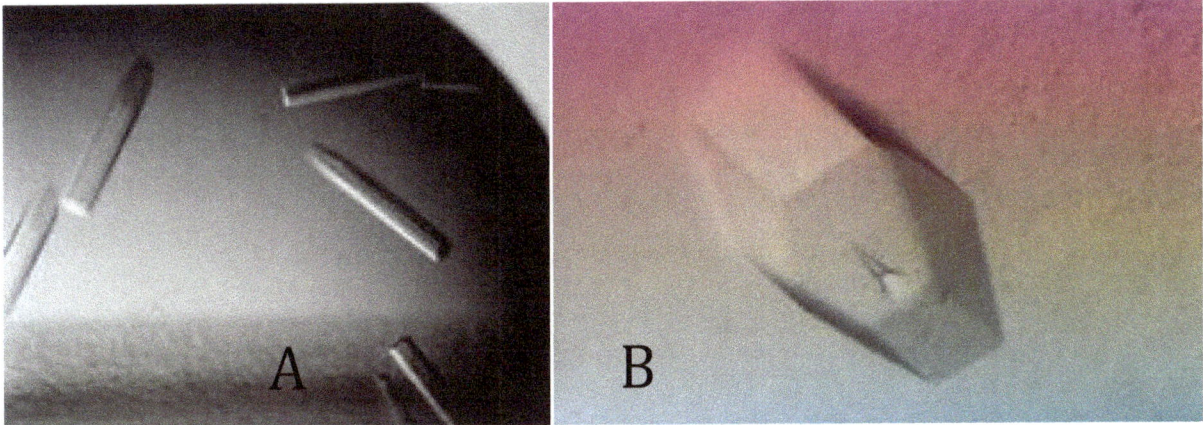

(A) Photograph of initial GroEL$_{E434K}$ crystals. Crystals were grown in a half height Greiner Crystal Ledge Plate. The longest crystal size was approximately 0.01 mm. (B) Photograph of an improved **GroEL$_{E434K}$ crystal.** Crystals were grown in a Linbro plate employing the hanging drop vapor diffusion method. Crystal size is 0.05 mm.

The salicylic acid effect on the tomato (*Lycopersicum esculentum* Mill.) sugar, protein and proline contents under salinity stress (NaCl)

Shahba, Zahra[1,2], Baghizadeh, Amin[2]*, Vakili Seid Mohamad Ali[3], Yazdanpanah Ali[2] and Yosefi Mehdi[1]

[1]Payame noor Najafabad University, Najafabad-Isfahan, Iran.
[2]International Center for Science, High Technology and Environmental Sciences, Kerman, Iran.
[3]Islamic Azad University - Jiroft Branch, Iran.

Plants growth is impressed by biotic and abiotic stress inversely. There are many reports about proteins change level in salinity stress. Leaves fill up more soluble sugar of glucose, fructose and proline with treatment of salicylic acid. In this research, tomato seeds planted in pots containing perlite were put in a growth chamber under controlled conditions of 27 ± 2 and 23 ± 2°C temperature, 16 h lightness and 8 h darkness, 15 Klux light intensity and 75% humidity; NaCl concentration of 0, 25, 50, 75 and 100 mM and salicylic acid concentration of 0, 0.5, 1 and 1.5 mM were used in the form of factorial experiment in a complete randomized design (CRD). Salinity increases the soluble sugar in leaf and root tissues, and salicylic acid decreases it. The leaf protein level decreased because of salinity effect, but salicylic acid could increase it. In the root, salinity increases protein, but salicylic acid with 1.5 mM concentration decreases it. Salinity increases the proline level in leaf and root, and salicylic acid did not significantly change in low salinity levels.

Key words: Salinity, stress, salicylic acid, tomato, proline, protein and sugar.

INTRODUCTION

Plants' growth and production are affected by natural stresses in the form of biotic and abiotic stresses, inversely. The abiotic stress causes loss of hundred million dollars annually, because of reduction and loss of products (Mahajan and Tuteja, 2005). Salinity is the most important limiting factor for crop production and it is becoming an increasingly severe problem in many regions of the world.

Plant's behaviorial response to salinity is complex, and different mechanisms are adopted by plants when they encounter salinity. The soil and water engineering methods increase farm production in the damaged soil by salinity, but achievement of higher purposes by these methods seems to be very difficult (Yokoi et al., 2002). The high salinity of the soil affected the soil penetration, decreased the soil water potential and finally caused physiological drought (Yusuf et al., 2007).The plants under salinity condition change their metabolism to overcome the changed environmental condition. One mechanisms utilized by the plants for overcoming the salt stress effects might be via accumulation of compatible osmolytes, such as proline and soluble sugar. Production and accumulation of free amino acids, especially proline by plant tissue during drought, salt and water stress is an adaptive response. Proline has been proposed to act as a compatible solute that adjusts the osmotic potential in the cytoplasm. Thus, proline can be used as a metabolic marker in relation to stress. The tomato is a major vegetable crop that has achieved tremendous popularity over the last century. It is grown in practically every country of the world - in outdoor fields, greenhouses and net houses.

Tomatoes, aside from being tasty, are very healthy as they are a good source of vitamins A and C. Compared with other plant, much less is known about the mechanism behind tomato plant response to salt stress.

*Corresponding author. E-mail: amin_4156@yahoo.com.

Salicylic acid is a plant phenol, and today it is in use as internal regulator hormone, because its role in the defensive mechanism against biotic and abiotic stresses has been confirmed. This research studies the salinity and salicylic acid effects on sugar, protein and proline contents of tomato which is sensitive to salinity.

MATERIALS AND METHODS

Planting

At first, the seeds were disinfected with hypocholorid sodium, and then 5 seeds were planted in each pot containing perlite and kept in a growth chamber under controlled conditions of $27 \pm 2°C$, $23 \pm 2°C$ temperatures, 16 h lightness, 8 h darkness, respectively, 15 Klux light intensity and 75% humidity. Then the pots were irrigated with deionized water and nutrient solution, with 6.5 pH every two day for one month. NaCl factor at 5 levels including 0, 25, 50, 75, 100 mM and salicylic acid treatment at 4 levels including 0, 0.5, 1 and 1.5 mM were used. The experiment was performed as factorial in the form of completely random plan (CRD Design) with 3 repetitions (60 pots). Salicylic acid treatments were sprayed on the leaves every two days for two weeks, and then different levels of salinity factor were used every two days with nutrient solution for 15 days.

The measurement of sugar content based on somogy 1952 method

0.05 g of fresh tissue of leaf and root was weighted by laboratory subtle scale (satrius) BP211D model with 0.0001 g accuracy. Each sample was grinded with 10 ml deionized water in a china mortar, then the mortar content was transferred to small container and located on a heater to boil. After that, the container contents were filtered by watman filter paper (number 1), for plant extraction. 2 ml of each extractionwas transferred to a test tube and 2 ml copper sulfate solution was added to each of the tube. Then, the tube caps were closed with cotton. Each of these tubes was kept in warm water bath with 100 °C temperature. In this term, CU^{2+} was reduced to CU_2O by monosacarido aldehid; here a brick red color was observed in the bottom of the test tube.

After cooling the pipes, 2 ml phosphomolibdic acid solution was added to them; after a moment, blue color appeared, and the test pipe was well shaken to spread the color within the test pipe. The solution absorption was in 600 nm, determined by spectrophotometer system, and then the sugar concentration was measured by using of standard curve. For spectrophotometer setting, a solution instead of plant extraction, which includes deionized water and the rest solution with sugar values, was measured and presented by using of relevant standard curve based on mg/g fw.

The measurement of protein concentration based on Bradford 1976 method

For protein extraction of root and leaf , one gram of each fresh tissue (leaf and root) was grinded in a chain mortar; it included 5 ml buffer Tris - HCl 0.05 M with pH=7.5. The obtained computable solution was transferred to centrifuge pipe and then, the samples were centrifuged by a refrigerator centrifuge for 25 min in 10000 g and 4°C. The obtained extraction was used for the measurement of protein solution concentration. Also, 0.1 ml protein extraction and 5 ml biore reagent were added to the test pipe, and vortexes quickly. After 25 min, their absorption was read by spectrophotometer system in 595 nm. The protein value was measured and presented by using of relevant standards curve based on mg/g fw.

Proline measurement method

0.02 g of root and fresh leaf tissue was grinded with 10 ml, 3% sulfosalicylic acid solution; the obtained extraction was centrifuged by using centrifuge napco 2028R model for 5 min in 10000 g. Then 2 ml of upper liquid was mixed with 2 mg ninhydrin reagent and 2 ml pure acetic acid; they were kept in hot water bath at 100°C for 1 h. After that for stopping all reactions, the pipes were cooled in ice bath, and then 4 ml Tollen's reagent was added, with the pipes well shaken. Separated layers were formed by fixing the pipes for 15 - 20 s. For measurement of proline concentration, the upper color layer of Tollens' reagent and proline were used. The absorption of some specific color material was determined through 520 nm, and the proline of each sample was obtained by using standard curve, based on mg/g fw.

Statistical analysis

In this study, the total number of experiments was done in different stages in completely randomized design with 3 repetitions; and the htest considers the reciprocal effect of Salicylic acid and salinity on different parameters as factorial. The levels of 0, 25, 50, 75, 100 mM of salinity were used and the levels of Salicylic acid were 0, 0.5, 1, 1.5 mM. The comparison of means was done with Duncan test to SPSS 12.0 software in probability level of 1%. For drawing graph, we used Excel 2003 software.

RESULTS

Leaf sugar

By high salinity concentrations, the level of sugar in leaf increases. Salicylic acid decreases the sugar at 0 and 25 mM salinity, which is related to 1 mM salicylic acid concentration. The highest reduction in the level of sugar in leaf in 50, 75, 100 mM salinity is related to 0.5 mM salicylic acid concentration. Salicylic acid balances in leaf sugar level by salinity control that decreases the sugar level due to salinity stress indirectly. The highest increasing sugar level at salinity is 100 mM and 0 mM salicylic acid concentration. The least sugar level is observed at 0 mM salinity and 1 mM salicylic acid (Figure 1).

Root sugar

According to Figure 2, at 0, 25 and 50 mM salinity level, salicylic acid decreases the sugar and at 75 and 100 mM NaCl concentrations, the sugar level increases at different salicylic acid concentrations. The balanced role of salicylic acid through salinity control is also observed at root sugar level, and the highest root sugar level is in 100 mM salinity and 0.5 mM salicylic acid concentration.

Leaf protein

According to Figure 3, salinity decreases the leaf protein

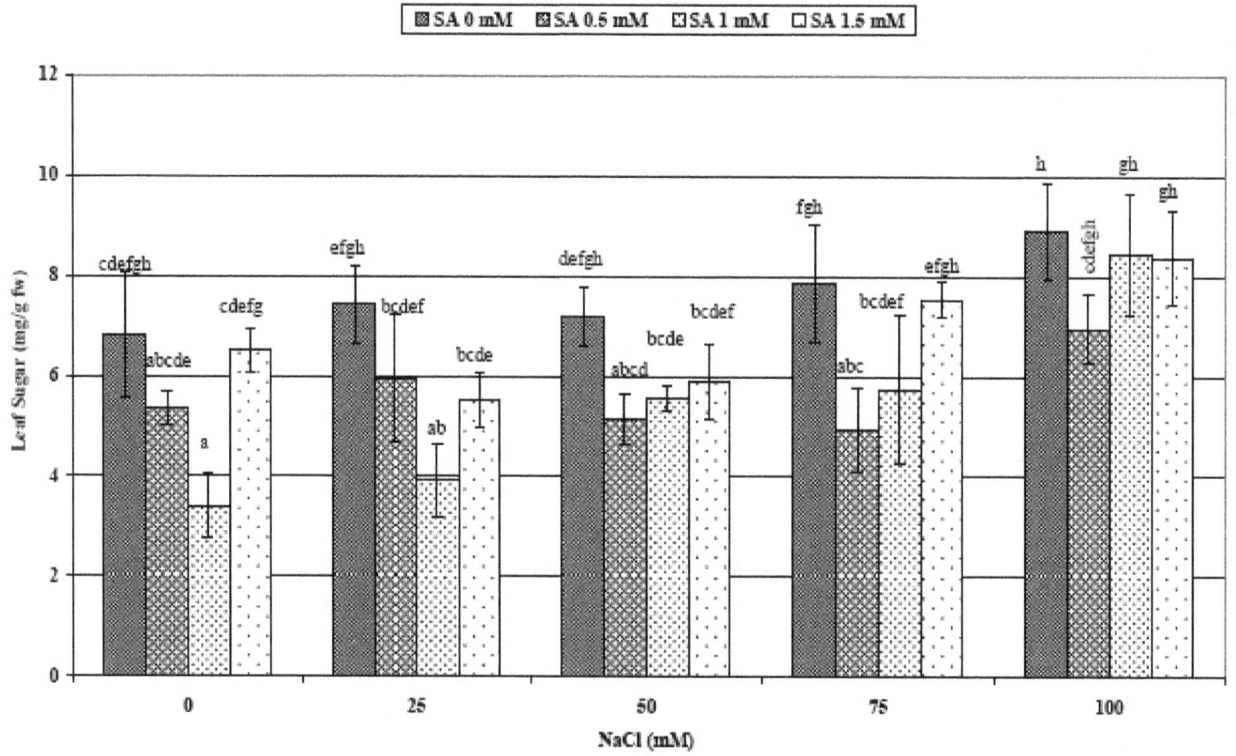

Figure 1. The effect of NaCl salinity concentrations and salicylic acid on the leaf sugar content.

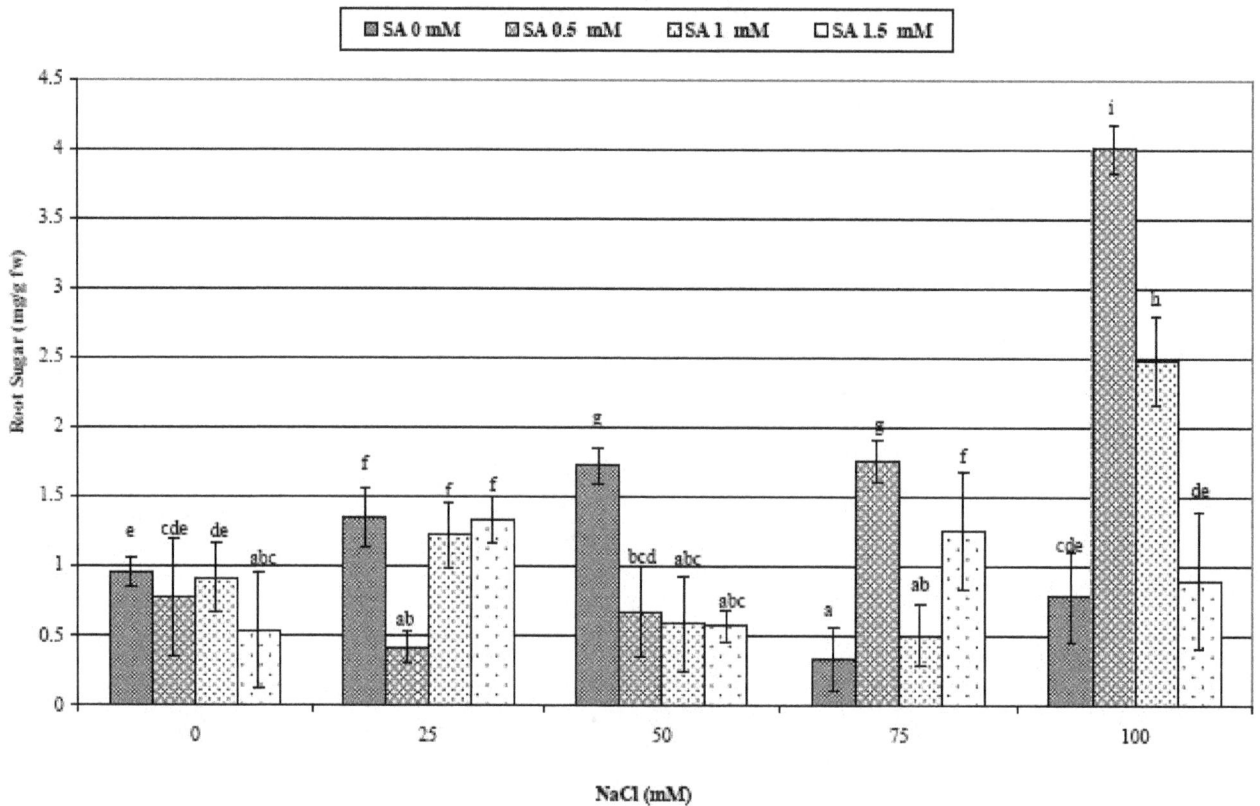

Figure 2. The effect of NaCl salinity concentrations and salicylic acid (SA) on root sugar content.

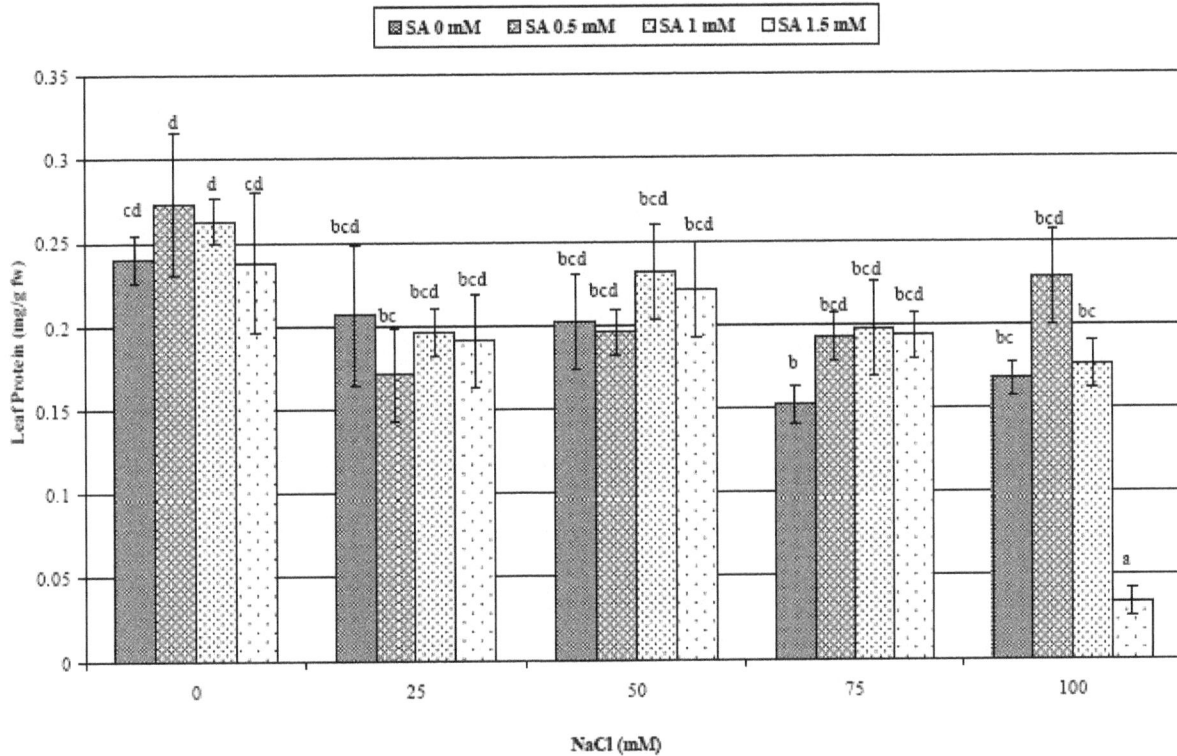

Figure 3. The effect of NaCl salinity concentrations and salicylic acid on leaf protein content.

level, which is more reduction at salinity high concentration (75 and 100 mM). Salicylic acid increases protein concentration by 50, 75 mM salinity levels. The least protein level is observed at 100 mM salinity level with 1.5 mM salicylic acid concentration, and also the highest level of it could be observed at 0 mM salinity level with 0.5 mM salicylic acid concentration.

It seems that there is negative interaction at high salinity concentration (100 mM) and salicylic acid treatment with high concentration (1.5 mM).

Root protein

All salinity levels except 100 mM level increases the protein. It seems that high defensive metabolism salinity makes trouble, but salicylic acid causes balance and decreases the protein level. The salicylic acid role could be observed at the balancing of high salinity concentration (Figure 4).

Leaf proline

Figure 5 shows that with increasing salinity level, proline increases, but decreases proline level by high salinity level at 75 and 100 mM, with use of salicylic acid treatment.

Root proline

Figure 6 shows that with increasing salinity, the root proline increases. But, there was no significant change atlow salinity level (0 - 25 mM), with use of salicylic acid treatment. With increasing salicylic acid concentration at 50, 70 and 100 mM salinity level, the proline decreases.

DISCUSSION

Sugar

The soluble sugar in oat organ plant (root and bud) increased with NaCl increasing (El-Tayeb, 2005). With salicylic acid, the leaves fill up more soluble sugar and proline (Szepesi, 2006). The increasing of photosynthesis carbohydrate is a signal for water deficiency tolerance. The high carbohydrate concentration with its role to reduce water potential helps to prevent oxidative losses and protein structure maintenance during water shortage. Also carbohydrates play a molecule role for sugar responsible genes that give different physiological response like defensive response and cellular expansion (Koch, 1996).

In one study on *prosopis alba*, salt stress increases soluble carbohydrate in the roots (Meloni et al., 2004).

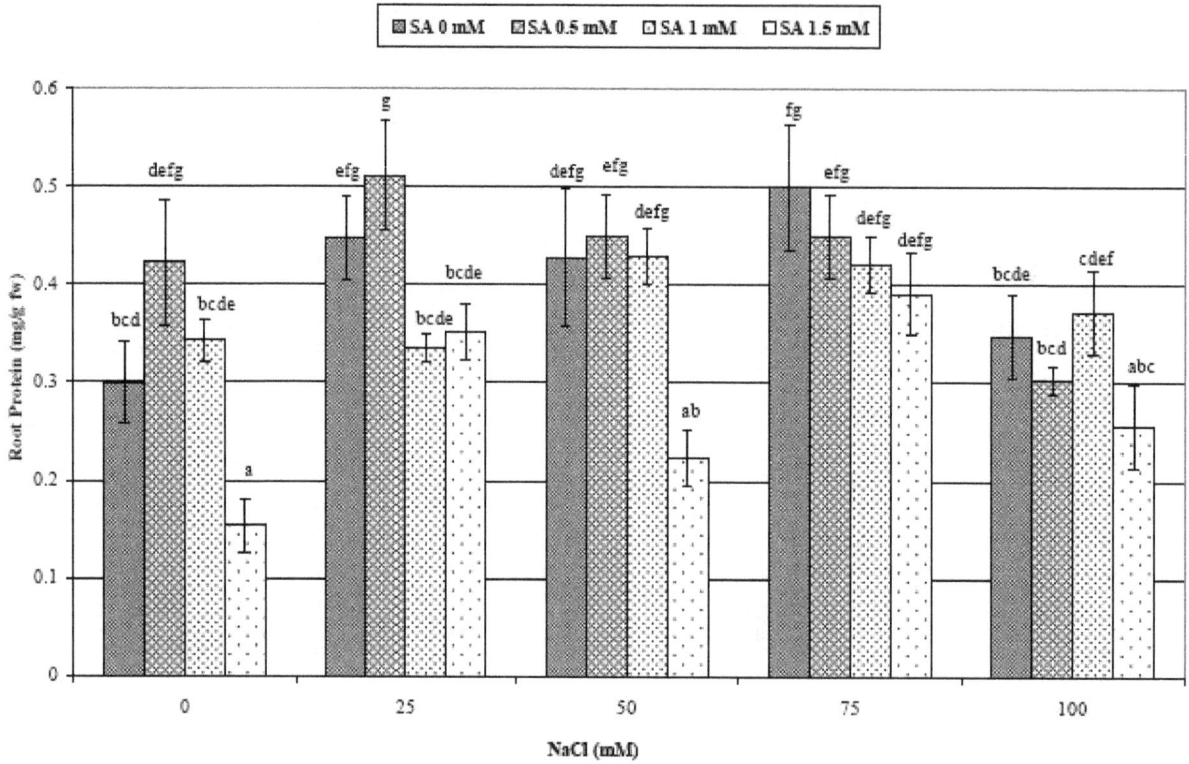

Figure 4. The effect of NaCl salinity concentrations and salicylic acid (SA) on the root protein content.

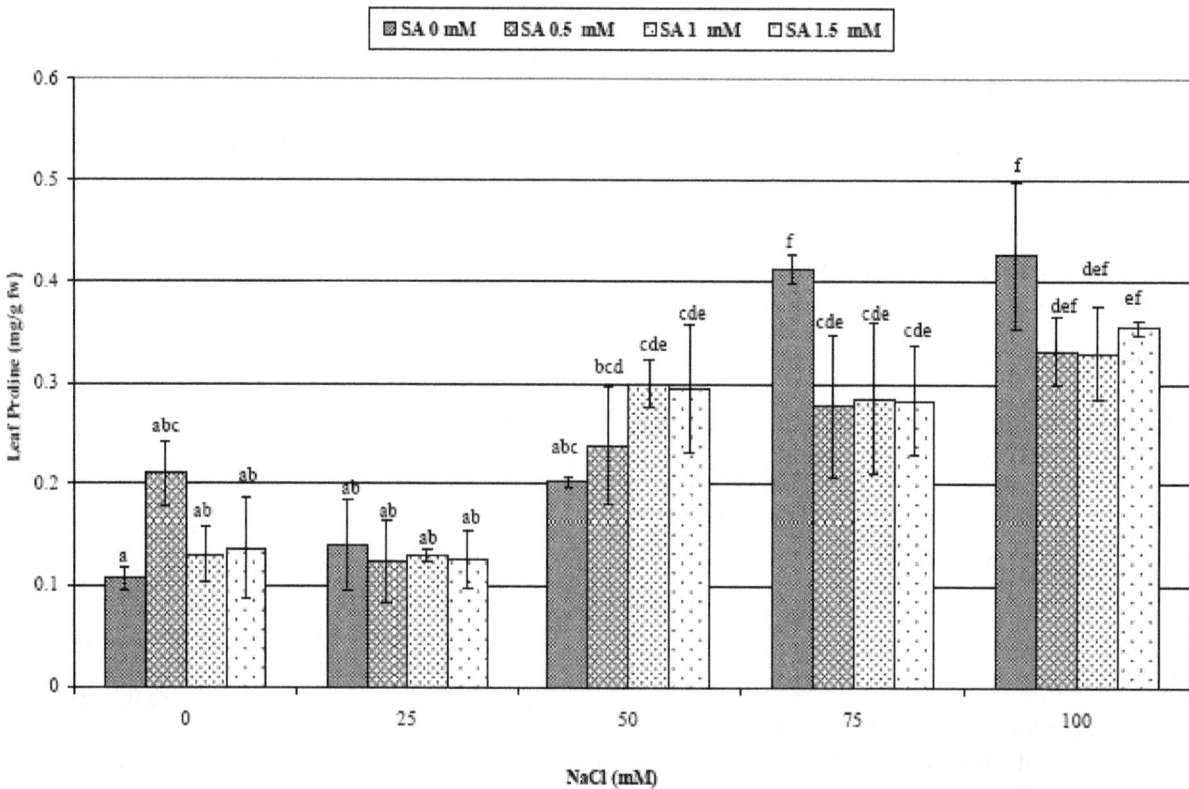

Figure 5. The effect of NaCl salinity and salicylic acid concentrations on leaf proline content.

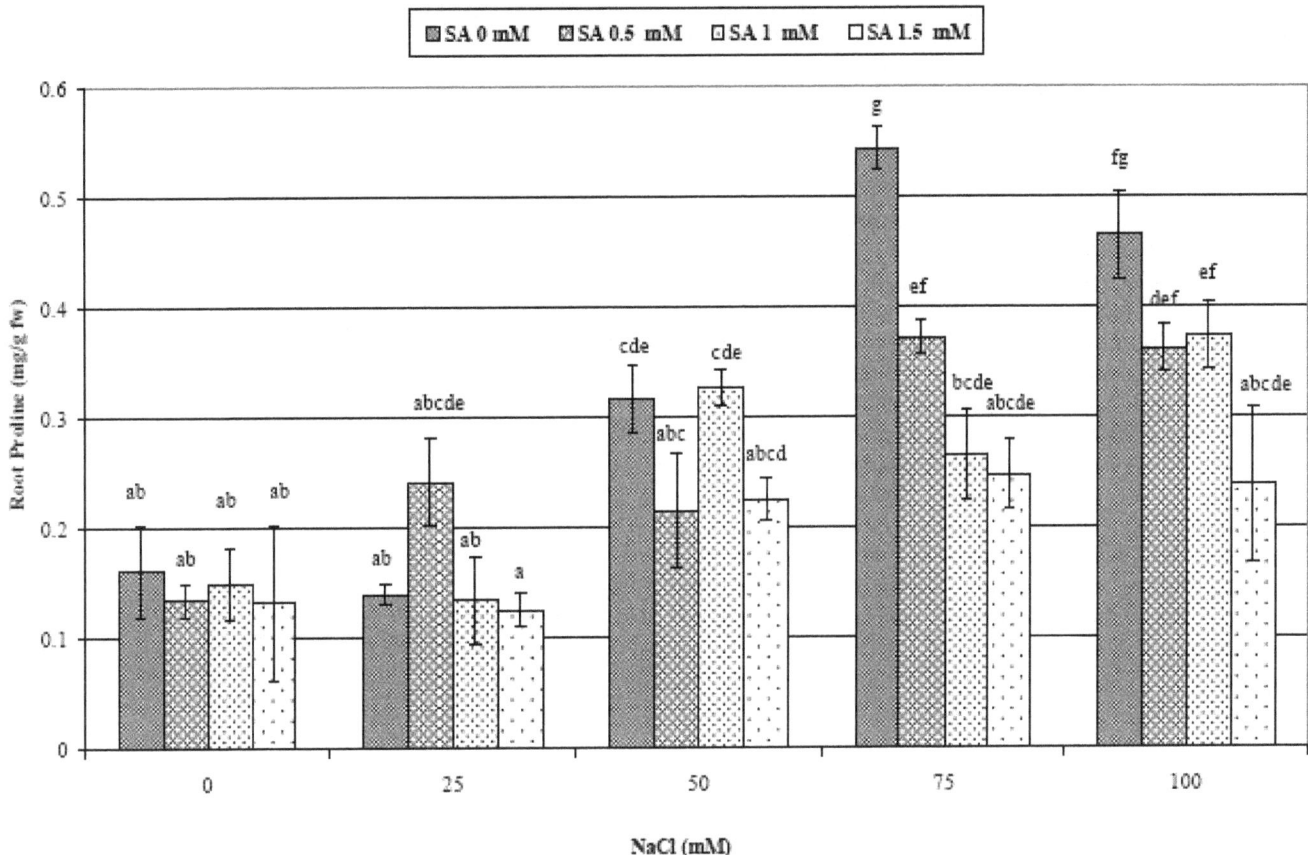

Figure 6. The effect of NaCl salinity and salicylic concentrations on root proline content.

In this research, salt stress increases leaf sugar, but salicylic acid decreases it. The same result was obtained in the root. Salicylic acid also causes balance in the sugar level at salinity stress condition. The increasing of induced glucose storage by salt stress is possible, that is for storage demand reduction of carbon or starch decomposition (Tattini, 1990). The increasing of all soluble carbohydrate in the root during salinity stress is effective on the balance against osmotic pressure. The plant cell for escaping from plasmolysis performance and creation during salt stress conditions should be changed and analyzed from macro molecule to micro molecule. Sucrose breaks down to glucose and fructose, and starch decomposition to glucose increases its osmotic pressure cell (Benbella, 1999). The use of salicylic acid could activate the consumption of soluble sugar metabolism by increasing osmotic pressure. It is supposed that salicylic acid treatment deranges the enzymatic system of polysacarid hydrolysis (Khodary, 2004). The salt stress in soybean varieties decreased sugar level (EL-Samad and Shaddad, 1997). In the accumulation of sugar at stress conditions, a protective mechanism enters the cell via sodium entry. So, some more of this kind of carbohydrate in cell area increases their membrane tolerance and

selectivity versus ion entry like sodium and chloride (Fernando et al., 2000).

Protein

There are many reports about increasing and decreasing of protein level in salinity stress. The soluble protein and free amino acid in barley organs (root and bud) increased with NaCl increasing. The study of maize plant and also all amino acids increased with salicylic acid (El Tayeb 2005). The increasing of amino acid in the plant tissue under stress is related to protein fraction (Hussein et al., 2007). In this research, it is written that the leaf protein level decreased by salt stress but salicylic acid could increase it. The cause of protein reduction at salinity condition is the prevention of nitrate reductase activity (Undovenko, 1971). The salt stress induced some changes on the protein of rice leaf shoots and roots, but not effective on leaf blade. The level of some protein decreases because of protein synthesis reduction (Kong-Ngern et al., 2005). Under high water stress, some plants produce materials with low molecular weight such as amino acid and polyamines, which reduce water potential

(Dantas et al., 2005). The plants produce some proteins in response to biotic and abiotic stresses, that some of these proteins deduct by phytohormones such as salicylic acid (Hussien et al., 2007). The increasing of nitrate reductase activity by salicylic acid depends on the material action with special inhibitors of nitrate (Ahmad et al., 2003). In this research, the salt stress increases the protein level in root, but salicylic acid decreases the protein levels.

The proteins at salt stress condition accumulate and act as osmotic regulator (Ahmad et al., 2003). The salinity stress deducts special protein in root and leaves of barley. Salinity stress increases amino acid content in wheat varieties (EL- Bassiouny and Bakheta, 2005). There are many reports about protein changes along with compatible stages that adapt the plants with changed environment (Kong-Ngern et al., 2005). The salinity stress interferes with nitrogen consumption and absorption. The salt stress condition could have effect on different stages of nitrogen metabolism, such as absorption, ionic reduction and protein synthesis (Meloni et al., 2004).

Proline

Proline is one of the most important compounds of plants defensive mixed action to salt stress. There are many reports about increasing proline under salt stress conditions in pretreatment wheat seedlings by salicylic acid and under salt conditions; the high level of ABA protected improves antistress activities (Sakhabutdinova et al., 2003; Inal, 2002).

According to the other studies, the proline accumulation by salicylic acid treatment increases in wheat, oat, bean and tomato, under oxidative stresses, (Tasgin et al., 2006). The more tolerant plants store more proline (Desnigh and Kanagaraj, 2007). In the present research, increasing salt increases proline in root and leaf, but salicylic acid at 0 and 25 mM of salt did not have a significant change on proline level. But, in high level of salt, the proline level decreases. Salinity and water shortage deduct the proline accumulation in seedlings which regulate osmotic pressure (Inal, 2002). High protection of abcissic acid in treated plants with salicylic acid and under salt stress increases proline and defensive proteins (Shakirova and Sakhabutdinova, 2003).

REFERENCES

Ahmad A, Fariduddin Q, Hayat S (2003). "Salicylic acid influences net photosynthetic rate, carboxylation efficiency, nitrate reductase activity and seed yield in *Brassica juncea*". Photosynthetica 41(2): 281-284.

Benbella M (1999). "Response of five sunflower genotypes (*Helianthus annus* L.) to different concentrations of sodium chloride". Helia. 22 (30):125-138.

Bradford MM (1976). "A rapid and sensitive method for the quantization of microgram quantities of protein utilizing the principle of protein-dye Binding". Anal. Biochem. 72:248-254.

Dantas BF, Ribeiro LDS, Aragao CA (2005). "Physiological response of cowpea seeds to salinity stress". Rev Bras. Sementes. 27: 144-148.

Desnigh R, Kanagaraj G (2007). "Influence of salinity stress on photosynthesis and ant oxidative systems in two cotton varieties". Gen. Appl. Plant Physiol. 33 (3-4):221-234.

El-Bassiouny HM, Bakheta MA (2005). "Effect of salt stress on relative water content, lipid per oxidation, polyamines, amino acids and ethylene of two wheat cultivars". Inter. J. Agric. Biol. 7: 363-365.

El-Samad HM, Shadad MA (1997). "Salt tolerance of soybean cultivars". Biol. Plant 39: 263-269.

El-Tayeb MA (2005). "Response of barley Gains to the interactive effect of salinity and salicylic acid". Plant Growth Regul. 45:215-225.

Fernando E, Cecilia P, Miriam B, Juan G, Gonzalez A (2000). "Effect of NaCl on germination, growth and soluble sugar". Bot. Acad. Sin. 12: 27-34.

Hussein MM, Balbaa LK, Gaballah MS (2007). "Salicylic Acid and Salinity Effects on Growth of Maize Plants". Res. J. Agric. Biol. Sci. 3(4): 321-328.

Inal A (2002). "Growth praline accumulation and ionic relations of tomato (*licopersicum esculentum* L.) as influence by NaCl and Na_2SO_4 salinity". Turk. J. Bot. 26:285-290.

Khodary SEA (2004). "Effect of Salicylic Acid on the Growth, Photosynthesis and Carbohydrate Metabolism in Salt Stressed Maize Plants". Int. J. Agric. Biol. 6(1):5-8.

Koch K (1996). "Carbohydrate-modulated gene expression in plants". Annu. Rev. Plant Physiol. Plant Mol. Biol. 47: 509-540.

Kong-Ngern K, Daduang S, Wongkham CH, Bunnag S, Kosittrakuna M, Theerakulpisuta P (2005). "Protein profiles in response to salt stress in leaf sheaths of rice seedlings". Science Asia 31: 403-408.

Mahajan S, Tuteja N (2005). "Cold, Salinity and Drought stresses: An overview". Biochem. Biophys. 444: 139-158.

Meloni DA, Gulotta MR, Oliva MA (2004). "The effects of salt stress on growth, nitrate reduction and proline and glycinebetaine accumulation in Prosopis alba". Plant Physiol. 16(1):39-46.

Sakhabutdinova AR, Fatkhutdinova DR, Bezrukova MV, Shakirova FM (2003). "Salicylic acid prevents the damaging action of stress factors on wheat plants". Bulg. J. Plant Physiol. Special Issue pp. 314-319.

Shakirova FM, Sakhabutdinova DR (2003). "Changes in the hormonal status of wheat seedlings induced by salicylic acid and salinity". Plant Sci. 164: 317- 322.

Somogy MA (1952). "A new reagent for determination of sugar". J. Biol. Chem. 160: 61-68.

Szepesi A (2006). "Salicylic acid improves the acclimation of *Lycopersicon esculentum* Mill. L. to high salinity by approximating its salt stress response to that of the wild species L. Pennellii". Acta. Biol. Szeged. 50(3-4):177.

Tasgin E, Atici O, Bantoglu NB, Popova LP (2006). "Effects of salicylic acid and cold treatment on protein levels and on the activities of antioxidant enzymes in the apoplast of winter wheat leaves". Phyto Chemistry 67:710-771.

Tattini M (1990). "Changes in nonstructural carbohydrates in olive leaves during root zone salinity stress". Physiol. Plant 98:117-124.

Undovenko GV (1971). "Effect of salinity of sobstrate on nitrogen metabolism of plants with different salt terance". Agro khimiya 3:23-31.

Yokoi S, Bressan RA, Hasegawa PM (2002). "Salt Stress Tolerance of Plants". JIRCAS Working Report pp. 25-33.

Yusuf M, Hasan SA, Ali B, Hayat S, Fariduddin Q, Ahmad A (2007). "Effect of salicylic acid on salinity induced changes in Brassica juncea". J. Integr. Plant Biol. 50(9): 1096-1102.

5

Virulence prediction model (virprob) using amino acid and dipeptide composition for human pathogens

S. B. Muley, V. Bastikar*, S. Bothe, A. Meshram and N. Roy

Department of Biotechnology and Bioinformatics, Padmashree Dr. D. Y. Patil University, Navi Mumbai, Maharashtra, India. – 400614.

Pathogenic bacteria that cause infectious disease are operated by various virulence mechanisms. Hence, it is important to develop a reliable system for predicting bacterial virulent proteins aiming at discovering novel drug/vaccine and for understanding virulence mechanisms in pathogens. On the basis of features like amino acid and dipeptide composition, it tried to identify the virulence potential in the given biological protein sequence of bacteria using statistical methods like regression analysis, which is of great use in the prediction strategies of the virulence protein. In this work a bacterial virulent protein prediction model, virprob, is proposed based on classifiers, where the features are extracted directly from the amino acid sequence of a given protein. It is a probabilistic model which predicts the virulence potential of the corresponding human pathogenic bacterial protein. An extensive evaluation according to a blind testing protocol, where the parameters of the system are calculated using the training set and the system is validated in independent dataset, has demonstrated the validity of virprob with 53.6% of accuracy. The statistical analysis method may increase the prediction accuracy when combined with machine learning techniques. The results of this analysis might help in rapidly advancing knowledge of infectious agents.

Key words: Virulence, proteins, virprob, pathogen.

INTRODUCTION

A pathogen is a biological agent that causes disease or illness to its host invading through several substrates and pathways. The majority of pathogens are harmless and sometimes beneficial; a few can cause infectious diseases such as the bacterium *Yersinia pestis*, which caused the Black Plague. Pathogenicity or virulence of a pathogen is its ability to cause disease. Virulence is the result of complex interplay between parasite and host. Only a small portion of the total population of microorganisms contains the attribute while the rest are harmless or even beneficial to humans and other animals. The pathogenic mechanisms of microorganisms causing diseases other than bacteria have been probed at the molecular level but many bacterial diseases are poorly understood. Discovering bacterial virulence factors is important in understanding bacterial pathogenesis and their interactions with the host, which may also serve as

novel targets for drug and vaccine development (Hsing et al., 2008).

Bioinformatics approaches and bacterial genomic data are being used to find new mechanisms of virulence, and eventually, targets for novel antimicrobials (Weinstock, 2000). Bioinformatics utilizes large databases of biological information with specific *in silico* tools to complement traditional wet laboratory-based biology (Murray, 1994). Detailed knowledge about the complete sequences of pathogen genomes provides wealth of information about the determinants of bacterial virulence. A large number of predicted proteins in these genomes are yet to be assigned any function and some of them could be virulent proteins. Due to diversity and complexity of virulence proteins, the computational tools for their interpretation, identification and characterization are still limited. Availability of accurate prediction methods for virulent proteins will enhance knowledge about bacterial virulence, annotations of (novel) virulent genes and development of novel antimicrobial targets. Similarity search methods like BLAST (Altschul et al., 1990) distinguish between virulent

*Corresponding author. E-mail: vabastikar@yahoo.co.in.

and non-virulent proteins with reasonable accuracy, but reasonable enough in cases where virulent proteins are evolutionarily distant and do not have significant sequence similarity to known virulent protein sequences. Several computational strategies have been proposed to deal with the problems of finding sequences with remote similarity and homology. PSIBLAST is one such algorithm, which aids in identification of remotely similar proteins (Altschul et al., 1997). Another reasonable method to overcome this limitation is the machine learning algorithms. Statistical methods like 'regression analysis' are also of great use in the prediction strategies of the virulence protein, which when combined with machine learning techniques may increase the prediction accuracy.

METHODOLOGY

Extraction of protein sequences

Virprob uses UNIPROT (Apweiler et al., 2004) and VFDB (Chen et al., 2005) as source of human pathogenic bacteria virulent proteins sequences. The sequence entries annotated as "Probable", "Putative", "By similarity", "Fragments" "Hypothetical", "Unknown" and "Possible" were removed. It yielded 406 annotated virulent protein sequences of human pathogenic bacteria referred as 'positive dataset'. For training with non-virulent protein sequences, we selected 350 annotated protein sequences of bacterial enzymes and other non-virulent proteins from UNIPROT database referred as 'negative dataset'. The non-virulent dataset sequences were mainly chosen from the bacterial proteomes, the virulent protein sequences of which are included in the positive dataset.

Reducing redundancy of sequences

For the refinement of dataset, we reduced similarity present between sequences. We used CD-HIT (Li and Godzik, 2006) to scale the redundancy in positive and negative dataset sequences so that no two sequences were more than 40% similar. CD-HIT yielded a non-redundant dataset of sequences, out of which 338 sequences were found to be virulent (positive dataset) sequences. Out of 338, we selected 296 sequences belonging to 12 different organisms: *Vibrio cholera, Staphylococcus epidermidis, Streptococcus pneumonia, Neisseria meningitides, Mycobacterium tuberculosis, Listeria monocytogenes, Helicobacter pylori, Haemophilus influenza, Escherichia coli, Clostridium perfringens, Brucella abortus* and *Bacillus anthracis* to make the training set. Hence, we used 296 positive sequences and 258 negative sequences from bacterial proteomes to complete our final non-redundant training dataset (554 sequences).

Generation of dataset for blind test

Sequences of a few organisms were excluded from the positive non-redundant training dataset to constitute a positive independent dataset. This was done to gauge the classifier prediction efficiency for the sequences of the organisms, which were not represented in the training dataset. Similarly, random non-virulent sequences from these organisms were included in the negative independent dataset. The dataset consists of 50 virulent and 50 non-virulent sequences from the following bacterial pathogens: *Bordetella pertusis, Legionella pneumophila, Mycoplasma pneumoniae* and

Yersinia pestis.

Protein features

Amino acid composition

Amino acid composition is the fraction of each amino acid in a protein. The fraction of all 20 natural amino acids was calculated using the following equation:

$$fraction\ of\ amino\ acid\ i = \frac{total\ number\ of\ amino\ acid\ i}{total\ number\ of\ amino\ acids\ in\ protein} \qquad \ldots\ldots\ldots (1)$$

i = 1 to 20

Dipeptide composition

Dipeptide composition was used to encapsulate the global information about each protein sequence, which gives a fixed pattern length of 400 (20 × 20). This representation encompassed the information about amino acid composition along local order of amino acid. The fraction of each dipeptide was calculated using following equation:

$$fraction\ of\ dep(i) = \frac{total\ number\ of\ dep(i)}{total\ number\ of\ all\ possible\ dipeptides} \qquad \ldots\ldots\ldots (2)$$

i = 1 to 400.

Binary logistic regression

Logistic regression model can be used for prediction of dependent variable on the basis of scale and/or categorical independents to rank the relative importance of independents, assess interaction effects and understand the impact of covariate variables. Binary logistic regression is a type of logistic regression model which is used when the dependent variable is of dichotomous type and the independent variables of any type.

The non-redundant dataset of 554 sequences was analyzed and the statistical model was generated using binary logistic regression. In logistic regression, we predict the probability of outcome variable Y occurring in a given known values of predictor variables (Xi). The logistic regression equation from which the probability of Y is predicted is given by equation:

$$P(Y) = \frac{1}{1 + e^{-(b_0 + b_1 X_1 + b_2 X_2 + \ldots + b_n X_n + \varepsilon_i)}} \qquad \ldots\ldots\ldots\ldots (3)$$

where, $P(Y)$ is the probability of Y occurring, e is the mathematical constant e = 2.7182 and $b_0, b_1, \ldots, b_n$ are the regression coefficient and ε_i is the error term.

The values of regression coefficient are estimated on the basis of dependent and independent variables which are used to predict the value of Y when X is given. The resulting value from the equation is a probability value ranging from 0 - 1. The value close to 0 indicates that Y is very unlikely to have occurred, and a value close to 1 indicates that Y is very likely to have occurred.

The values are estimated using maximum-likelihood estimation, which selects coefficients that make the observed values most likely to have occurred. So, we tried to fit a model to data that allows us to estimate values of the outcome variable from known values of the predictor variable or variables. In the present study, the dependent variable is of dichotomous type, the value 1 indicates virulent protein and 0 indicates non-virulent protein. The independent variables comprise 20 variables for amino acid composition and 400 variables for dipeptide composition.

The data is analyzed using Forward LR method of Logistic Regression in SPSS. There are several methods that can be used in logistic regression. Stepwise methods are defensible when used in situations in which no previous research exists to base hypotheses for testing, and in situations in which causality is not of interest and the researcher merely wishes to find a model to fit his data (Andey,????). In the present hypothesis, there is no prior research present to defend. For this analysis Forward: LR method of regression was used in which the computer begins with a model that includes only a constant and then adds single predictors into the model based on the value of the score statistic. The variable with the most significant score statistic is added to the model and continues until none of the remaining predictors have a significant score statistic (the cut-off point for significance being .05). At each step, the computer also examines the variables in the model to see whether any should be removed. In the Forward: LR method, the current model is compared to the model when that predictor is removed. If this removal makes a significant difference as to how well the model fits the observed data, then the computer retains that predictor (because the model is better when predictor is included). However, when the predictor removal makes little difference to the model, the computer rejects that predictor (Andey, 2005).

RESULTS AND DISCUSSION

Variables used in the equation

The same table is divided into three tables. In the table, first is the constant and rest are Amino acid or Dipeptide composition (comp). For example, ALcomp indicates - Alanine Leucine Composition. "Standard amino acid abbreviations are used."

The results in Table 1, gives the estimates for the values of regression coefficients, that is, in column B, Wald statistic and other statistics of the desired Logistic Regression Model. The B statistic and the corresponding significance p-value, test the significance of each of the independent variable in the model. If the p-value is less than 0.05 then the independent variable is significant in the model. The independent variables included in the equation of the final model have statistical significance (p-value) less than 0.05.

The B-values are the coefficient values that we would replace in logistic regression equation (3) to establish the probability of the protein virulence potential. The Exp (B) statistic gives us the change in odds. If the value is greater than, then it indicates that as the predictor increases, the odds of the outcome occurring also increases. Conversely, a value less than 1 indicates that as the predictor increases, the odds of the outcome occurring decrease. The Wald statistic tells us whether a variable is a significant predictor of the outcome or not (Andey, 2005).

The classification in Table 2 shows that, for the current data set, the Model accuracy for the classification is almost 83%. Virprob was validated with the independent dataset. The accuracy of prediction achieved is 53.6%. The dataset consists of 50 virulent and 50 non-virulent sequences from the following bacterial pathogens: *B. pertusis, L. pneumophila, M. pneumoniae* and *Y. pestis*.

Conclusion

Discovering virulence factors is important in understanding bacterial pathogenesis and their interactions with the host, which may also serve as novel targets in drug and vaccine development. On the basis of features like amino acid and dipeptide composition, we tried to identify the virulence potential in the given biological protein sequence. In Forward: LR method, logistic regression will move forward while dropping non-significant variables. Same approach can be used for other organism specific bacteria. Statistical method is a pro-step for machine learning technique and the output of this method can be used for machine learning.

In summary, probabilistic models for biological sequences (DNA and proteins) are frequently used in bioinformatics. We describe statistical tests designed to detect the order of dependency among elements of the sequence and to select the most appropriate probabilistic model for an experimental biological sequence. We demonstrate how one can estimate virulence potential in a particular protein that is human pathogenic bacteria by applying binary logistic regression analysis using SPSS. This is one of the approach through which we can define the probability of the particular protein being virulent.

Description of virprob

The programs runs on Windows operating system with Framework. NET 2.0 or higher. To demonstrate the functionality of virprob, we take protein sequences from the blind test dataset. First, we take raw virulent protein sequence of *M. pneumonia* (Dallo et al., 1990) from positive independent dataset and enter it into the textfield and click PREDICT. The output is result of two step process. In first step, it calculates the Amino Acid and Dipeptide Composition and substitutes them with corresponding regression coefficients values in the equation 3. P(Y) is calculated. In final step, P(Y) value is compared with the fixed threshold values. Since the P(Y) value of the entered protein exceeds the threshold of Strong Virulent Potential, the output is displayed as "PROTEIN HAS STRONG VIRULENCE POTENTIAL", as shown in Figure 1. The protein is responsible for the pathogenecity of the organism.

Now we take another raw virulent protein sequence of *M. pneumonia* (Himmelreich et al., 1996) from positive independent dataset. Click CLEAR DATA, it will clear the earlier data from the program. Enter the virulent raw

Table 1. Estimates for the values of regression coefficients.

Variables	Regression coefficients(B)	Standard error	Wald statistics	Significance Value (P value)	Exp (B)
Constant	7.376	1.703	18.764	0.000	0.001597
ALcomp	-0.752	0.231	10.6:09	0.001	0.472
APcomp	1.011	0.390	6.725	0.10	2.749
AVcomp	0.556	0.331	2.822	0.93	1.743
CAcomp	- 2.365	0.916	6.665	0.010	0.094
CGcomp	-2. 508	0.767	10.693	0.001	0.061
DAcomp	-1.189	0.322	13.672	0.000	0.304
DGcomp	-1.002	0.344	8.488	0.004	0.367
DPcomp	- 1.335	0.474	7.928	0.005	0.263
EQcomp	1. 518	0.396	14.791	0.000	4.565
FCcomp	2.380	1.107	4.623	0.032	10.804
FLcomp	1.741	0.440	15.666	0.000	5.706
FTcomp	1.994	0.474	17.728	0.000	7.346
FYcomp	-2.233	0.637	12.307	0.000	0.107
GQcomp	-1.339	0.416	10.367	0.001	0.262
HQcomp	- 2.765	0.787	12.342	0.000	0.063
IDcomp	-0.958	0.390	6.021	0.014	0.384
ILcomp	1. 085	0.325	11.118	0.001	2.960
KRcomp	-0.938	0.395	5.626	0.018	0.391
KVcomp	-0.888	0.321	7.316	0.007	0.420
KVcomp	-2.795	0.999	7.829	0.005	0.061
LFcomp	-1.947	0.379	26.413	0.000	0.143
MDcomp	-2.157	0.662	10.941	0.001	0.116
MRcomp	-2.414	0.691	16.672	0.000	0.089
NKcomp	-1.286	0.402	10.237	0.001	0.276
NWcomp	2.706	1.150	5.536	0.019	14.965
PScomp	-1.399	0.471	8.826	0.003	0.247
QCcomp	-4.945	1.129	19.185	0.000	0.007
QFcomp	1. 631	0.561	8.456	0.004	0.196
QMcomp	-2.049	0.722	8.058	0.005	0.129
QNcomp	-1. 027	0.626	3.814	0.051	0.366
QRcomp	-1.369	0.641	6.309	0.012	0.267
QTcomp	-2.293	0.490	21.896	0.000	0.101
RLcomp	0. 846	0.276	9.424	0.002	2.330
RVcomp	1.309	0.397	10.876	0.001	0.270
SNcomp	-1.204	0.486	6.133	0.013	0.300
THcomp	-2.135	0.843	11.043	0.001	0.118
TQcomp	1. 477	0.499	8.739	0.033	4.376
TScomp	-0.921	0.385	5.721	0.017	0.398
TWcomp	2.723	1.294	4.428	0.035	16.229
VAcomp	-1. 808	0.334	29.316	0.000	0.164
VTcomp	1.187	0.342	12.021	0.001	3.276
WNcomp	-2.285	1.042	4.810	0.028	0.102
WScomp	-3.900	1.021	14.607	0.000	0.020
YLcomp	-1.685	0.614	7.537	0.006	0.185
E	-0.430	0.077	31.174	0.000	0.651
G	-0.259	0.077	11.360	0.001	0.772
N	0. 217	0.092	5.533	0.019	1.242
S	0. 466	0.088	28.244	0.000	1.594
W	0.547	0.204	7.166	0.007	1.727

Table 2. Classification of data set.

	Predicted		
	Non Virulent	**Virulent**	**Percentage**
Non virulent	215	43	83.3
Virulent	48	248	83.8
Overall percentage	-	-	83.6

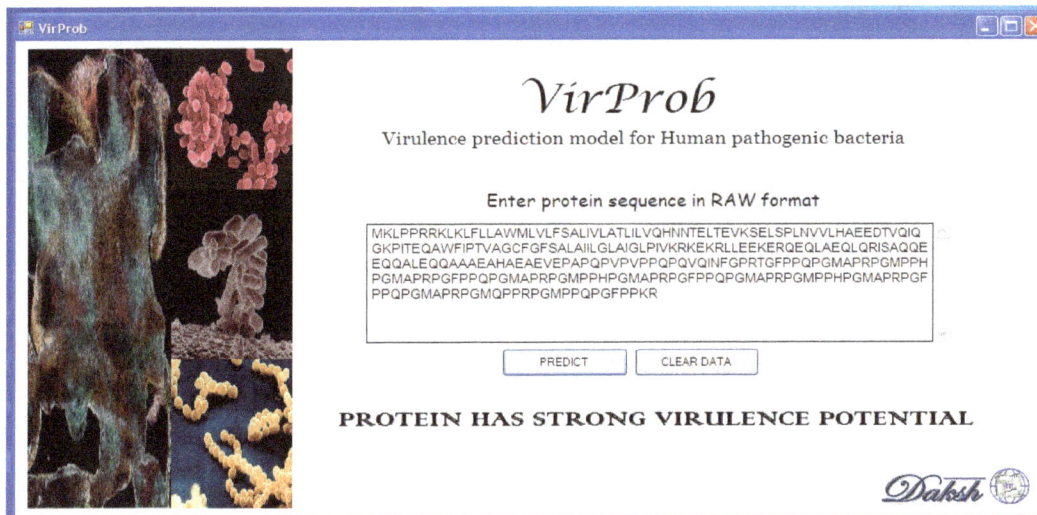

Figure 1. First raw virulent protein sequence of *M. pneumonia.*

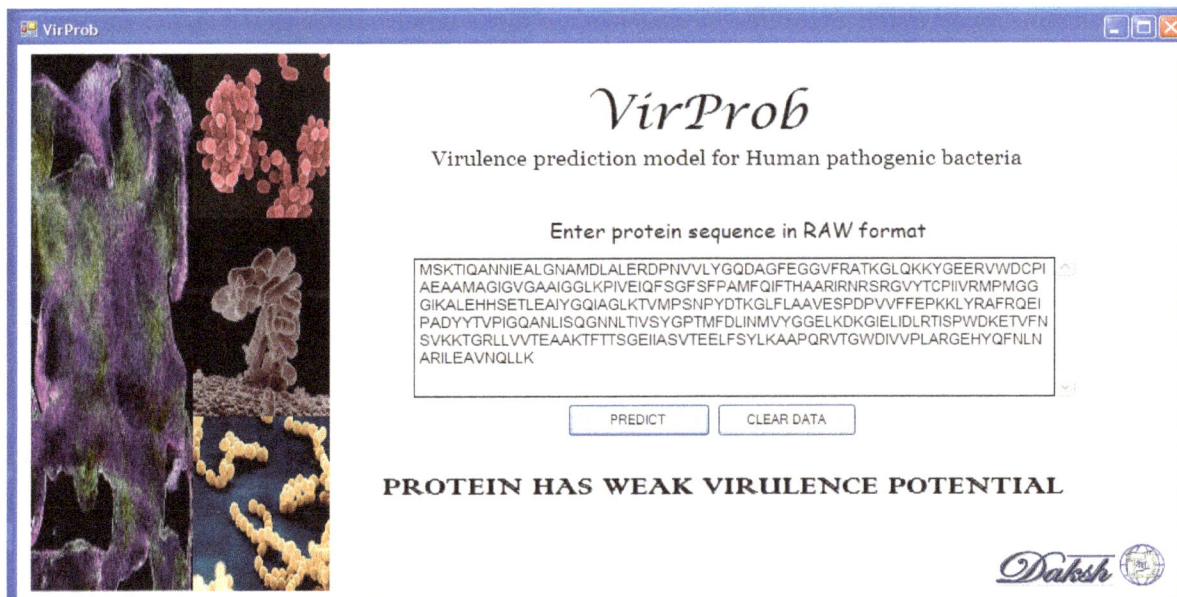

Figure 2. The second raw virulent protein sequence of *M. pneumonia.*

sequence and click PREDICT. The output is "PROTEIN HAS WEAK VIRULENCE POTENTIAL" (Figure 2). The reason is that the P(Y) value for this sequence is greater than the threshold of Virulence Potential but less than that of Strong Virulent Potential. The protein somewhere helps in the virulence but not completely a virulent protein.

Now we take raw non-virulent protein sequence of *M.*

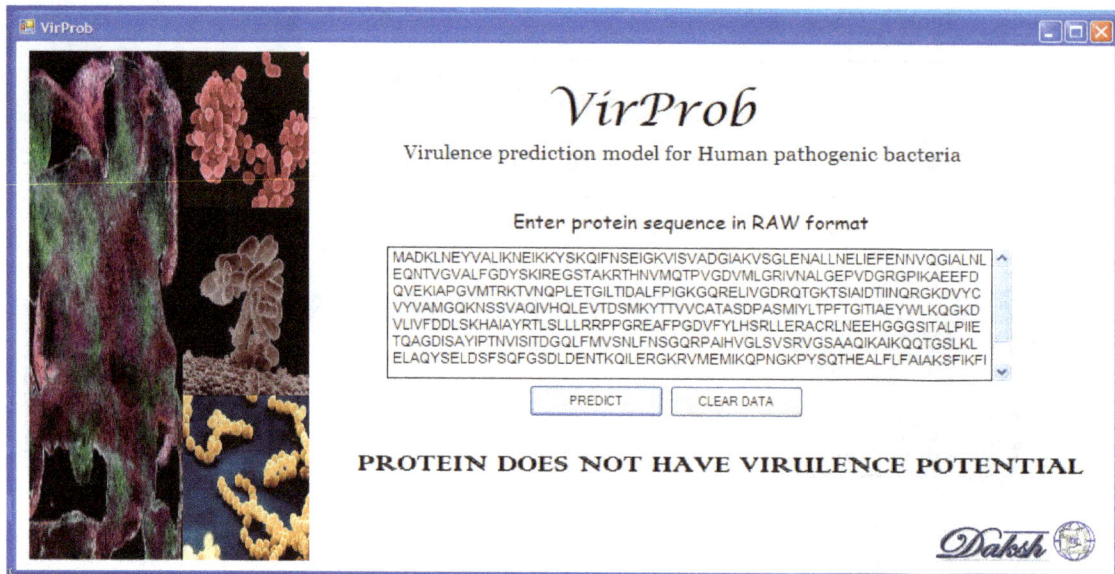

Figure 3. Raw non-virulent protein sequence of *M. pneumonia*.
The threshold values are as follows:
No virulence potential = less than 0.5
Virulence Potential = 0.5 to less than 0.8
Strong Virulence Potential = greater than 0.8

pneumonia (Himmelreich et al., 1996) from negative independent dataset. Click CLEAR DATA, it will clear the earlier data from the program. Enter the non-virulent raw sequence and click PREDICT. The output is "PROTEIN DOES NOT HAVE VIRULENCE POTENTIAL" (Figure 3). The reason is the P(Y) value of this sequence does not exceed the threshold value of Virulence Potential and is part of the organism's normal biological function.

REFERENCE

Altschul SF, Gish W, Miller W, Myers EW, Lipman DJ (1990). Basic local alignment search tool, J. Mol. Biol. 215:403-410.

Altschul SF, Madden TL, Schaffer AA, Zhang J, Zhang Z, Miller W, Lipman DJ (1997). Gapped BLAST and PSI-BLAST: a new generation of protein database search programs. Nucleic Acids Res., 25: 3389-402.

Andey Field (2005). "Discovering statistics using SPSS" Second Edition

Chen L, Yang J, Yu J, Yao Z, Sun L, Shen Y, Jin Q (2005). VFDB: a reference database for bacterial virulence factors. Nucleic Acids Research, 33: D325-D328

Dallo SF, Chavaoya A, Baseman JB (1990). Characterization of the gene for a 30-kilodalton adhesion-related protein of Mycoplasma pneumoniae.

Himmelreich R, Hilbert H, Plagens H, Herrmann R (1996). Sequence analysis of 56 kb from the genome of the bacterium Mycoplasma pneumoniae comprising the dnaA region, the atp operon and a cluster of ribosomal protein genes.

Himmelreich R, Hilbert H, Plagens H, Pirkl E, LiB C, Herrmann R (1996). Complete sequence analysis of the genome of the bacterium Mycoplasma pneumoniae.

Hsing JU. Andrew WU, Wan HJ, Michael PJ (2008). Discovery of virulence factors of pathogenic bacteria. Curr Opin Chem Biol. 12(1): 93-101.

Murray RP (1994). Bioinformatics and drug discovery, Curr. Opin. Biotechnol., 5: 648-653.

Rolf A, Amos B, Cathy H, Wu, Winona CB (2004). Brigitte Boeckmann, Serenella Ferro, Elisabeth Gasteiger, Hongzhan Huang, Rodrigo Lopez, Michele Magrane, Maria J. Martin, Darren A. Natale, Claire O'Donovan, Nicole Redaschi and Lai-Su L. Yeh: UniProt: the universal protein knowledgebase Nucleic acids research, 2004 - Oxford University Press.

Soccaronan M (2004). 1 Contact Information, K. Koscaronmelj2, N. Mariniccaron-Fiscaroner3 and L. Vidmar: A prediction model for community-acquired Chlamydia pneumoniae pneumonia in hospitalized patients. Infection, Aug., 32(4): 204-9.

Weinstock GM (2000). Genomics and bacterial pathogenesis, Emer. Infect. Dis., 6: 496-504.

Weizhong Li, Adam G (2006). CD-hit: a fast program for clustering and comparing large sets of protein or nucleotide sequences. Bioinformatics, 22(13): 1658-1659.

6

Epigenetic regulation of PGC1 α in human type 2 diabetes

Y. Dhanusha Yesudhas

Department of Bioinformatics, Karunya University, Coimbatore India. E-mail: dhanusha2504@gmail.com

Type 2 diabetes mellitus (T2DM) is the most common metabolic disease in the world, reaching epidemic proportions. PPARGC1A mRNA expression is reduced in islets from patients with diabetes 2 and it is influenced by both genetic and epigenetic factor. The epigenetic modification, results as a two fold increase in DNA methylation of the PGC1A promoter of diabetes. This two fold increase in DNA methylation is due to the DNA methyl transferase1 (DNMT1) enzyme. In the present study, methylation activity of DNMT 1 enzyme was inhibited with suitable methylation inhibitors which led to the decrease of the hypermethylation of PGC 1 alpha protein. The interaction study of this modified DNMT1 with PGC 1 alpha leads to the new way of drug discovery in type 2 diabetes.

Key words: Type 2 diabetes mellitus (T2DM), methyl transferase1 (DNMT1), epigenetic modification.

INTRODUCTION

Epigenetics is the study of inherited changes in phenotype (appearance) or gene expression caused by mechanisms other than changes in the underlying DNA sequence. Epigenetic modifications, specifically like DNA methylation and histone modification are the key events regulating the process of normal human development (Reik et al., 2001). Type 2 diabetes (T2DM) is characterised by chronic hyperglycaemia as a result of impaired pancreatic beta cell functions and insulin resistance in peripheral tissues that is skeletal muscle, adipose tissue and liver (Yahli et al., 2005). The transcriptional coactivator peroxisome proliferator activated receptor gamma coactivator-1 alpha (protein PGC-1α; gene PPARGC1A) is an important factor regulating the expression of genes for oxidative phosphorylation and ATP production in target tissues through coactivation of nuclear receptors (Ling et al., 2008). It was previously shown that the expression of PPARGC1A and a set of genes involved in oxidative phosphorylation are reduced in skeletal muscle from patients with type 2 diabetes. Furthermore, a common

polymorphism, Gly482Ser, in the PPARGC1A gene has been associated with increased risk of type 2 diabetes and an age-related reduction in muscle PPARGC1A expression (Yunhua et al., 2003). Obesity, reduced physical activity and ageing are well known risk factors for type 2 diabetes (T2DM) (Puigserver and Spiegelman, 2003). However, all individuals exposed to an affluent environment do not develop the disease. One likely reason is that genetic variation modifies individual susceptibility to the environment. However, the environment could also modify genetic risk factors by influencing expression of a gene by DNA methylation or histone modifications. Cytosine residues occurring in CG dinucleotides are targets for DNA methylation and gene expression is usually reduced when DNA methylation takes place at a promoter. Whether DNA methylation influences gene expression in target tissues for type 2 diabetes (T2DM) and thereby the pathogenesis of the disease remains to be demonstrated (Puigserver et al., 2003).

Sequencing method is used to determine the pattern of methylation. Treatment of DNA with bisulfate (salt) converts cytosine residues to uracil, but leaves 5-methyl-cytosine residues unaffected. Thus, bisulfite treatment introduces specific changes in the DNA sequence that depend on the methylation status of individual cytosine residues, yielding single-nucleotide resolution information

---

Abbreviations: PGC1A, Peroxisome proliferator-activated receptor gamma coactivator 1-alpha (PPARGC1A); **DNMT1,** DNA methyl transferase 1; **T2DM,** type 2 diabetes.

about the methylation status of a segment of DNA. Various analyses can be performed on the altered sequence to retrieve this information (Long-Cheng and Rajvir, 2002).

The present study investigated: (1) DNA methylation analysis of PGC1 alph; (2) the modification of the methylation activity of DNMT by methylation inhibitors and (3) the inhibition of the hypermethylation of PGC1 α by modified DNMT1 (protein-protein interaction study).

MATERIALS AND METHODS

Protein Structure

The ttructure of the enzyme DNMT1 and the PGC 1 alpha protein was analyzed by SAVS server. the results are shown in Figure 1a and b.

DNA methylation analysis

DNA methylation involves the addition of a methyl group to the 5TH position of the cytosine pyrimidine ring or the number 6 nitrogen of the adenine purine ring. The process takes place in the so called CpG islands, located in the promoter of the eukaryotic genes. The methylation analysis was done using MethPrimer software. Meth-Primer software uses bisulfite sequencing method to determine the pattern of methylation. Treatment of DNA with bisulfate (salt) converts cytosine residues to uracil, but leaves 5-methylcytosine residues unaffected. Thus, bisulfite treatment introduces specific changes in the DNA sequence that depend on the methylation status of individual cytosine residues, yielding single-nucleotide resolution information about the methylation status of a segment of DNA. Various analyses can be performed on the altered sequence to retrieve this information (Long-Cheng and Rajvir., 2002); the sequence of the gene was pasted which is methylated, the job was submitted, and the results were gotten as original sequences in the first row and the bisulfate sequences in the next row and the CpG sites are marked as ++ and the unmethylated cytosine residue was converted as uracil, and the methylated cytosine remained as same (http://www.urogene. org/methprimer/index1.html).

Modification of methylation activity of DNA methyl transferase1 (DNMT1)

Inhibition of DNMT1 was done, because it is responsible for the hyper methylation of PGC1 alpha (it transfers the methyl group to the cytosine residue) which causes the type 2 diabetes (T2DM). The inhibitor selection for this DNMT1 was based on the literature search and in that some of the inhibitor compounds are in the phase 1 and phase 2 trail (Goffin and Eisenhauer, 2002). The molecular properties of these inhibitors were analyzed using various softwares. Inhibition was done with the help of docking tools (Schrondiger and autodock). This docking was first done using Autodock and the top six compounds were selected as the input for Schrödinger docking (Akio et al., 2009). Compounds selected for inhibition of DNMT1 were: analogs of azacitidine; hydralazine analogs; decitabine analogs; procainamide analogs and zebularine analogs

Inhibition of DNMT1 using Schrödinger

Glide searches for favorable interactions between one or more

typically small ligand molecules and a typically larger receptor molecule usually a protein. Each ligand must be a single molecule, while the receptor may include more than one molecule: a protein and a cofactor. GLIDE can be run in rigid or flexible docking modes; the later automatically generates conformation for each input ligand. The combination of positions and orientation of the ligand relative to the receptor, along with its conformation in flexible docking which is referred to as a ligand pose. The ligand poses that GLIDE generates pass through a series of hierarchical filters that evaluate the ligand interaction with the receptor.

Poses that pass these initial screens enter the final stage of the algorithm, which involves evaluation and minimization of a grid approximation to the OPLS-AA nonbonded ligand–receptor interacttion energy. Final scoring is then carried out on the energy-minimized poses. Schrödinger's proprietary GLIDE Score multi ligand scoring function is used to score the poses. If Glide Score was selected as the scoring function, a composite Emodel score is then used to rank the poses of each ligand and to select the poses to report to the user. Emodel combines Glide Score non-bonded interaction energy, and for flexible docking, the excess internal energy of the generated energy conformation.

Inhibition of PGC1 Alpha with modified DNMT1

From Protein Data Bank, PDB files were taken. DNMT1 PDB id is 3PT9, length of the protein is 873 residues, resolution is 2.5 Å. PGC 1 Alpha PDB id is 1XB7, length of the protein is 247 and the resolution is 2. 5 Å. Based on the resolution and the SAVS server result the PBD files were taken. The Protein - Protein docking was done using ClusPro server.

RESULTS

DNA methylation Analysis

The CpG site was identified from the MethPrimer software. Results were observed. Original sequences in the first row and the bisulfate sequences in the next row and the CpG sites are marked as ++ and the unmethylated cytosine residue was converted as uracil, and the methylated cytosine remained as same; from that we can find the methylated and the unmethylated molecules (Figure 2).

Inhibition using Autodock

The best ligand structures with their binding energy and the interaction of these inhibitors obtained from autodock is shown in Table 1. The best 2 ligand structure and its interaction with the target protein are shown in Figures 3 and 4.

Inhibition using Schrödinger Glide

Glide offers the full range of speed vs. accuracy options, from the high-throughput virtual screening (HTVS) mode for efficiently enriching million compound libraries, to the standard precision (SP) mode for reliably docking tens to hundreds of thousands of ligand with high accuracy, to

PROCHECK

Ramachandran Plot
3CS8

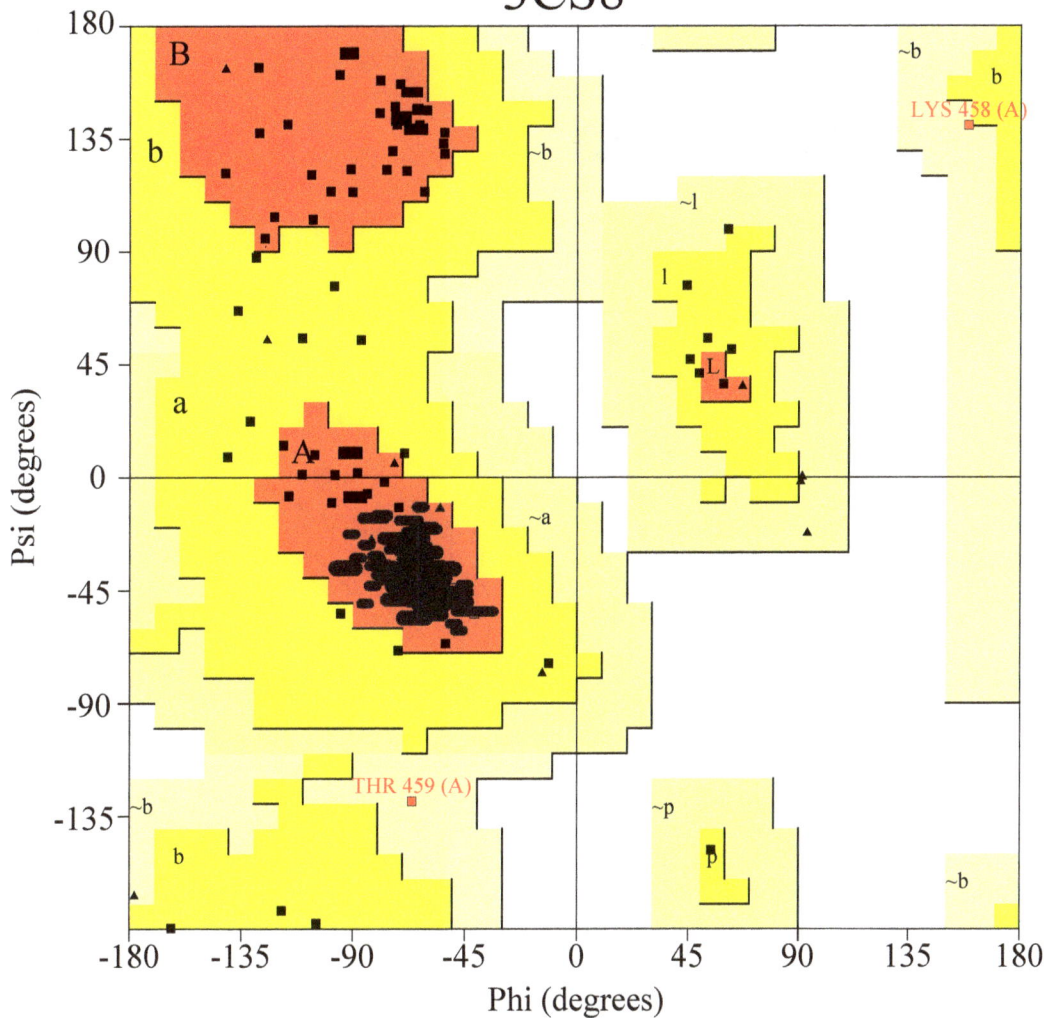

Plot statistics

Residues in most favoured regions [A,B,L]	228	90.8%
Residues in additional allowed regions [a,b,l,p]	21	8.4%
Residues in generously allowed regions [~a,~b,~l,~p]	2	0.8%
Residues in disallowed regions	0	0.0%
	----	------
Number of non-glycine and non-proline residues	251	100.0%
Number of end-residues (excl. Gly and Pro)	4	
Number of glycine residues (shown as triangles)	13	
Number of proline residues	13	

Total number of residues	281	

Based on an analysis of 118 structures of resolution of at least 2.0 Angstroms
and R-factor no greater than 20%, a good quality model would be expected
to have over 90% in the most favoured regions.

3CS8_01.ps

Figure 1 a. PGC1 Alpha. Resolution=2.4; length =275; secondary structure: 62% helical (14 helices; 172 residues):
4% beta sheet (4 strands; 12 residues); active site residues, CYS 285, TYR 473, SER 225, **THR 229.**

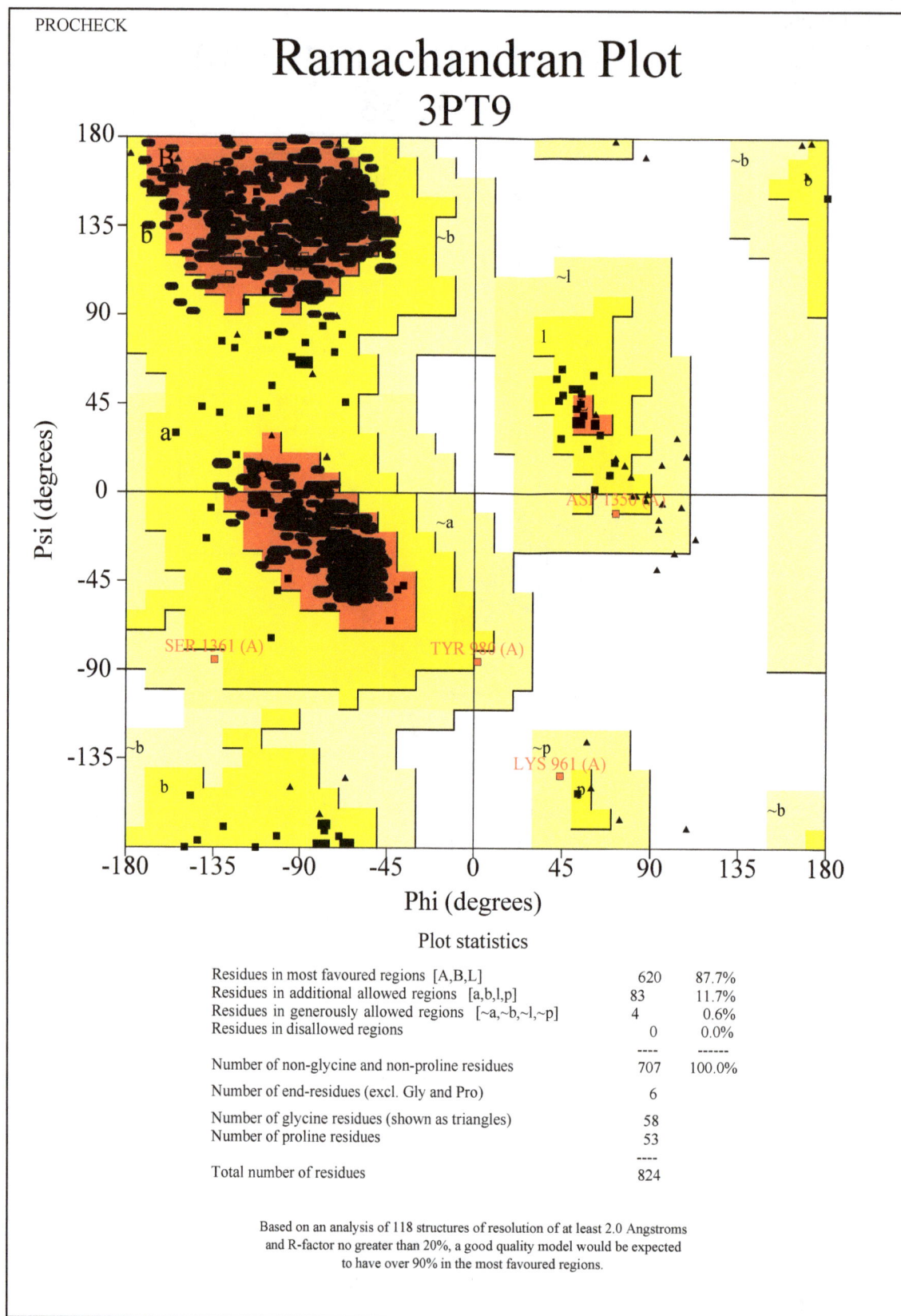

Figure 1b. **DNMT1** Resolution =2.5 Å; length = 873 residues; secondary structure 21% helical (30 helices; 188 residues); 25% beta sheet (52 strands; 220 residues); active site residues ARG1187, ARG1181, VAL1189, GLN1212, **SER A1149, GLY A 1150, LEU1189.**

MethPrimer result

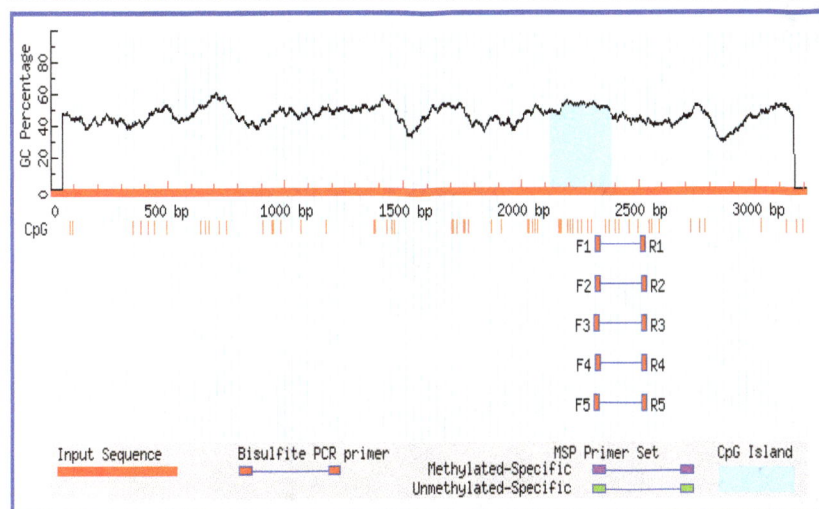

```
3061  TGTGTGCTTGGTTTAGGGGAAGTATGTGTGGGTACATGTGAGGACTGGGGGCACCTGACC
      |||||:|||||||||||||||||||||||||||:|||||||||:||||||:|::|||::
3061  TGTGTGTTTGGTTTAGGGGAAGTATGTGTGTGGGTATATGTGAGGATTGGGGGTATTTGATT


3121  AGAATGCGCAAGGGCAAACCATTTCAAATGGCAGCAGTTCCATGAAGACACGCTTAAAAC
      ||||||++:|||||:||||||::||||:|||||::|||||||::||||||:|++:||||||:
3121  AGAATGCGTAAGGGTAAATTATTTTAAATGGTAGTAGTTTTATGAAGATACGTTTAAAAT


3181  CTAGAACTTCAAAATGTTCGTATTCTATTC
      :|||||:||:|||||||||++|||||:|||||:
3181  TTAGAATTTTAAAATGTTCGTATTTTATTT


********************************************************************
*  Explanations                                                    *
*------------------------------------------------------------------*
*  Upper row: Original sequence                                     *
*  Lower row: Bisulfite modified sequence                          *
*             (For display, assume all CpG sites are methylated)   *
*  ++        CpG sites                                             *
*  ::::      Non-CpG 'C' converted to 'T'                          *
*  >>>>>>    Left  primer                                          *
*  <<<<<<    Right primer                                          *
*                                                                  *
********************************************************************
```

Figure 2. MethPrimer Output.

the extra precision (XP) mode where further elimination of false positives is accomplished by more extensive sampling and advanced scoring, resulting in even higher enrichment. XP does more extensive sampling than SP. This SP and XP uses rigid docking only; only the ligands structure got moved and docked to protein in different conformations.

The top six ligand molecules obtained from the auto-dock result was given as the input to the glide package and the XP and SP docking was done; the docking score and the glide energy were noted.

Induced fit docking

The active site geometry of a protein complex depends heavily upon conformational changes induced by the bound ligand. The Induced Fit protocol begins by docking the active ligand with Glide. In order to generate a diverse ensemble of ligand poses, the procedure uses reduced van der Waals radii and an increased Coulomb-vdW cut-off, and can temporarily remove highly flexible side chains during the docking step. For each pose, a Prime structure prediction is then used to accommodate

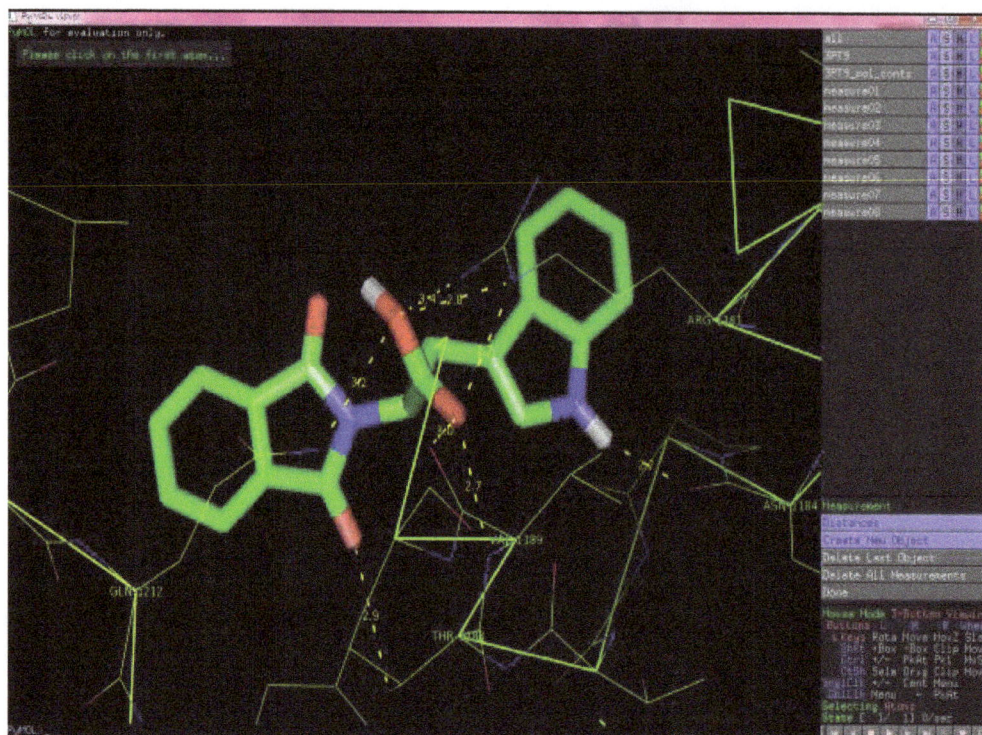

Figure 3. SGS 107 ligand interaction with DNMT1. The best ligand structure which shows the binding energy of **-6.16** with their interaction was viewed using PyMol viewer and the interaction showed are (ARG 1187) O-H...O=2.8, (ARG 1181) N-H...O=2.7, (ARG 1181) N-H...O=3.4, (VAL 1189) N-H...O=3.0, (GLN 1212) N-H...O=3.2, (GLN 1212) N-H...O=3.4, (THR 1188) O-H...O=3.3.

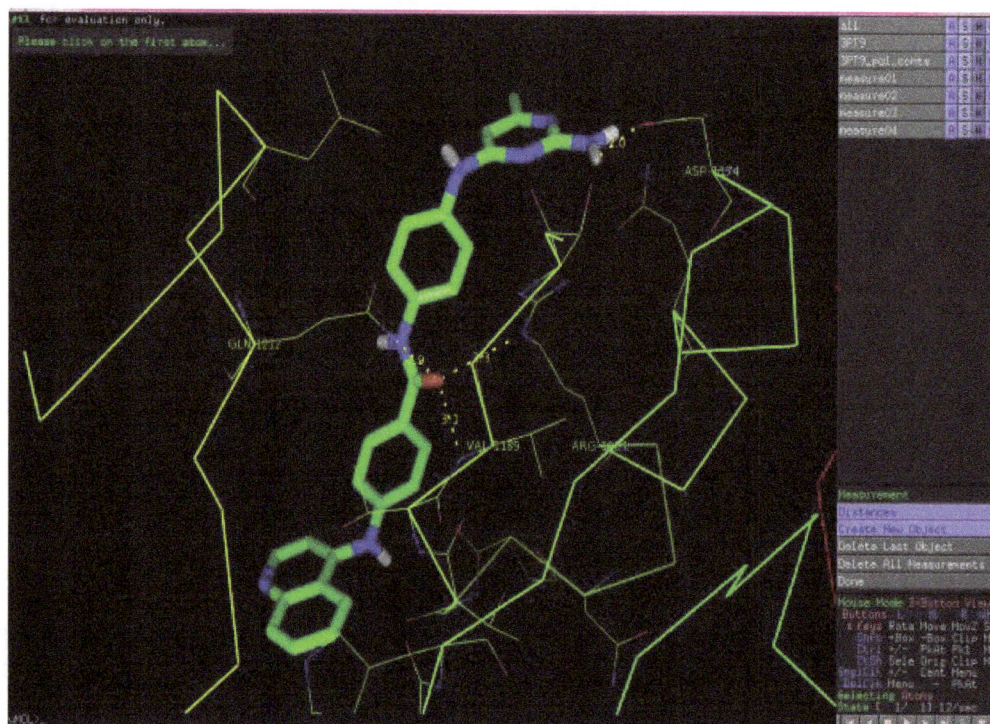

Figure 4. Zebularine ligand interaction with DNMT1The best ligand structure which shows the binding energy of **-5.73** with their interaction was viewed using PyMol viewer. And the interaction shows are (GLN1212) N-H...O=2.9 (VAL1189) N-H...O=3.1, (ARG1181) N-H...O=3.3.

Figure 5. Schrödinger result .The best compound selected from the Schrödinger result was the zebularine molecule and it had the binding energy of -32.8633. The ligand structure shows the interaction of (LEU1189) O-H...O=3.4, (ARG1181) N-H...O=2.9, (GLN1212) N-H...O=3.2, and shows the docking Score of -5.1374.

the ligand by reorienting nearby side chains. These residues and the ligand are then minimized. Finally, each ligand is re-docked into its corresponding low energy protein structures and the resulting complexes are ranked according to GlideScore. Accuracy is ensured by Glide's superior scoring function and Prime's advanced conformational refinement. Simply, we can call it as flexible docking. Both ligand and the protein molecule will move and docked in different conformations. The screened compounds were selected and the induced fit docking was done, and the results are shown below (Figure 5).

Protein-protein docking

Macromolecular docking is the computational modelling of the quaternary structure of complexes formed by two or more interacting biological macromolecules. Protein - protein complexes are the most commonly attempted targets of such modeling. So this protein - protein docking or interaction has the effect to change one protein's function or structure or some of the properties. Based on this technique, the docked DNMT1 protein gets again docked with the PGC 1 alpha protein which may change the hypermethylation process of PGC 1 alpha Protein (Marcotte et al., 1999).

So this protein-protein interaction work was carried out by ClusPro Server.

ClusPro Server

The user can input the PDB codes of the crystal structures of their choice in the receptor and ligand fields, as well as any chain identifiers that they would like to use. Once the PDB files have been uploaded to the server, they are processed into the input files necessary for DOT or ZDOCK, as well as CHARMM minimized for 100 steps with a constrained backbone. The minimized PDBs are then imported to a supercomputer for the running of DOT/ZDOCK, filtering, and clustering. The ClusPro output is shown in the Figure 6.

DISCUSSION

Epigenetics is the study of inherited changes that occur other than DNA sequences level which means the changes does not occur in the sequence. Our lifestyles, food habit, and the environment all these factor make these changes. The epigenetic factors are DNA methylation and histone modification. The major disease caused by this factor is cancer and type 2 diabetes. The DNA hyper methylation of PGC1 alpha protein is responsible for causing the type 2 diabetes (T2DM). This hyper methylation is due to DNMT1 enzymes. Inhibiting the activity of DNMT1 leads to the inhibition of hyper methylation.

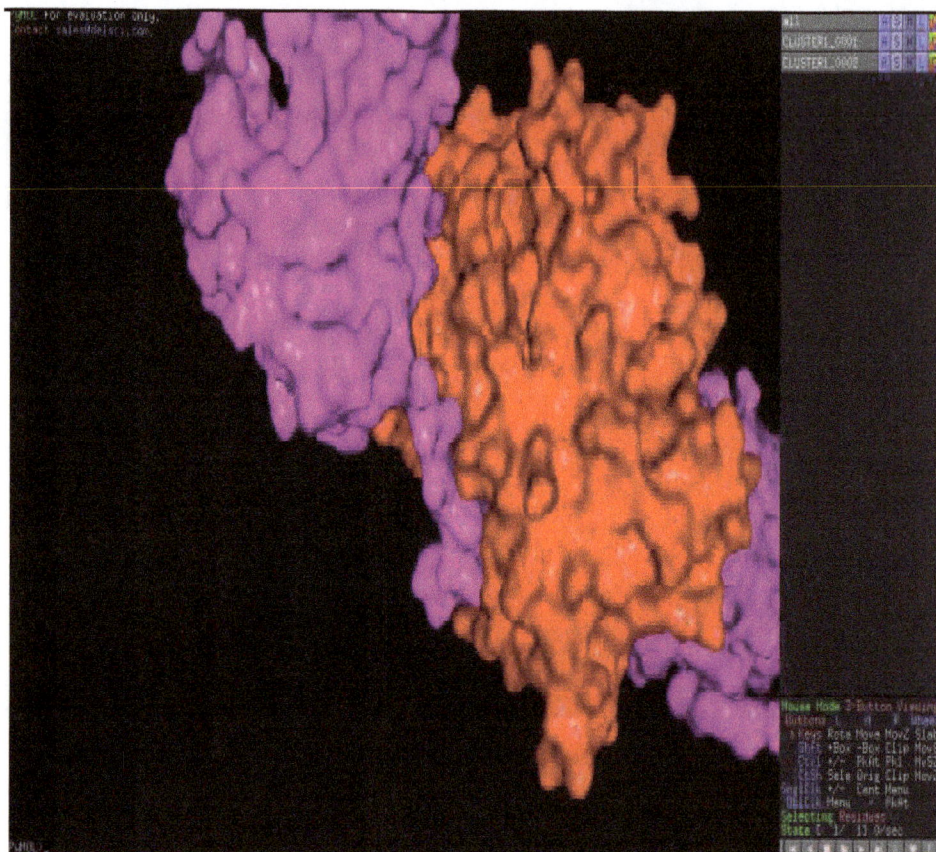

Figure 6. Modified DNMT1 docked with PGC 1 alpha. The two proteins had the interaction with the residues of ARG1187, ARG1181, VAL1189, GLN1212, SER A1149, and SER 225, PHE 226, THR 229, GLY 258, ARG 280. Magenta color = DNMT1; Red = PGC 1 alpha. The ligand molecules are showing the interaction with the active site residues and it has the binding region as in the methylasetransferase binding domain.

The inhibition was done by docking methods.

The selection of protein for docking studies is based upon several factors like, it should contain a co-crystal ligand, structure should be determined by X-ray diffraction, and resolution between 2.0-2.5 Angstroms, and out of the 10 entries of DNMT1, 3PT9 was taken for docking analysis (based on the Ramachandran plot statistics) as it showed 620 most favored regions, 83 in additionally allowed region and none of the residue in disallowed regions. The 17 small molecules are obtained by the literature search, and its structure is drawn using chemsketch and its properties are analyzed using Molinspiration tool. Based on the docking results that are based on docking score and interaction, the best inhibitor molecule was found out and it has the binding score more than the native PDB ligand. The ligand molecules show the interaction with the active site residues and it has the binding region as in the methylasetransferase binding domain. This domain is responsible for catalyzing the reaction of C-5 cytosine methylation in DNA to produce C5-methylcytosine. So inhibition of this DNMT1 has the ability to decrease the hypermethylation of proteins. The docked DNMT1 protein can inhibit the

methylation property of PGC 1 alpha. Protein - Protein docking was done between docked DNMT1 and PGC 1 alpha protein, which has the interacting region of ARG1187, ARG1181, VAL1189, GLN1212, SER 1149, and SER 225, PHE 226, THR 229, GLY 258, ARG 280; has the same active site residues as inhibited DNMT1. So, we can clearly say that the probability of inhibiting the activity of PGC 1 alpha protein is more in this Protein-protein docking. This approach is used to find the new way of drug discovery for type 2 diabetes.

REFERENCES

Akio K, Tibor A, Rauch b, Ivan T, Hsun T (2009). "Insulin Gene Expression Is Regulated by DNA Methylation" *PLoS ONE.*, **9**: 6953.

Goffin J, Eisenhauer E (2002). DNA methyltransferase inhibitors—state of the art. Ann. Oncol. 13: 1699–1716.

Ling C, Del Guerra S, Lupi R, Ronn T, Granhall C, Luthman H, Masiello P, Groop G Del S (2008). Epigenetic regulation of PPARGC1A in human type 2 diabetic islets and effect on insulin secretion. Diabetol. 51: 615–622.

Long-Cheng L, Rajvir D (2002). MethPrimer:designing primers for methylation PCRs. Bioinformatics 18:1427-1431.

Marcotte E, Pellegrini M, RiceD, YeatesT, Eisenberg D (1999). Detecting protein function and protein-protein interactions from genome sequences. Science 285: 751-753

Reik W, Dean W, Walter J (2001). Epigenetic reprogramming in mammalian development. Science 10:1089-93.

Puigserver P, Spiegelman B (2003). Peroxisome proliferator-activated receptor-gamma coactivator 1 alpha (PGC-1 alpha): transcriptional coactivator and metabolic regulator. Endoc. Rev. 24: 78–90.

Puigserver P, Rhee J, Donovan J, Walkey C, Yoon J, Kitamura Y, Altomonte J, Accili D, Spiegelman B (2003). Insulin-regulated hepatic gluconeogenesis through FOXO1-PGC-1alpha interaction. Nature 423: 550–555.

Sun C, Zhang F, Ge X (2007). "SIRT1 improves insulin sensitivity under insulin-resistant conditions by repressing PTP1B". Cell Metab., 6: 307–19.

Yahli L, Barbara M, Roger D, Kornberg D (2005). Chromatin remodeling by nucleosome disassembly in vitro. PNAS 103: 3090-3093.

Yunhua L, Clifton B, Oluf P, Leslie B (2003). A Gly482Ser Missense Mutation in the Peroxisome Proliferator-Activated Receptor γ Coactivator-1 Is Associated With Altered Lipid Oxidation and Early Insulin Secretion in Pima Indians. Diabetes 5: 895-898.

7

Using electrical impedance tomography in following up skin conductivity change for different sonophoresis conditions

Mamdouh M. Shawki* and Abdel-Rahman M. Hereba

Bio-Medical Physics Department, Medical Research Institute, Alexandria University, Egypt.

Sonophoresis is the using of ultrasound waves to increase the entrancement of genes and drugs surpassing the skin barrier. Many mechanisms have been described to illustrate the sonophoresis mode of action; some describe changes in skin resistance during sonophoresis. Electrical impedance tomography (EIT) technique uses voltage measurement through a group of electrodes and reconstructs the data to a conductivity picture. The aim of this work is to evaluate the possibility of using EIT in order to detect changes in mice skin conductivity under pulsed ultrasound waves with powers of 1 and 3 W/cm^2 and times of exposure of 2, 4, and 6 min. EIT system was designed and performed locally. The mice skin was obtained from ten albino mice. Immediately after the skin part was exposed to certain sonophoretic condition, its complex biological impedance was measured using the EIT device. The resultant pictures then were analysed using special software. The results indicate that EIT can be used as a good technique for skin conductivity scan. The data analysis shows that as the ultrasound power increases, the skin conductivity increases. However, there is no significant decrease in skin impedance with time in the same ultrasound power at the time range used.

Key words: Electrical impedance tomography (EIT), sonophoresis, skin conductivity.

INTRODUCTION

Transdermal drug delivery offers several advantages, particularly for those drugs that when taken orally are lost due to gastrointestinal and liver metabolism (first-pass effect) (Hadgraft and Guy, 1989). However, because of the presence of the stratum corneum, the outermost layer of the skin is generally impermeable, especially to large and/or hydrophilic molecules such as peptides and proteins (Schaefer and Redelmeier, 1996). To increase the transdermal drug delivery rate to therapeutically significant levels, various chemical and physical approaches have been investigated, including chemical enhancers (Johnson et al., 1997), electric fields (iontophoresis (Takasuga et al., 2011) and electroporation (Prausnitz et al., 1993), and ultrasound waves (sonophoresis) (Polat et

al., 2011). Sonophoresis is a process that exponentially increases the absorption of topical compounds (transdermal delivery) into the epidermis, dermis and skin appendages. Sonophoresis occurs because ultrasound waves stimulate micro-vibrations within the skin epidermis and increase the overall kinetic energy of molecules making up topical agents. It is widely used in hospitals to deliver drugs through the skin by mixing them with a coupling agent (gel, cream, ointment) that transfers ultrasonic energy from the ultrasound transducer to the skin (Pahade et al., 2010).

Cavitation, temperature elevation, acoustic streaming and convective dispersion are believed to be factors that may substantially contribute to the efficacy of ultrasound. Cavitation effects are generally recognized to be the dominant mechanism of sonophoretic enhancement. These effects are directly related to the observed drop in the electrical resistance of the skin. Due to significant hindrance of the ionic transport through the lipid bilayers,

*Corresponding author. E-mail: mamdouh971@hotmail.com.

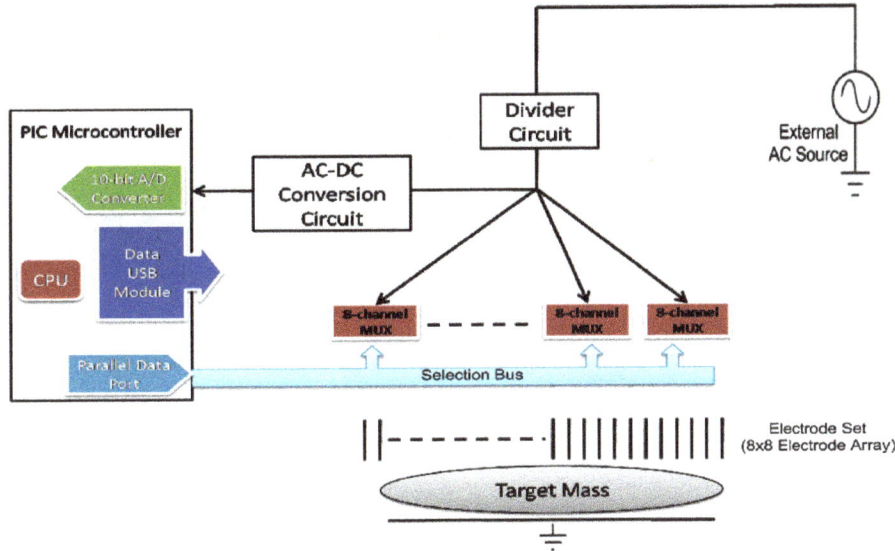

Figure 1. System block diagram.

the electrical resistance of the skin is very high and it has been used to verify that the structural integrity of skin samples is within the normal range. However, the continuous monitoring of skin electrical resistance during ultrasound application can provide insight into the mechanisms governing the enhancement of mass transfer through this biological barrier (Bagshaw et al., 2003). Electrical impedance tomography (EIT) is a noninvasive technique whereby images of the conductivity within a body can be reconstructed from voltage measurements made on the surface (Cheneyy et al., 1993). Electrodes are attached to the body in a fashion similar to electroencephalography (EEG), and each measurement is typically made from a combination of four electrodes, two to inject current and two to sample the resulting voltage distribution (Costaa et al., 2009).

For a particular pair of current injection electrodes, several voltage measurements are taken, the injection pair is then switched, and the process repeated to produce one complete data set. A reconstruction algorithm is used to relate the measured voltages to the conductivity within the body (Metherall et al., 1996). In EIT, the hardware usually will be set to measure with a sampling rate of 10 to 25 per second and for a period of 30 to 60 s. Thus, each measurement session consists of several hundreds of single tomograms. The pixel values in the tomograms approximate local resistivity. The pixel values equal relative impedance change (dimensionless unit), a value that is derived either from a baseline measurement or from parts of the actual measurement itself (Bodenstein et al., 2009).

Tang et al. (2001) illustrated the relationship between permeability of a hydrophilic permeant (P_{diff}) through a skin, and the skin conductivity (σ) by the following equations:

$$P_{diff} = C * \sigma / \Delta x \tag{1}$$

$$C = (kT/ 2z^2 f c_{ion} e_0) \times (D^\infty_p H \lambda_p) / (D^\infty_{ion} H \lambda_{ion}) \tag{2}$$

Here: Δx is skin thickness, C is constant, z is the electrolyte valence, F is the Faraday constant, c_{ion} is the electrolyte molar concentration, e_0 is the electronic charge, K is the Boltzmann constant, and T is the absolute temperature. D^∞_{ion} and D^∞_p are the diffusion coefficients at infinite dilution for ion and permeant respectively. $H(\lambda_{ion})$ and $H(\lambda_p)$ are the diffusion hindrance factors for ion and permeant respectively.

Substituting skin resistivity $(R) = \Delta x / \sigma$ (3)

$$Log\ P_{diff} = Log\ C - Log\ R \tag{4}$$

The aim of this study is to evaluate the role of EIT technique in the detection of the change in skin conductivity due to different sonophoresis powers and times of exposures, which in turn according to equation (4) gives direct relation with solute permeability.

MATERIALS AND METHODS

Constructing of electrical impedance tomography system

The device parts

The system was constructed locally at the Bio-Physics Department, Medical Research Institute, Alexandria University. The system block diagram is illustrated in Figure 1.

Data acquisition system: It is a microcontroller-based system responsible for data collection and interfacing; the system is based

Figure 2. Device main panel view.

on a Microchip PIC high performance microcontroller (part no: PIC18F4550). The data acquisition process is performed by applying an external AC voltage to a simple voltage divider circuit where the applied voltage is divided between a constant value resistor and the target mass impedance at a specific electrode.

Voltage measurement: The AC voltage across the electrode is converted to an equivalent DC voltage through a simple AC-DC conversion circuit (full wave rectifier + RC filter). The analog DC voltage is then converted to a digital value using the integrated analog to digital converter module. The A/D converter has 10-bit resolution of the digital result, 5 V of reference voltage and voltage sensitivity of 4.89 mV. These values guarantee an accurate measurement of the analog DC voltage. The value of the measured DC voltage is indicative of the magnitude of the bioelectrical impedance.

Electrode multiplexing: The electrode set (64 electrodes) is multiplexed using a set of 8-to-1 channel multiplexers, so that the voltage across each electrode is measured independently. The multiplexing system is then a set of 8 multiplexers; each set is targeting 8 electrodes. The microcontroller activates one of the multiplexers and then sends the selection data (for the channel selection within the same multiplexer) to gain the access to one electrode while disabling the other 63 ones. The analog voltage across the selected electrode is converted to the corresponding DC value using the AC/DC converter circuit. The DC voltage is measured and then converted to a digital value using the 10-bit A/D converter module.

Data delivery: The digital value for each voltage is transmitted via the built-in USB module. This digital value is then used to generate the tomographic image on the target computer.

Image reconstruction system

It is a software-based system responsible for data handling and tomographic image generation. The developed software is split into two main interconnected components; Data acquisition software (VB.NET based) and image generating software (MATLAB based).

Data acquisition software: This part of the software (namely: Bio-Image Scanner V2.0) was developed using (Microsoft Visual Basic .NET) to perform, control the hardware data acquisition system via the computer's USB HID (Human Interface Device), collect, and store data elements corresponding to each member of the electrode set. The software sends an asynchronous message to the data acquisition system via USB, forcing it to start its operation (multiplex, read, send, etc.). The data acquisition system responds

with a stream of data (64 units) representing the data from the electrode set, then the software waits for this data stream to store it and export it to the last stage (image generating software).

Data exporting mechanism: On the read of a new data unit, the software saves this unit in an ASCII-formatted text file using the .NET file system capabilities. When the data stream ends, the software starts the image generating executable (plot_impedance.exe) via a simple communication with the windows shell.

Image generation software: This part of the software (namely: plot_impedance.exe) was developed (using MATLAB 7.5) to perform reordering of imported data elements, performing linear data interpolation, and generating a tomographic image using a predefined color code. The software reads the text file that was created by the data acquisition software and translates the text contents into "double" format, creates a matrix of integers with the size of 8x8 (each matrix element corresponds to an electrode), then performs 2D linear data interpolation between the data elements resulting into a new matrix with the size of (700x700) data elements, and finally the (700x700) 2D matrix is plotted using filled contours to output an image.

Device features and general description

The device has an internal power supply unit and operates on 220V AC; it has 2 BNC connectors, one for the ground-connected metal slab and another for the AC input from a standard function generator. The device main panel view is shown in Figure 2.

Device interface: The Bio-image scanner uses a full speed USB 2.0 interface for communications and data transfer. The device is connected to the USB hub using a USB cable.

Operation setup: The output of the function generator was adjusted to 9V Sine wave and the frequency range (typically 100 KHz) was chosen. The output resolution was optimized for 9V AC input, so that any change may cause the output image to have a bad color resolution, both BNC connectors of the AC input and the GND metal Slab were connected, the device to the USB hub using a USB cable was connected, and then the device was powered on, so that the "Bio Image Scanner v2.0" program could be run.

Sonophoresis on mice skin

Ten albino mice aged 8 to 10 weeks and weighing between 20 to 25 g (obtained from animal house, Medical Research Institute,

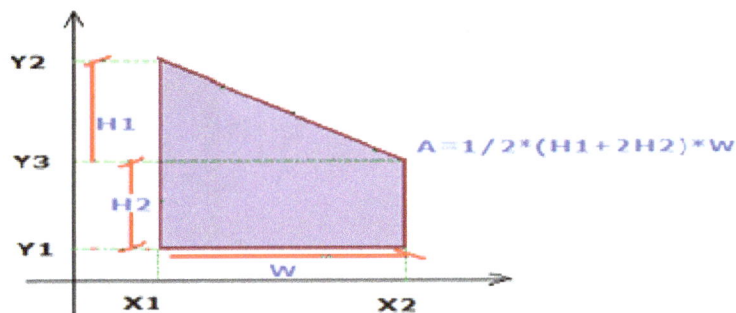

Figure 3. The idea of polygon area calculation.

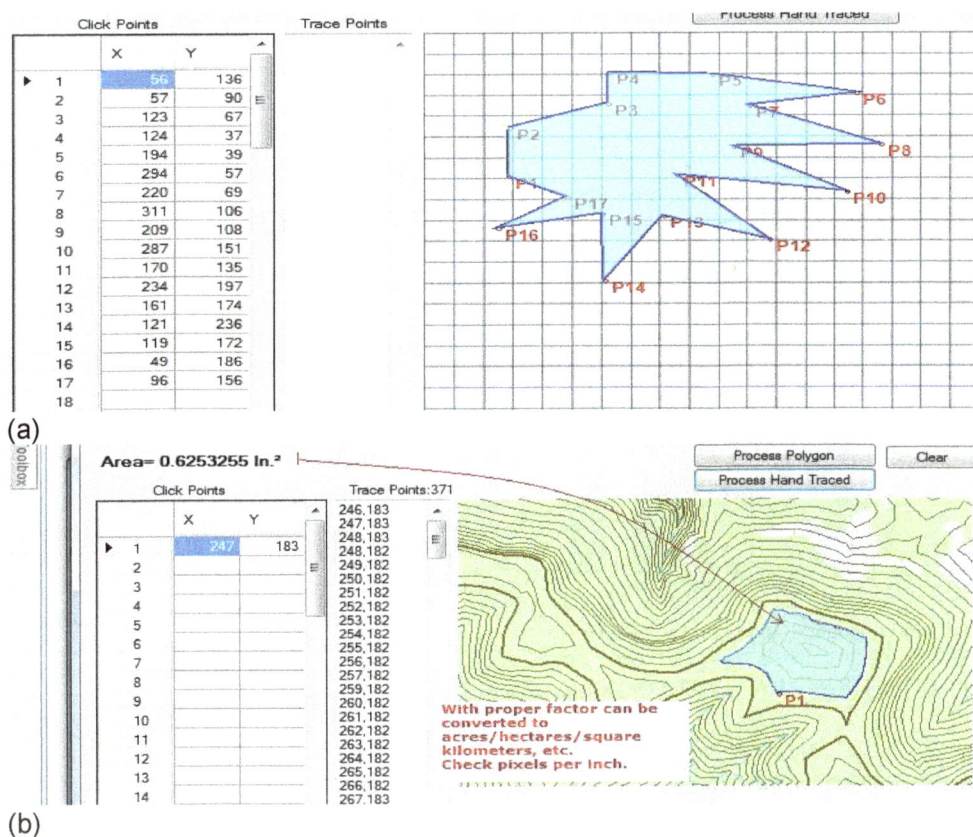

Figure 4. (a) and (b) Shows the steps of area of polygon program operation.

Alexandria University) were anesthetized and sacrificed and the whole skin was excised and sliced into equal pieces using a scalpel approximately 3 cm in length, 3 cm in width, and 8 mm in thickness. 35 skin samples were obtained of approximately equal volumes. Experiments were conducted within 90 min from animal sacrifice with the tissue stored in 0.9% NaCl at room temperature until use. The skin samples were randomly classified into seven groups; each of five samples. Control group was not exposed to ultrasound waves. Three groups were exposed to pulsed ultrasound waves (Ultrasonic Therapeutic Apparatus CSL-1, made in China) 800 KHz, 1 W/cm^2 for 2, 4, and 6 min, respectively. The remaining three groups were exposed to the same ultrasonic apparatus but at 3 W/cm^2 for 2, 4, and 6 min, respectively. Immediately after the exposure time, each sample was examined using the electrical impedance tomography device.

Designing special program for area calculation of images of irregular shapes

Private program was designed by a specialist to determine the area of the different irregular shapes in each picture which was called area of polygon program. The programming language is C-Sharp, and the development enrolment is visual studio.net 2008. The idea of obtaining the area is to move from point to point on a path and use trapezium areas. These areas are the shadow under the line that connects two consecutive points on a Cartesian chart. The idea is shown in Figure 3. An irregular area needs plenty more points in the path to be defined, but the process is the same. The creation of the path depends on the type of shape. On the program, the picture box was clicked on and the array list was filled. Then the data grid appeared when creating the path for a polygon, as shown in Figure 4.

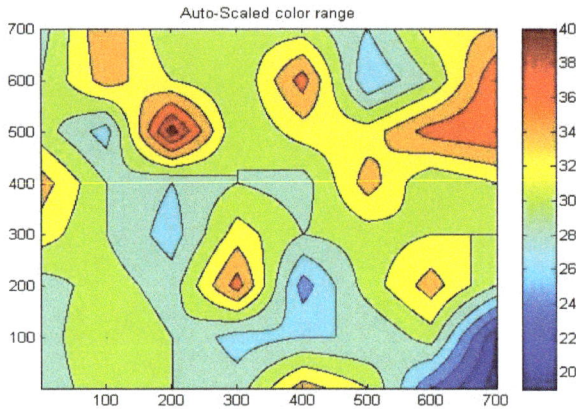

Figure 5. Electrical bio-impedance tomography picture for control skin: 2D linear data interpolation with the size of (700x700) data elements.

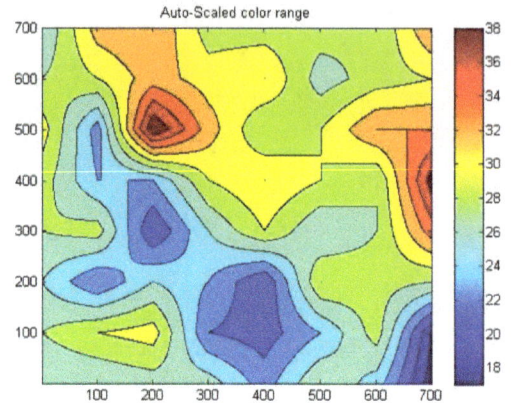

Figure 7. Electrical bio-impedance tomography picture for skin exposed to US, 1 W/cm^2, for 4 min: 2D linear data interpolation with the size of (700x700) data elements.

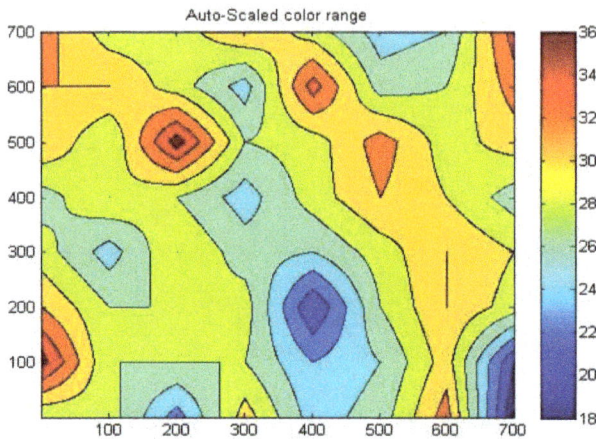

Figure 6. Electrical bio-impedance tomography picture for skin exposed to US, 1 W/cm^2, for 2 min: 2D linear data interpolation with the size of (700x700) data elements

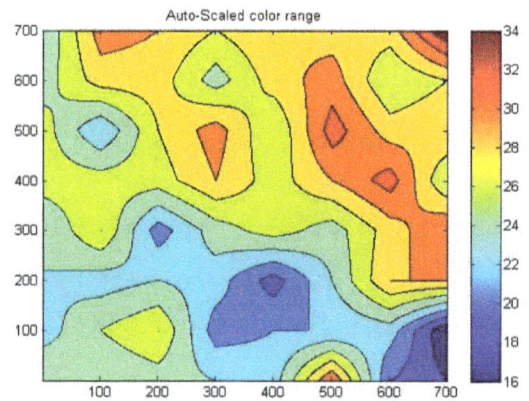

Figure 8. Electrical bio-impedance tomography picture for skin exposed to US, 1 W/cm^2, for 6 min: 2D linear data interpolation with the size of (700x700) data elements.

Complex bioelectrical impedance calculation

The complex impedance was calculated as a result of each image colors area processing (using area of polygon software) by calculating the area percentage of each color regarding to the total skin EIT picture area then multiply the value in the corresponding impedance. The total impedance of the sample is the sum of the whole picture parts impedances. Analysis for five samples (pictures) from each group was done. Impedance averages and standard deviations were calculated for each group. One-way analysis of variance (ANOVA) was used to compare each variable in the different studied groups. For all statistical comparisons, a value of $p < 0.05$ was considered significant.

RESULTS

Figures 5 to11 show examples of bioelectrical impedance tomography pictures for each examined group with color legend for the impedance. Color area calculation done by area of polygon software is placed on Figure 11 as an

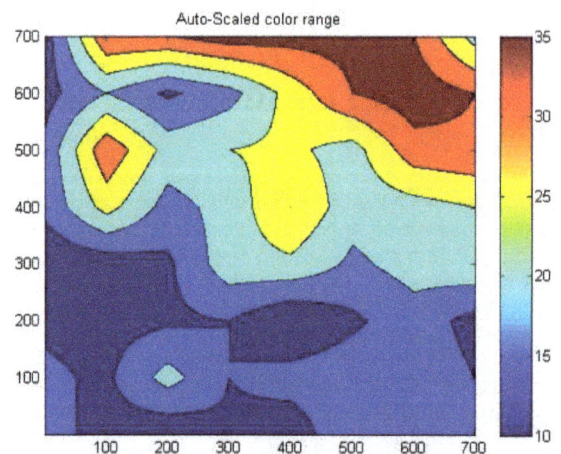

Figure 9. Electrical bio-impedance tomography picture for skin exposed to US, 3 W/cm^2, for 2 min: 2D linear data interpolation with the size of (700x700) data elements.

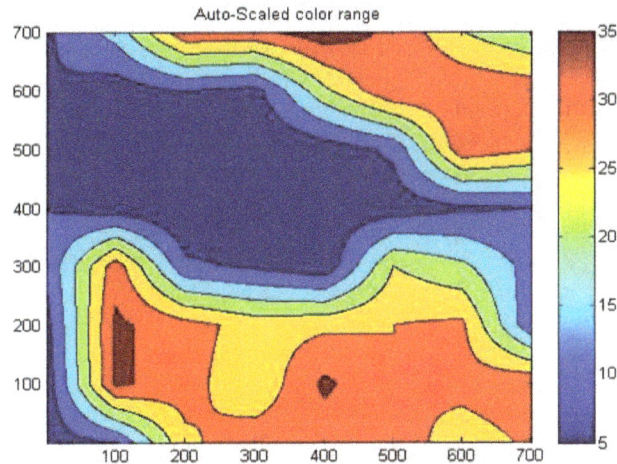

Figure 10. Electrical bio-impedance tomography picture for skin exposed to US, 3 W/cm^2, for 4 min: 2D linear data interpolation with the size of (700x700) data elements.

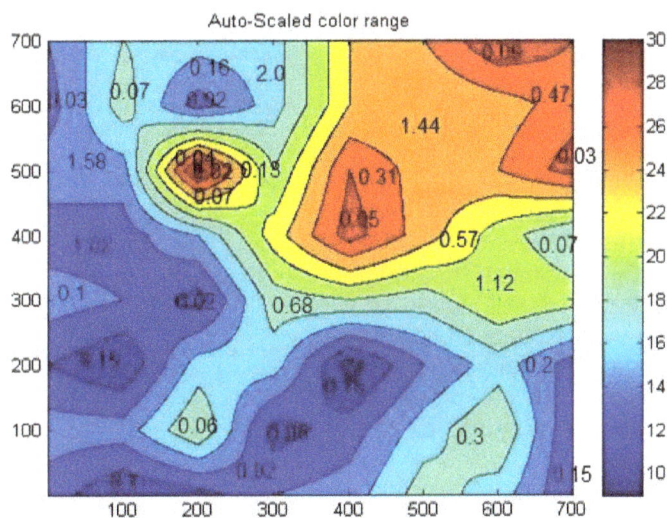

Figure 11. Electrical bio-impedance tomography picture for skin exposed to US, 3 W/cm^2, for 6 min. Area distribution in (inches)^2 obtained by area of polygon program is illustrated.

Table 1. Statistical analysis (averages and standard deviations) of the bio-impedance results.

Group	Control	US*, 2min	US*, 4min	US*, 6min	US**, 2min	US**, 4min	US**, 6min
Bioelectrical impedance	30.60±1.81	27.37±1.05	27.02±1.11	26.67±1.45	18.54±0.62	17.88±0.88	17.57±0.95

US*: Groups exposed to ultrasound 1 W/cm^2. US**: Groups exposed to ultrasound 3 W/cm^2.

example. Table 1 summarizes the statistical analysis of the impedance results.

DISCUSSION

There are a variety of medical problems for which it would be useful to know the time-varying distribution of electrical properties inside the body. High-conductivity materials allow the passage of both direct and alternating currents; high-permittivity materials allow the passage of only alternating currents. Both of these properties are of interest in medical applications, because different tissues have different conductivities and permittivities (Cancel et al., 2004). The application of ultrasound (sonophoresis) has been shown to increase the low skin permeability for various drugs (Sarheed and Abdul Rasool, 2011). The mechanisms and applications of sonophoresis have been reviewed before (Polat et al., 2010). Cavitation, the growth and collapse of gas bubbles, is generally recognized to be the dominant mechanism of sonophoresis. Cavitation is thought to disorder the lipid bilayers in the outermost layer of the skin, the stratum corneum, creating mass transfer pathways and thus increasing the diffusion coefficient of solutes.

However, cavitation alone cannot account for the total mass transfer enhancement observed. Several mechanisms seem to contribute to this transport phenomenon, among them, structural changes caused by cavitation (Smith, 2007), thermal effects (Kalluri and Banga, 2011), mixing in the liquid phase (Levy et al., 1989) and acoustic streaming through hair follicles and sweat ducts (Tachibana and Tachibana, 1993). The electrical resistance of the skin is a good instantaneous indicator of the structural properties of the skin (Allenby et al., 1961). Due to significant hindrance of the ionic transport through the lipid bilayers, the electrical resistance of the skin is very high and, in most previous studies, it has been used to verify that the structural integrity of skin samples is within the normal range. So, the continuous monitoring of skin electrical resistance during ultrasound application can provide insight into the mechanisms governing the enhancement of mass transfer through this biological barrier. Tang et al. (2001) indicated that in the context of the porous pathway hypothesis, when the transport of a permeant occurs via pure diffusion, log P_{diff} versus log R should exhibit a linear behavior with a slope of -1. So, following up the skin electrical properties through different sonophoresis conditions will guide to a prediction of the optimum permeability conditions, and that was achieved through our usage of electrical impedance tomography. It was concluded from Figures 5 to 11 and Table 1 that EIT showed the difference in skin conductivity between control and sonophoretic groups. The complex impedance of the control group was significantly decreased due to sonophoretic exposure either for 1 or 3 W/cm^2. The P-values between control and groups exposed to 1 W/cm^2 for 2, 4, and 6 min were 0.0087, 0.0055, and 0.0053 respectively, while they were less than 0.0001 for 3 W/cm^2 at the same time intervals.

It was also noticed that there was no significant impedance decreases within the same power at the studied time intervals; the P-values at 1 W/cm^2 were 0.6223, 0.4074, 0.6795 as a comparison between 2, 4, 6 min respectively, and they were 0.2076, 0.0922, 0.6795 for 3 W/cm^2 at the same time intervals. These results may be because the used time intervals were narrow and more investigation in a wider range may be performed in the future. At each corresponding time of exposure, there was very high significant decrease in skin impedance at 3 W/cm^2 than 1 W/ cm^2 (all P-values were less than 0.0001) and that may be because higher powers cause higher

thermal effect, and due to the relatively higher coherence of thermal effect with cavitation in power of 3 W/cm^2 more decrease in skin impedance occurred. These results support the facts that EIT can be used to detect the impedance (or conductivity) change in the skin under sonophoresis exposure, which in turn can give the optimum conditions of exposure, as well as the idea that sonophoresis decreases skin resistance which in turn increases the permeability of solutes to the skin which has many bio-medical applications.

ACKNOWLEDGEMENT

The authors thank Mr. S. M. Irshad Hassan, Lecturer of computer engineering, Al Baha College of Sciences, Kingdom of Saudi Arabia, for designing the area of polygon software.

REFERENCES

Allenby A, Schock C, Tees TF (1961). The effect of heat and organic solvents on the electrical impedance and permeability of excised human skin. Br. J. Dermatol., 81: 31–62.
Bagshaw P, Liston D , Bayford H, Tizzard A, Gibson P, Thomas A, Sparkes K, Dehghani H, Binnie D, Holder S (2003). Electrical impedance tomography of human brain function using reconstruction algorithms based on the finite element method. NeuroImage, 20: 752–764.
Bodenstein M, Matthias D, Markstaller K (2009). Principles of electrical impedance tomography and its clinical application. Crit. Care Med., 37: 713-724.
Cancel M, Tarbell M, Jebria A (2004). Fluorescein permeability and electrical resistance of human skin during low frequency ultrasound application. JPP. 56: 1109–1118.
Cheneyy M, Isaacsony D, Newell C (1993). Electrical Impedance Tomography. SIAM. Review, 41: 85–101.
Costaa LV, Limab R, Amato BP (2009). Electrical impedance tomography, Curr. Opin. Crit. Care, 15: 18–24.
Johnson ME, Blankschtein D, Langer R (1997). Evaluation of solute permeation through the stratum corneum: Lateral bilayer diffusion as the primary transport mechanism. J. Pharm. Sci., 86: 1162-1172.
Hadgraft J, Guy RH (1989). Transdermal drug delivery, New York, Marcel Dekker. Reference to a chapter in an edited book : Developmental issues and research initiatives.
Kalluri H, Banga AK (2011). Transdermal delivery of proteins, AAPS. Pharm. Sci. Tech., 12: 431-441.
Levy D, Kost J, Meshulam Y, Langer R (1989). Effect of ultrasound on transdermal drug delivery to rats and guineapigs. J. Clin. Invest., 83: 2074–2078.
Metherall P, Barber DC, Smallwood RH, Brown BH (1996). Three dimensional electrical impedance tomography. Nature, 380: 509–512.
Pahade A, Jadhav VM, Kadam VJ (2010). Sonophoresis: An Overview. Int. J. Pharm. Sci. Rev. Res., 3: 24-32.
Polat BE, Blankschtein D, Langer R (2010). Low-frequency sonophoresis: application to the transdermal delivery of macromolecules and hydrophilic drugs. Expert Opin. Drug Deliv., 7: 1415-1432.
Polat BE, Hart D, Langer R, Blankschtein D (2011). Ultrasound-mediated transdermal drug delivery: mechanisms, scope, and emerging trends. J. Control Release, 152: 330-348.
Prausnitz MR, Bose VG, Langer R, Weaver JC (1993). Electroporation of mammalian skin: A mechanism to enhance transdermal drug delivery. Proc. Nat. Acad. Sci., 90: 10504-10508.
Sarheed O, Abdul Rasool BK (2011). Development of an optimised application protocol for sonophoretic transdermal delivery of a model hydrophilic drug. Open Biomed. Eng. J., 5: 14-24.

Schaefer H, Redelmeier TE (1996). Skin barrier. Reference to a chapter in an edited book: principles of percutaneous absorption, K. Basel.

Smith NB (2007). Perspectives on transdermal ultrasound mediated drug delivery. Int. J. Nanomed., 2: 585-594.

Tachibana K, Tachibana S (1993). Use of ultrasound to enhance the local-anesthetic effect of topically applied aqueous lidocane. Anesthesiology, 78: 1091–1096.

Tang H, Mitragotri S, Blankschtein D, Langer R (2001). Theoretical Description of Transdermal Transport of Hydrophilic Permeants: Application to Low-Frequency Sonophoresis. J. Pharm. Sci., 90: 545-568.

Takasuga S, Yamamoto R, Mafune S, Sutoh C, Kominami K, Yoshida Y, Ito M, Kinoshita M (2011). *In-vitro* and *in-vivo* transdermal iontophoretic delivery of tramadol, a centrally acting analgesic. J. Pharm. Pharmacol., 63:1437-1445.

8

The use of a continuity equation of fluid mechanics to reduce the abnormality of the cardiovascular system: A control mechanics of the human heart

L. S. Taura[1], I. B. Ishiyaku[2] and A. H. Kawo[3]*

[1]Department of Physics, Bayero University, Kano, Nigeria.
[2]Department of Physics, Gombe State University, Gombe, Nigeria.
[3]Department of Biological Sciences, Bayero University, Kano, Nigeria.

The paper is aimed at presenting the differential equations for the cardiovascular system with the help of continuity equation of fluid mechanics to reduce the abnormality of the rate of blood flow and variation of blood volume in different parts of the system. The equations are used to explain the Frank-Starling mechanism, which plays an important role in the maintenance of the stability of the distribution of blood in the system. This is a reasonable approach based on mathematical considerations as well as being further motivated by the observations that many physiologists cite optimization as a potential influence in the evolution of biological systems. We present a model as an application in the provision of a basis for developing information on steady state relations and also to study the nature of the controller and key controlling influences. The model further provides an approach for the study of complex physiological control mechanisms of the cardiovascular system and possible pathways of interaction between the cardiovascular and respiratory control systems. The study also provides an easy way for students of both physics and mathematical sciences, with no previous knowledge of human physiology, to understand the basic systems in cardiovascular concept.

Key words: Continuity equations, fluid mechanics, cardio-vascular system.

INTRODUCTION

In this paper, the concept of fluid mechanics are shown to be useful for the understanding of cardiovascular mechanics, exemplifying the fact that physics can play a fundamental role in the investigation of phenomena in human physiology. To properly understand the application of fluid mechanics (Falkovic, 2011) to human cardiovascular system (Hugh et al., 2005), it is necessary to make the assumption that the cardiovascular system is a closed system and also to describe the cardiovascular system itself (Dwivedi and Dwivedi, 2007). The cardiovascular system is a closed tabular system in which the blood is propelled by a muscular heart via two circuits: the pulmonary and systemic that consist of arterial, capillary and venous components (Dwivedi and Dwivedi, 2007; Mohammadali et al., 2009). It is a system that keeps life pumping through the body with its complex pathways of veins, arteries and capillaries (Mohammadali et al., 2009; West, 2008). The cardiovascular system consists of the heart, blood vessels and the circulatory system (Taylor et al., 1997). On the other hand, the heart is a muscular organ that provides a continuous blood circulation through the cardiac cycle. The heart is divided into four chambers: the two upper chambers called the left and right auricles and the two lower chambers called the left and right ventricles. There is a thick wall of muscles separating the right side and the left side of the heart called the septum. The heart consists of two large veins, the superior/anterior and the inferior/posterior vanae cavae, which bring red deoxygenated blood from

---

*Correspondence author. E-mail: ahkawo@yahoo.com.

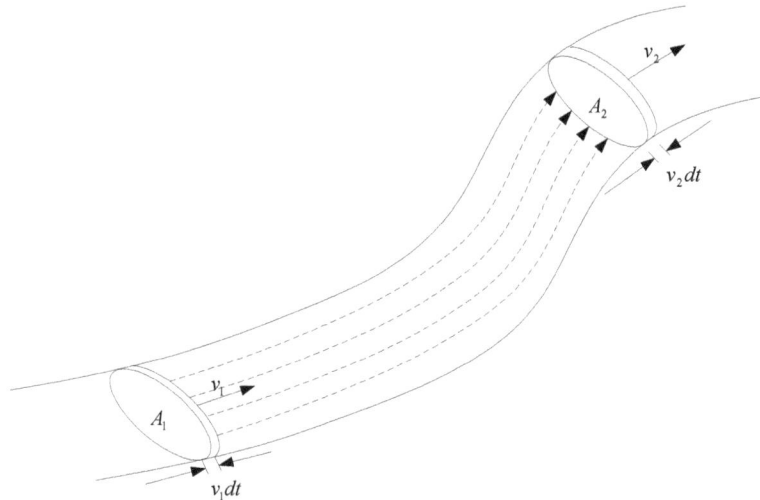

Figure 1. Diagrammatic representation of a flow tube with changing cross-sectional area. If the fluid is incompressible, the product Av has the same value at all points along the tube.

the various parts of the body with the exception of the lungs. The superior is located near the top of the heart while the inferior is located just beneath the superior (Taylor et al., 1997). The main function of the heart is therefore to pump blood round the tissue through systolic and diastolic processes triggered by the spontaneous discharge (Mathias and Andrew, 2011). These can be explained through electrical and mechanical activities of the heart (Massey and Ward-Smith, 2005; Falkovic, 2011). These processes are therefore interesting not only for physics students, but also for students of the biological and medical sciences, because they illustrate the possibility of formulating theories in these areas; hence the justification for the present study.

AN OVERVIEW OF CONTINUITY EQUATION

One may need to know the idea behind the continuity equation of fluid mechanics before discussing it with respect to cardiovascular system. The mass of a moving fluid doesn't change as it flows. This leads to an important quantitative relationship called the continuity equation. Zemasky (2005) considers a portion of flow tube between two stationary cross sections with areas A1 and A2 (Figure 1). The fluid speeds at these sections are v_1 and v_2, respectively. No fluid flows in or out across the sides of the tube because the fluid velocity is tangent to the wall of every point on the wall. During a small time interval dt, the fluid at A1 moves a distance $v_1 dt$, so a cylinder of fluid with height $v_1 dt$ and volume $dv_1 = A_1 v_1 dt$ flows into the tube across A1. During this

same interval, a cylinder of volume $dv_2 = A_2 v_2 dt$ flows out of the tube across A2. Let's first consider the case of an incompressible fluid so that the density ρ has the same value at all points. The mass dm_1 flowing into the tube across A1 in time dt is $dm_1 = \rho A_1 v_1 dt$. Similarly, the mass dm_2 that flows out across A2 in the same time is $dm_2 = \rho A_2 v_2 dt$. In steady flow the total mass in the tube is constant, so $dm_1 = dm_2$ and $\rho A_1 v_1 dt = \rho A_2 v_2 dt$ or, continuity equation, incompressible fluid:

$$A_1 v_1 = A_2 v_2 \qquad (1.1)$$

The product $A v$ is the volume flow rate dv/dt, the rate at which volume crosses a section of the tube; *volume flow rate*

$$\frac{dv}{dt} = A v \qquad (1.2)$$

The mass flow rate is the mass flow per unit time through a cross section. This is equal to the density (ρ) times the volume of flow rate (dv/dt)

Equation (1.1) shows that the volume flow rate has the same value at all points along any tube. When the cross

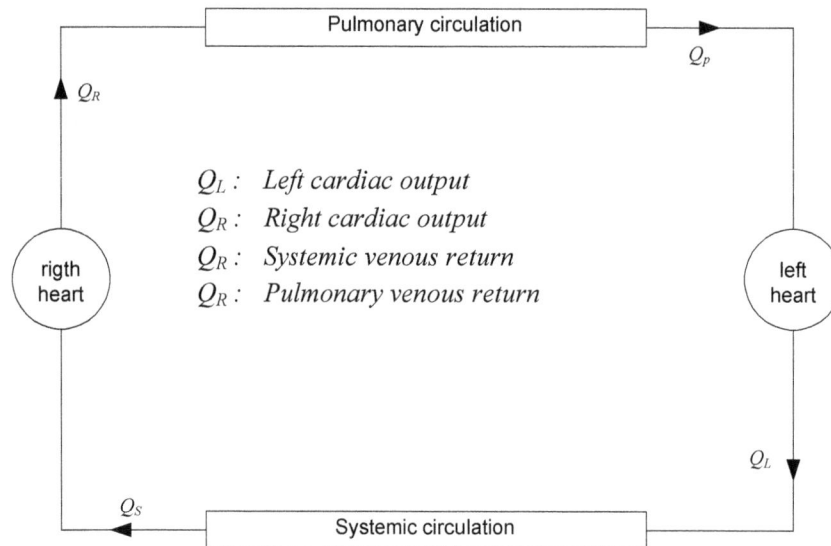

Figure 2. The cardiovascular system.

section of a flow tube decreases, the speed increases, and vice versa.

We can now generalize Equation (1.1) for the case in which the fluid is not incompressible. If ρ_1 and ρ_2 are densities at sections 1 and 2, then, the continuity equation, compressible fluid:

$$\rho_1 A_1 v_1 = \rho_2 A_2 v_2 \tag{1.3}$$

Cardiovascular system

Cardiovascular system consists of a double pump, the heart, and two distinct circulatory systems, i.e., systemic and pulmonary. The heart is divided into two parts, which are called the left and right hearts. Each has two chambers, the atrium and the ventricle, which periodically contract and relax. The relaxation and contraction are synchronized such that when the atrium is contracting, the ventricle is relaxing and vice-versa. The atrium receives and stores blood during the ventricular contraction, and blood flows from the atrium to the ventricle pumped by each heart per unit time is called the cardiac output (Williams and Ganong, 2005). The flow of blood through the system circulations depends on the contraction of the left ventricle, whereas the right heart drives blood through the pulmonary circulation, where blood is oxygenated and CO_2 is disposed. A system of valve ensures that blood flows in the direction as shown in Figure 2.

In Figure 2, Q_L is the cardiac output, Q_R right cardiac output, Q_S systemic nervous return and Q_P pulmonary

venous return respectively.

The cardiovascular system is regulated by both internal and external factors. The effectiveness of the regulation is manifested by the remarkable stability of the system. External factors including nervous activity and chemical substance called hormones, can affect the cardiac performance. There is an internal control mechanism, called the Frank–Starling mechanism that plays an essential role in the maintenance of the balance between the right and the left ventricular outputs and in the distribution of blood between the systemic and pulmonary circulation. A mathematical analysis of this control mechanism is essential for the thorough understanding of cardiovascular mechanics.

MATERIALS AND METHODS

Differential equation for the cardiovascular system

Cardiovascular mechanics is based on the continuity equation and on the momentum equations for blood flow throughout the system. Because the cardiovascular system is very complex, idealizations and approximations are necessary. However, there are numerous mathematical models, which differ in the way the momentum equations are taken into account. In simple models, the momentum equation are taken into account by phenomenological equations that relate the pressure, volume, and the flow of blood in different parts of the system, and by parameters that depends on the blood viscosity and on the geometry and elastic properties of the vascular beds. In more complex models, Navier – Stokes equations are applied to the blood flow and numerically integrated for the assumed boundary conditions (t and t + T). In this paper, only the continuity equation was considered. This is because it is sufficient to explain a control mechanism of blood flow in cardiovascular system.

Let $V_R^i(t)$ be the instantaneous volume of blood contained in the right heart at time t. The continuity equation can be written as

$$d\left[V_R^i\left(t\right)\right]/\,dt = Q_S^i\left(t\right) - Q_R^i\left(t\right) \tag{1.4}$$

Where $Q_R^i\left(t\right)$ is the systemic or right venous return, that is, the instantaneous rate of blood flow (in liters per minute) from the systemic circulation into the right atrium, and $Q_R^i\left(t\right)$ is the instantaneous right cardiac output. The superscript i denotes instantaneous values.
Also, the instantaneous volume of blood in the systemic circulation, in the left heart, and in the pulmonary circulation satisfies

$$dV_S^i\left(t\right)/\,dt = Q_L^i - Q_S^i\left(t\right) \tag{1.5}$$

$$\frac{dV_L^i\left(t\right)}{dt} = Q_P^i\left(t\right) - Q_L^i\left(t\right) \tag{1.6}$$

$$\frac{dV_P^i\left(t\right)}{dt} = Q_R^i\left(t\right) - Q_P^i\left(t\right) \tag{1.7}$$

Where: $Q_L^i\left(t\right)$ and $Q_P^i\left(t\right)$ are the instantaneous left cardiac output and pulmonary venous return respectively. From (1.4) - (1.7), it can be deduced that:

$$d\left(V_L^i + V_R^i + V_S^i + V_P^i\right)dt = 0 = \frac{dV}{dt}, \tag{1.8}$$

which, expresses the conservation of total blood volume in the cardiovascular system.
In steady state, the instantaneous left cardiac out is a periodic function of time and can be written as

$$Q_L^i\left(t\right) = Q_L + f\left(t\right) \tag{1.9}$$

Where: Q_L is the average value of the left cardiac output, given by:

$$Q_L = \left(\frac{1}{T}\right)\int_t^{t+T} Q_L^i\left(t\right)dt \tag{1.10}$$

which, is always constant in time and its value is one after integration. The periodic function $f\left(t\right)$ has an average value equal to zero; T is the cardiac period.
It is observed that during transient phenomena, all the physiological variables in Equations (1.4) - (1.7) are not exactly periodic functions of time, and consequently, their average values in a given cardiac cycle are not necessarily equal to the corresponding values in another cycle. Thus, during transient phenomena, the average values of physiological variables depend on time.

For transient phenomena, during which the physiological quantities have averages that are slowly varying functions of time, Equations (1.4) - (1.7) yield:

$$\frac{dV_R}{dt} \cong Q_S - Q_R \tag{1.11}$$

$$\frac{dV_S}{dt} \cong Q_L - Q_S \tag{1.12}$$

$$\frac{dV_L}{dt} \cong Q_P - Q_L \tag{1.13}$$

$$\frac{dV_P}{dt} \cong Q_R - Q_P \tag{1.14}$$

In this approximation, the form of the equations for the average quantities is almost the same as that for the instantaneous ones. In the steady state, the time derivation on the left-hand side of Equations (1.11) - (1.14) are equal to zero, and hence, all rates of blood flow on the right-hand side have the same value.
It is important to note that there must exists a control mechanism that maintains for long time a balance between the left and right cardiac output, because without such a mechanism the distribution of blood in the system would be unstable. For example, if the right cardiac output remained larger than the left one for a long time, an abnormal accumulation of blood in the pulmonary circulation would occur, whereas the systemic circulation would gradually be emptied of blood. The control mechanism that prevents this problem and maintains the stability of the blood distribution is known as the Frank–Starling mechanism.

The ventricular function curve

It is noted that in 1914, Starling used a canine heart – lung preparation to demonstrate that increasing the stretch on the ventricle of the mammalian heart during relaxation increases the pressure developed during contraction, as depicted by Frank in 1895 where he used a ventricle of the frog heart to demonstrate a similar relationship. Starling found experimentally that there is a relation between cardiac output and right atrial filling pressure. The latter variable determines the degree of filling of the ventricle and may be regarded as a measure of the average blood volume in the heart. The experimental data obtained by Starling shows that the cardiac output first increases and then decreases when the right atrial filling pressure rises (William, 2005). The second part of the relation is called "the descending limb of the starling curve" and has been controversial. Because the right atrial filling pressure is a measure of the average volume of blood contained in the heart, Starling's experimental results can be expressed by saying that cardiac outputs are functions of the average volume of blood contained in the respective hearts. The relations $Q_L = Q_L\left(v_L\right)$ and $Q_R = Q_R\left(v_R\right)$ are called ventricular functions or cardiac output.
When subjects stand in the upright position, the ventricle clearly operates on the ascending limb of its function curve in a relatively steep region, and consequently, fluid administration can markedly

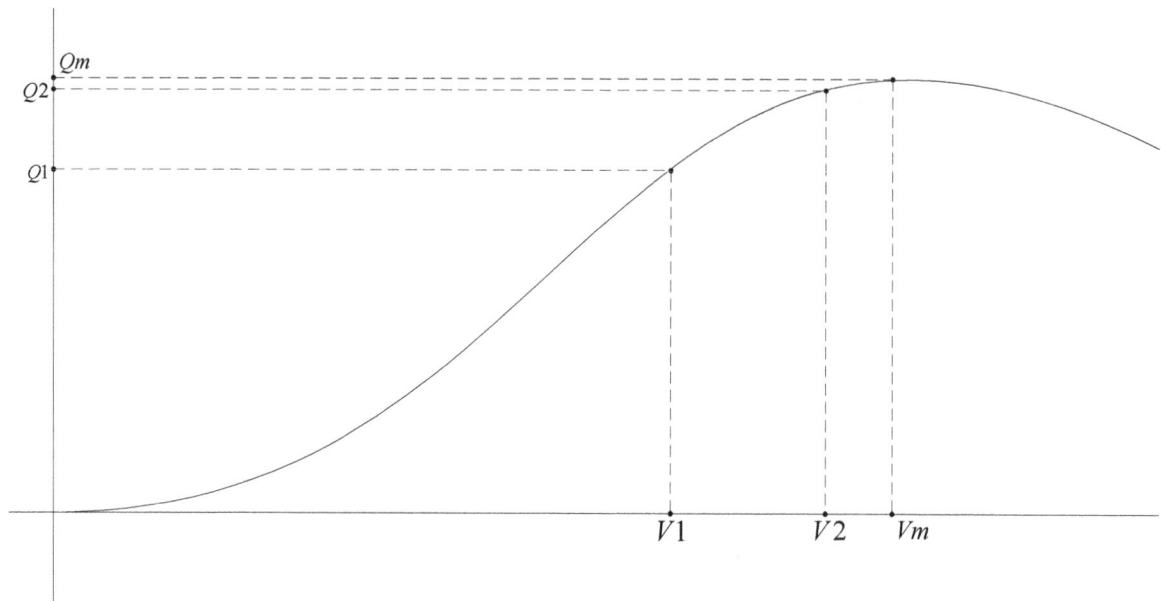

Figure 3. Ventricular function Q denotes cardiac output and v is blood volume in the heart.

enhance pump function. In contrast, in the supine position, in which the subject lies float on hither back, the ventricle operates closer to the maximum of its function curve. Attempts to increase the filling volume lead to an increase in filling pressure, but only to modest improvement in ventricular performance (Uehara and Sakane, 2002).

RESULTS AND DISCUSSION

The experimental observations show that the ventricular function is non-linear as illustrated in Figure 3, in which Q represents Q_L or Q_R, and v represents v_L or v_R. For simplicity, only one function is shown in Figure 4, but it should be noted that the two ventricular functions are not exactly the same (Uehara and Sakane, 2002).

Relationship between the Frank–Starling mechanism and differential equations of the cardiovascular system

The Frank–Starling mechanism or Starling's law of the heart states that the stroke volume of the heart increases in response to an increase in volume of blood filling the heart (the end diastolic volume). The increase volume of blood stretches the ventricular wall, causing cardiac muscle to contract more forcefully (the Frank–Starling mechanism). The stroke volume may also increase as a result of greater contractility of the cardiac muscle during exercise, independent of the end-diastolic volume. The Frank–Starling mechanism appears to make its greatest contribution to increasing stroke volume at lower work

rates, and contractility has its greatest influence at higher rates (Costanzo, 2007). This allows the cardiac output to be synchronized with the venous return, arterial blood supply and humeral length without depending upon external regulation to make alterations. In addition, the Frank-Starling mechanism plays an essential role in the maintenance of the balance between the right and the left ventricular outputs and in the distribution of blood between the systemic and pulmonary circulation (Uehara and Sakane, 2002; Mathias and Andrew, 2011).

Before using Equations (1.11) - (1.14) to explain the control mechanism, it is interesting to consider one quantitative explanation that can be found in physiology textbooks (McGeon, 1996). It was reported that the most important intrinsic mechanism involved in the control of cardiac output is usually referred to as Starling's law of the heart, or the Frank–Starling mechanism, after the two physiologist who first described it. Starling's law helps explain two important features of cardiac function, namely that cardiac output equals venous return and that the average outputs from the two ventricles are equal. If venous return suddenly rises above ventricular output, blood will accumulate in the ventricle, increasing the end diastolic volume. Starling's law predicts that this will lead to an increase in both stroke volume and cardiac output. Until a new state is reached in which cardiac output equals venous return again. Because the output from one ventricle is responsible for the venous return to the other side of the heart in the intact circulation, this mechanism will also ensure that the cardiac output from the two ventricles remains equal. For example, if the cardiac output from the left ventricle increases, this will increase

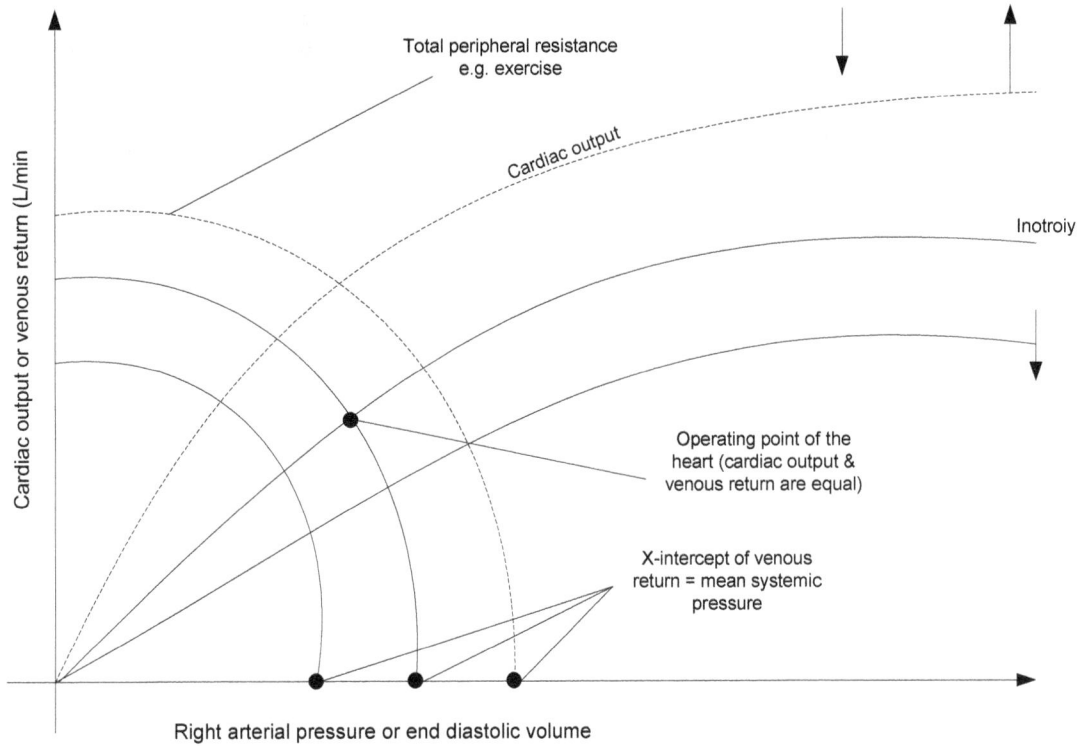

Figure 4. The Frank–Starling law of the heart. The three curves illustrate that shift along the same line indicate a change in preload, while shifts from one line to another indicate a change in after load or contractility.

right venous return and right ventricular output will rise as a consequence.

The above quantitative explanation can be substantiated by a discussion based on Equations(1.11) - (1.14), and on the hypothesis that both hearts work on the ascending part of their ventricular function curves. Equations (1.11) - (1.14) yield

$$\frac{dv_L}{dt} = -Q_L\left(v_L\right) + Q_R\left(v_R\right) - \frac{dv_P}{dt}, \qquad (1.15)$$

$$\frac{dv_R}{dt} = Q_L\left(v_L\right) - Q_R\left(v_R\right) - \frac{dv_S}{dt}. \qquad (1.16)$$

The blood volume distribution in the cardiovascular system depends on body posture relative to the gravitational field. For example, the average blood volumes in the pulmonary circulation and in the heart in the supine position are larger than in the upright position, because gravity induces redistribution of blood volume in the system (Grodins, 2001). Therefore, in a movement from sitting to supine position, there is a transient behaviour of the system during which the blood is redistributed between the pulmonary and systemic circulation according to Equations (1.15) and(1.16).

A complete analysis of transient cardiovascular phenomenon would require the derivation of additional differential equations to describe the response of the pulmonary and systemic circulation to the perturbation of the system. In Equations (1.15) and (1.16) the circulation response is represented by the time derivatives, $\frac{dv_P}{dt}$ and $\frac{dv_S}{dt}$. These quantities depends on the blood viscosity and on the geometry and elastic properties of the vascular beds, which are constituted by arteries and veins (McGeon, 1996). The number of required equations depends on the number of components of the pulmonary and systemic systems.

If we assume that the vascular beds are rigid, Equations (1.15) and (1.16) can be reduced to:

$$\frac{dv_L}{dt} = -Q_L\left(v_L\right) + Q_R\left(v_R\right), \qquad (1.17)$$

$$\frac{dv_R}{dt} = Q_L\left(v_L\right) + Q_R\left(v_R\right). \qquad (1.18)$$

These equations contain only physiological quantities

related to the heart, the active element of the system, and are useful in explaining the role of both hearts in the control mechanism. Mathematically, Equations (1.17) and (1.18) represent the core of the Frank – Starling control mechanism. The average cardiac outputs are implicit time functions with time derivatives given by:

$$\frac{dQ_L}{dt} = \left(\frac{dQ_L}{dv_L}\right)\left(\frac{dv_L}{dt}\right)$$

$$= \left(\frac{dQ_L}{sv_L}\right)\left[Q_R\left(v_R\right) - Q_L\left(v_L\right)\right] \tag{1.19}$$

$$\frac{dQ_R}{dt} = \left(\frac{dQ_R}{dv_R}\right)\left(\frac{dv_R}{dt}\right)$$

$$= \left(\frac{dQ_R}{sv_R}\right)\left[Q_L\left(v_L\right) - Q_R\left(v_R\right)\right] \tag{1.20}$$

Equations (1.19) and (1.20) yield:

$$\frac{d\left(Q_L - Q_R\right)}{dt} = -\left(Q_L - Q_R\right)\left[\frac{dQ_L}{dv_L} + \frac{dQ_R}{dv_R}\right]$$

$$= -\left(Q_L - Q_R\right)\varphi/T \,, \tag{1.21}$$

Where:

$$\left(Q_L - Q_R\right) = T\left[\frac{dQ_L}{dv_L} + \frac{dQ_R}{dv_R}\right] \tag{1.22}$$

If, due to a perturbation of the system, the left cardiac output becomes larger thank the right one, both outputs will vary in time and the difference between them will vary according to Equation (1.21). If both hearts work on the ascending part of the ventricular function curve, the derivatives $\frac{dQ_L}{dv_L}$ and $\frac{dQ_R}{dv_R}$ will be positive and, in this case, Equation (1.21) shows that the difference between left and right cardiac output will decrease in time until the steady state is again restored. The positive value of the φ

expresses the fact that the Frank – Starling mechanism is effective in restoring the steady state of the cardiovascular system. Furthermore, for positive values, the larger the value of φ, the faster the steady state is again attained. Therefore, φ, measures the effectiveness of the Frank – Starling mechanism.

If both hearts work on the descending limb of the ventricular function curve, the derivatives $\frac{dQ_L}{dv_L}$ and $\frac{dQ_R}{dv_R}$ will be negative and consequently, dv_L, φ, will also be negative. In this case, Equation (1.21) shows that the difference between left and right cardiac output will increase with time and the Frank–Starling mechanism would be completely exhausted as a control mechanism, a fact expressed by the negative value of φ. Therefore, both hearts cannot work on the descending limb of the ventricular function curve.

In the linear approximation, in which the time variations of $\frac{dQ_L}{dv_L}$ and $\frac{dQ_R}{dv_R}$ are neglected, Equation (1.21) has the solution:

$$Q_L\left(t\right) - Q_R\left(t\right) = \left[Q_L\left(0\right) - Q_R\left(0\right)\right]\exp\left[-\left(\varphi/T\right)t\right],$$

Which, shows that for φ> 0, the transient duration is inversely proportional to φ.

The above conclusions concerning the Frank–Starling mechanism were deduced without assuming a particular form for the ventricular function. This generality is important because the ventricular function (and φ) varies from person to person, and from moment to moment, depending on the individual's physical conditions. In spite of these variations, the control mechanism is able to maintain the stability of the system.

If we assume, for simplicity, that the left and right ventricular functions have the same mathematical form, we can write $Q_L\left(v_L\right) = Q\left(v_L\right)$ and $Q_R\left(v_R\right) = Q\left(v_R\right)$. Thus, for a steady state, Equation (1.22) reduces to:

$$\varphi\left(v\right) = 2T\left(\frac{dQ}{dv}\right) \tag{1.24}$$

Because of the variability of φ, its numerical value is used to compare the cardiovascular system of different individuals, or of the same person under different physical and health conditions as an example, it is interesting to compare the congestive failing heart with

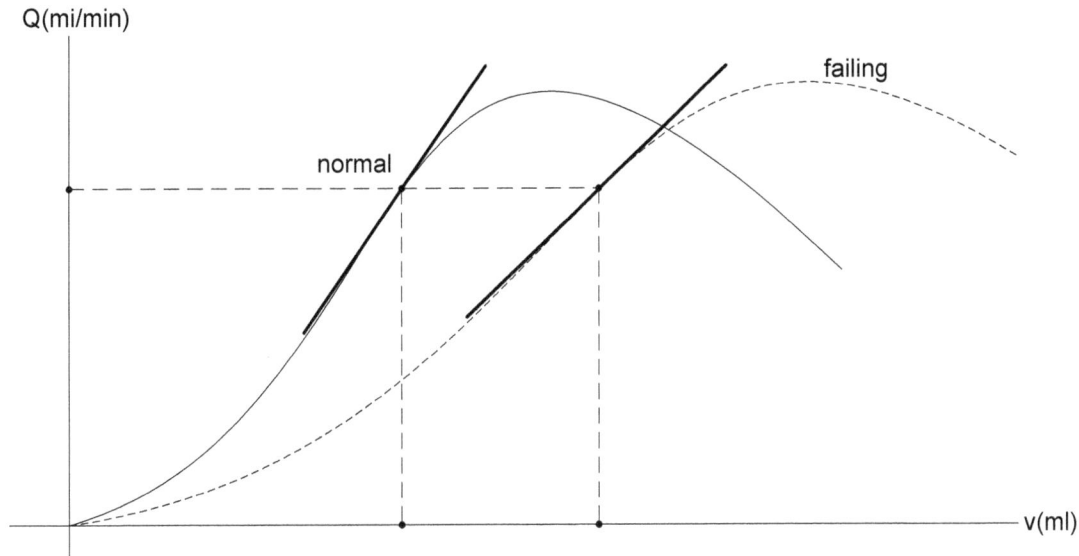

Figure 5. Ventricular function curve for the normal and failing heart.

the normal one. In the former, the cardiac muscle is weaker than the normal one. Research has shown that the ventricular function curve of the failing heart is depressed in comparison to the normal heart curve, and the failing heart works on a relatively flat part of the ventricular function curve. Because, of cardiovascular adaptive abilities, it is possible for cardiac output to be nearly normal, even though the cardiac muscle is severely diseased. However, to maintain normal cardiac output, the heart must dilate.

These observations can be expressed mathematically by saying that $\dfrac{dQ}{dt}$ for the failing heart is smaller than the corresponding value for the normal heart, and consequently, φ, for the failing heart is small compared to the normal value. Figure 5 shows the ventricular function curve of the failing heart in comparison to the normal curve.

In Equations (1.15) and (1.16), the derivatives $\dfrac{dv_P}{dt}$ and $\dfrac{dv_S}{dt}$ represent the response for the pulmonary and systemic circulation to a perturbation of the system and depend on the geometry and elastic properties of the vascular beds and on the blood viscosity. Equations (1.17) and (1.18) were derived by assuming that $\dfrac{dv_P}{dt} = \dfrac{dv_S}{dt} = 0$, which neglect the elastic properties of the vascular beds (Grodins, 1999).

The first mathematical description of the complete cardiovascular system was published by Grodins in 1966 where it was reported that the cardiovascular system manifests mechanical self-regulation. His model includes the elastic properties of vascular beds, and satisfies the conditions that $\dfrac{dQ_L}{dv_L}$ and $\dfrac{dQ_R}{dv_R}$ must be positive.

The mechanical self-regulation of the cardiovascular system is based on the fact that the left and right hearts are connected in series, and both hearts work on the ascending part of their ventricular function curve.

Blood volume re-distribution

Since the heart operates like two pumps connected in series through the pulmonary and systemic circulations (Figure 2), its optimum for redistribution of blood volume in the system is described below (Figures 6 to 10): Consider a situation in which the subject moves from sitting to supine position. The cardiovascular system will show a transient behaviour during which blood will be redistributed between the systemic and pulmonary circulation (Dwivedi and Dwivedi, 2007; West, 2008). As noted, a complete analysis of the transient behaviour requires the derivation of additional differential equations. Although, Equations (1.15) and (1.16) do not constitute a complete set, interesting results can be derived from them as follows:

$$\frac{d\left(v_L - v_R\right)}{dt} = -2Q_L + 2Q_R - \frac{d\left(v_P - v_S\right)}{dt}, \quad (1.25)$$

If we integrate Equation(1.25), we have:

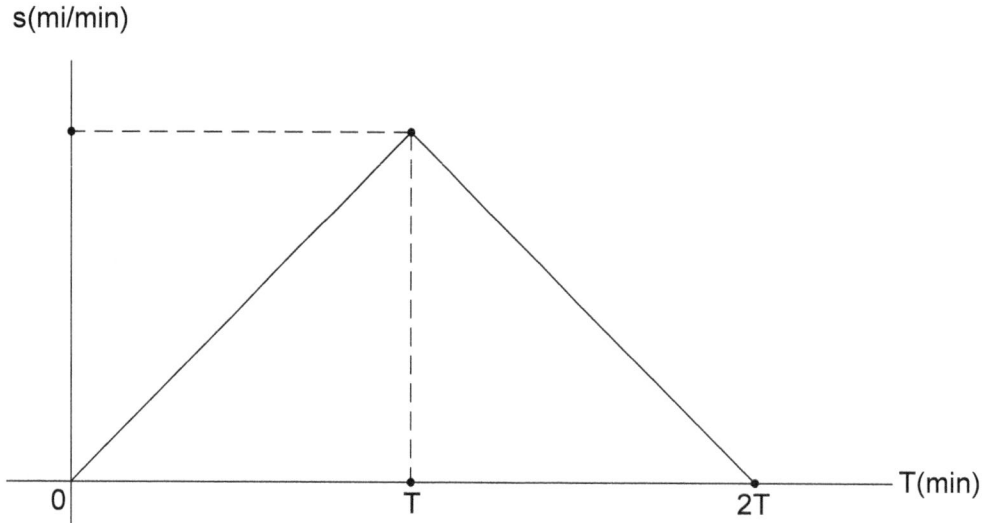

Figure 6. The time function s (t). T is the cardiac period.

$$\int_{t_I}^{t_F} \left(Q_R - Q_L \right) dt = \frac{1}{2}\left[v_P\left(t_F\right) - v_P\left(t_I\right) - v_S\left(t_F\right) + v_S\left(t_I\right) \right]$$

$$= \frac{1}{2}\left[\Delta v_P - \Delta v_S \right]$$

(1.26)

Where: t_I and t_F denotes the initial and the final instants of blood volume redistribution respectively (Figures 6 -10). For a movement from sitting to supine position, the integral in Equation (1.26) is positive because the average blood volume in the pulmonary (systemic) circulation in the supine position is larger (smaller) than in the sitting position. Therefore, during the redistribution of blood (Figure 8) between the pulmonary and systemic circulation, the function $Q_R\left(t\right)$ must necessarily be different from $Q_L\left(t\right)$, $Q_R\left(t\right)$ and $Q_L\left(t\right)$ depend on how the subject move between the initial and final position, but the integral $\int_{t_I}^{t_F}\left(Q_R - Q_L\right)dt$ depends only on variation of the volume of blood in the pulmonary and systemic circulation, as shown by Equation (1.26). The average value of the difference between $Q_R\left(t\right)$ and $Q_L\left(t\right)$, during a transient of duration $\Delta = t_F - t_I$ is given by:

$$\left[Q_R - Q_L \right]_{av} \Delta = \frac{1}{2}\left[\Delta v_P - \Delta v_S \right]$$

(1.27)

The duration of the transient depends on how quickly or slowly the subject moves from one position to another. The faster the movement, the smaller the value of Δ. Hence, Δ cannot equal zero because as $\Delta \to 0$, $\left[Q_R - Q_L \right]_{av} \to \infty$, which is physiologically impossible. Hence, Δ must satisfy the condition $\Delta \geq \Delta_{min} > 0$, where Δ_{min} is the minimum transient duration, which occurs when $\left[Q_R - Q_L \right]_{av}$ is a maximum. It is interesting to note that this result was derived from the fact that in Equation(1.26), the integral depends only on the distribution of blood corresponding to the initial and final steady states, that is, it does not depend on how the subject moves from one position to another. Thus, for a movement from sitting to supine position, there is a minimum transient duration that depends on the characteristic of the heart vascular beds, as well as on blood viscosity, so that it is a measure of the Δ_{min} effectiveness of the cardiovascular system, considered as a whole, in restoring the steady state that was perturbed by the movement of the subject. Additional equations are necessary for a theoretical estimate of Δ_{min}, but it can be experimentally determined. The arterial blood pressure shows a transient behaviour when the subject moves from sitting to supine position, so that Δ_{min} can be experimentally determined by observing this transient behaviour for subject movements at different speeds (Figures 6 to 10).

Because the physics of the cardiovascular system is

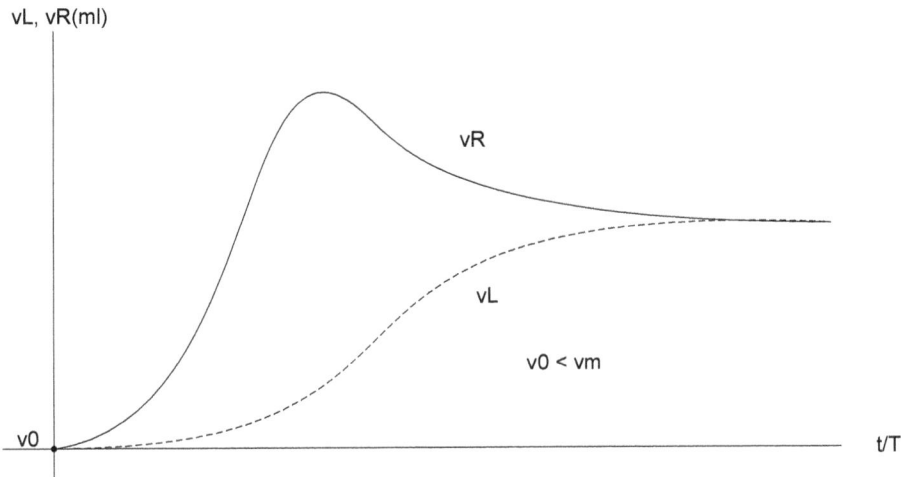

Figure 7. Blood volumes in the left and right heart, v_L and v_R, as time functions, initial volume v_0 smaller than the volume v_m corresponding to the maximum of the ventricular function.

not discussed in physics textbooks, all the idealizations and approximations are clearly stated in this paper, so that it will be useful for teaching one way of doing physics. Due to its great complexity and idealized representation of the cardiovascular system is necessary for deriving differential equations for the blood flow. The idealized representation illustrated in Figure 2 is very simple, but it is sufficient for the derivation of blood flow in the system. It would be interesting to compare the idealized representation in Figure 2 with the real anatomy of the cardiovascular system to stressing the drastic simplification used in our treatment.

The reason for taking the average of Equations (1.4) – (1.7) is to simplify the mathematical discussion, which is much simpler in terms of average quantities. For example, in the steady state, the time derivatives of the instantaneous blood volumes in the left and right heart, $\dfrac{dv_L^i}{dt}$ and $\dfrac{dv_R^i}{dt}$, are functions of time, whereas for the average blood we have:

$$\frac{dv_L}{dt} = \frac{dv_R}{dt} = 0$$

The averaging process, which in this case was accomplished by approximation, represents a great simplification of the mathematical description.

Equations (1.17) and (1.18) were derived from Equations (1.15) and (1.16) assuming that $\dfrac{dv_S}{dt} = \dfrac{dv_P}{dt} = 0$. This assumption described the limiting

case in which the elastic properties of the vascular beds are neglected. The study of "Limiting cases" was considered to be one of the most useful and educational things we can do with any equation. In our treatment, the study of the limiting case made possible an analytical discussion of the Frank–Starling mechanism in which it was not necessary to assume a particular form for the ventricular functions, a generality that is important considering that the ventricular function varies from person to person, and from moment to moment.

It is also noted that the output from one ventricle is responsible for the venous return to the other side of the heart (Mc. Geon's 1996). This statement is expressed mathematically by Equations (1.12) and (1.14), which more precisely show that the venous return $Q_S\left(Q_P\right)$ is related not only to cardiac output $Q_L\left(Q_R\right)$, but also to the time derivative of the volume of blood in the systemic (pulmonary) circulation $\dfrac{dv_S}{dt}\left(\dfrac{dv_P}{dt}\right)$. In the steady state, and also in the limit in which vascular beds are assumed to be infinitely rigid, these times derivative are equal to zero, and consequently, $Q_L = Q_S$ and $Q_P = Q_R$. If the systemic venous return Q_S suddenly rises above the right cardiac output Q_R, blood will accumulate in the heart, as can be seen from Equation (1.11) with respect to the relationship between cardiac output and blood volume in the heart, this increase of blood volume in the right heart will lead to an increase in the right cardiac

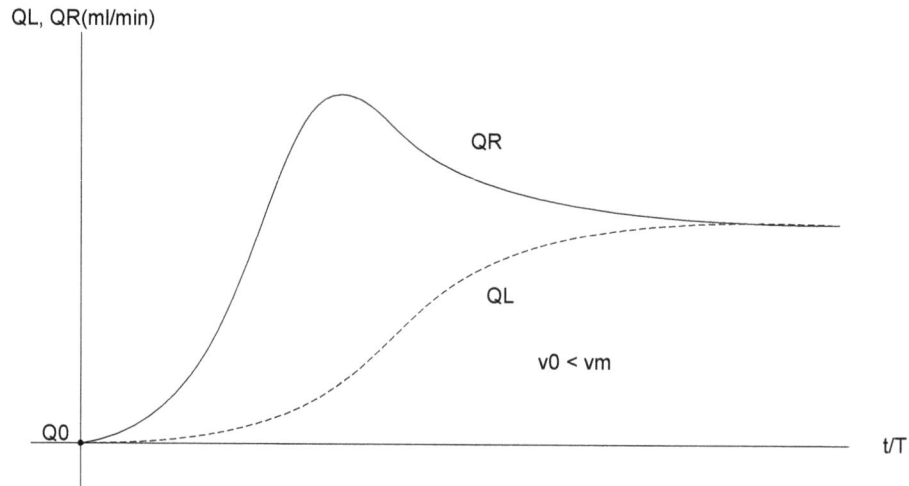

Figure 8. Left and right cardiac outputs Q_L and Q_R, as time functions; $v_0 < v_m$ (initial heart operating on the ascending part of the ventricular function curve).

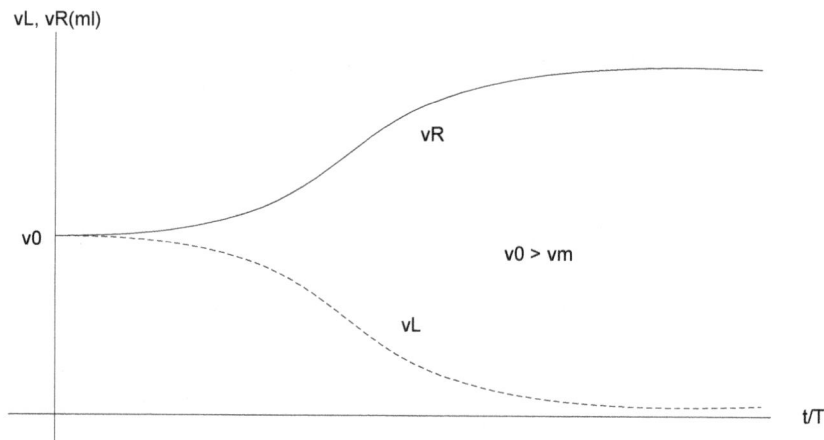

Figure 9. Blood volumes in the left and right heart, v_L and v_R, as time functions, initial volume v_0 larger than the volume v_m corresponding to the maximum of the ventricular function.

output until a new steady state is reached in which the cardiac output Q_R equals the nervous return Q_S again.

The experimental observation that there is a relation expressed by the ventricular function between cardiac output and the volume of blood in the heart is essential for explaining the auto-regulation of the system. Because there is as yet no theory that derives the ventricular function from first principles, this function was included in Equations (1.15) and (1.16) as expressing experimental data. The inclusion of an empirical relation illustrates the formation of a phenomenon of logical theory that contains elements based on experimental observation that are

waiting for a more fundamental explanation (Uehara and Sakane, 2002).

Conclusions and recommendations

The present work describes the application of the continuity equation to the rate of blood flow and variation of the volume of blood in different parts of the cardiovascular system in conjunction with physiological observations has provided a set of differential equations that are useful for clarifying the essential points of the Frank–Starling mechanism. The basis hypothesis for

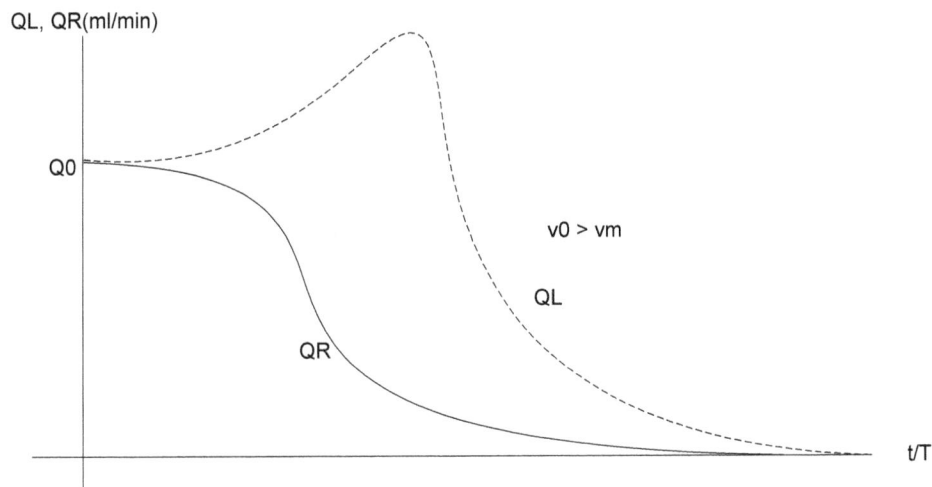

Figure 10. Left and right cardiac outputs Q_L and Q_R, as time functions; $v_0 < v_m$ (initial heart operating on the ascending part of the ventricular function curve)

explaining this mechanism is that the heart operates like two pumps connected in series through pulmonary and systemic circulation (Figure 2). There is a relationship expressed by the ventricular function between cardiac output and blood volume contained in the heart. Both hearts work on the ascending part of their respective ventricular function curves; and the total blood volume in the cardiovascular system is constant. There is need for extension of this work in order to derive more differential equations that will explain different complex models used in cardiovascular system.

REFERENCES

Costanzo LS (2007). Physiology. Williams and Wilkins Edition, pp. 81.

Dwivedi G, Dwivedi S (2007). History of medicine: Shuruta – the Clinician – Teacher par Excellence. Indian J. Chest Dis. Allied Sci. 49: 243-244.

Falkovic G (2011). Fluid mechanics: A short course for physicists. Cambridge University press, UK, pp. 107.

Grodins F (1999). Integrative cardiovascular physiology: A mathematical synthesis of cardiac and blood vessel hemodynamic. Rev. in Biol. Sci., 34: 93–116.

Hugh D, Young R, Freedman A (2005). Sears and Zemansky's University Physics, pp. 549-550.

Massey B, Ward-Smith J (2005). Mechanics of fluids. 8th edition, pp. 415.

Mathias H, Andrew LH (2011). Fluid-Structure interaction in internal physiological flows. Ann. Rev. Fluid Mech., 43:141-162.

McGeon JG (1996). Physiology. Churchill Livingstone publishers, New York, pp. 44.

Mohammadali M, Shojaa R, Shane T, Marios L, Majid K, Farid A, Aaron A, Cohen G, Tubbs RS, Loukas M, Khalil M, Alakbarli F, Cohen-Gardon AA (2009). Vasovagal syncope in the Canon of Avicenna: The first mention of carotid artery hypersensitivity. Int. J. Cardiol., 134(3): 297-301.

Taylor DJ, Green NPO, Stout GW (1997). Biological Science. Third edition. Cambridge University press limited, UK, pp. 984.

Uehara M, Sakane, KK (2002). Physics of the cardiovascular system: An intrinsic control mechanism of the human heart. Am. J. Physics 71(4): 338–344.

West JB (2008). Ibn Al-Nafis: The pulmonary circulation and the Islamic golden age. J. Appl. Physiol. 105(6):1877-1880.

Williams F, Ganong MD (2005). Review of medical physiology. 22nd edition, pp. 572–578.

Zemasky SG (2005). Engineering fluid mechanics. New Delhi publishing house, India, pp. 145.

Effect of the tannoid enriched fraction of *Emblica officinalis* on α-crystallin chaperone activity under hyperglycemic conditions in lens organ culture

P. Anil Kumar, P. Yadagiri Reddy, P. Suryanarayana and G. Bhanuprakash Reddy*

[1]National Institute of Nutrition, Biochemistry, Jamai Osmania, Tarnaka, Hyderabad, AP 500007 India.

Chaperone-like activity (CLA) of α-crystallin is known to be compromised in diabetic conditions and associated with cataract formation. Protecting α-crystallin CLA may help in delaying and/or preventing cataracts. In this study, we employed a lens organ culture model to study the effect of hyperglycemia on the CLA of α-crystallin and investigated the protective effect the tannoids of *Emblica officinalis* had on the CLA of α-crystallin. Goat lenses were treated with 30 mM glucose with or without an aqueous extract of *E. officinalis* tannoids (25 or 50 µg/ml) for 12 days. Cataract development due to hyperglycemia was monitored and a lens soluble protein profile was analyzed using HPLC. α-crystallin fractions from cultured lenses were isolated by gel filtration; CLA, hydrophobicity and structural confirmation of α-crystallin were assessed using light scattering methods. Culturing the lenses with 30 mM glucose resulted in the development of cortical cataracts and the formation of high molecular weight aggregates. α-crystallin isolated from lens incubated in hyperglycemic conditions displayed a significant decrease in CLA. Co-culturing lenses with glucose and tannoids normalized the altered crystallin profile, preserved α-crystallin CLA and prevented cataract formation. This suggests that tannoids may mitigate hyperglycemia mediated manifestations to α-crystallin thereby preventing cataract formation. Tannoids of *E. officinalis* prevented the loss of α-crystallin CLA and cataract formation in lens organ culture. Thus, lens organ culture can be employed to investigate the pharmacological potential of compounds that modulate α-crystallin CLA and consequently delay or prevent cataractogenesis.

Key words: α-Crystallin, chaperone-like activity, hyperglycemia, lens organ culture, tannoids.

INTRODUCTION

Cataracts, which result from the loss of transparency of normally crystalline eye lens is mainly due to the disruption of the micro architecture of the lens. Human cataract formation is a debilitating eye disease that afflicts millions worldwide. Cataracts account for an estimated 16 million cases of blindness worldwide, with approximately half of all cases originating from Africa and Asia (Brian and Taylor, 2001; Congdon et al., 2003). Besides aging, diabetes mellitus (DM) is a significant etiological factor that contributes to the early onset of cataract formation. Studies indicate that hyperglycemia and the duration of diabetes increase the risk of development of a cataract (Brian and Taylor, 2001; Congdon et al., 2003; Ughade et al., 1998). Considering the magnitude of DM worldwide including India (King et al., 1998; Wild et al., 2004) diabetic- induced cataracts may pose a major problem in the management of blindness. Hence, medical treatment for cataract prevention or for slowing down the progression of cataracts is a highly desired alternative and discovering an effective medical treatment for cataracts is likely to have a global impact on eye health. Crystallins are the major structural proteins in the eye lens accounting for up to 90% of the total soluble protein (Harding, 1991; Horwitz, 2003; Bloemendal et al., 2004). There are three distinct families of crystallins: α-, β- and γ-crystallins, whose structure, stability and short-range interactions are thought to contribute to lens transparency (Harding, 1991; Horwitz, 2003; Bloemendal et al., 2004). However, during aging and in clinical conditions, such as

*Corresponding author. E-mail: geereddy@yahoo.com

diabetes, crystallins undergo extensive modifications due to increased oxidative stress and altered intraocular milieu.

Oxidatively challenged, unfolded crystallins are vulnerable to disulphide cross links, and prone to the formation of high molecular weight (HMW), water-insoluble aggregates. α-crystallin exists as an heterooligomer of approximately 800 kDa with two subunits, αA and αB occurring in a stoichiometry of 3:1 (Horwitz, 2003; Kumar and Reddy, 2009; Srinivas et al., 2008). α-crystallin has been shown to function like a chaperone in suppressing protein aggregation under various conditions (Horwitz, 2003; Kumar and Reddy, 2009), so as to protect other lens proteins from the adverse effects of heat, chemicals and UV irradiation. Hence, α-crystallin is instrumental in maintaining the transparency of the lens with its chaperone-like activity (CLA) (Horwitz, 2003; Kumar and Reddy, 2009). Nevertheless, α-crystallin CLA gets compromised under certain conditions. Decreased α-crystallin CLA and the appearance of α-crystallin subunits in HMW fractions are associated with the formation of cataracts. It is believed that preserving α-crystallin's CLA could be a potential means either to prevent or delay the incidence of cataracts (Kumar and Reddy, 2009). Earlier, we showed that compounds that prevent the loss of α-crystallin CLA could delay diabetic cataract formation in experimental rats (Kumar et al., 2005, 2009). We have also reported that the tannoid-enriched fraction of Emblica officinalis inhibits aldose reductase in vitro and prevents hyperglycemia-induced lens opacification in organ culture and streptozotocin-induced diabetic cataracts in rats (Suryanarayana et al., 2004, 2007).

In the present study, we have investigated the effect of tannoid principles of E. officinalis to prevent hyperglycemia-induced lens opacification and their efficacy to prevent the loss of CLA of α-crystallin in lens organ culture.

MATERIALS AND METHODS

ANS (8-anilinonaphthalene-1-sulphonic acid), TC-199 medium (#M3769), glucose, fructose, penicillin, streptomycin, sodium azide, sodium chloride, sodium bicarbonate, EDTA and Tris were purchased from Sigma Chemical Co (St. Louis, MO). Molecular weight markers were from Bio-Rad (Hercules, CA); Sephacryl S-300HR was from Amersham Biosciences (Uppsala, Sweden). TSK 3000 HPLC column from Tosoh (Tokyo, Japan). Tannoid-enriched extract of E. officinalis was obtained in the form of a standardized mixture from Indian Herbs Research and Supply Company (Saharanpur, India; U. S. patent # 6,124,268). The relative proportions of the different tannoids in the standardized extract are as follows: emblicanin A and B, 35 to 55%; punigluconin, 4 to 15%; pedunculagin, 10 to 20%; rutin, 3%; and gallic acid, 1% (Suryanarayana et al., 2004, 2007).

Lens organ culture

Eye balls from six to eight months old goats were obtained from a local abattoir and were dissected within 2 h post mortem. Dissections were performed under a flow hood by an anterior approach. Lenses devoid of any damage were cultured with their anterior surface up. The integrity of the lenses was assessed by measuring the protein content of 20 µl aliquot of the conditioned medium at 2 h after cultivation (Tumminia et al., 1994; Moghaddam et al., 2005). Quantification of the proteins in conditioned medium was performed by the Lowry method. Lens culture medium is composed of TC-199, L-glutamine, sodium bicarbonate, 1% antibiotics (penicillin/streptomycin 100 units/ml). The medium was prepared under sterile conditions adjusting the pH (7.4) and osmolarity (298 ± 2 mOSm/kg). Individual lenses were incubated in 5 ml TC-199 medium at 37°C and 5% CO_2 with 30 mM of glucose. The high glucose concentration was chosen according to published reports (Zigler and Hess, 1985; Suryanarayana et al., 2004). A stock solution of tannoids was added to the medium to produce final concentrations of 25 µg or 50 µg/ml. Lenses incubated in the medium containing 30 mM fructose were treated as osmotic controls (Tumminia et al., 1994; Kamiya and Zigler, 1996). The medium was changed every 48 h and lenses were maintained in culture for 12 days and observed for development of opacification. Visual inspection was conducted by placing lenses on a transparent glass slide with grids.

Photographing of the grid through the lenses was performed with a digital camera (Sony, DSC-P7). After 12 days of culture, lenses were homogenized in a buffer containing 25 mM Tris, 100 mM NaCl, 0.5 mM EDTA, and 0.01% NaN3, pH 8.0 (TNEN buffer). The homogenate was centrifuged at 10,000 xg for 30 min at 4°C and the supernatant was collected as lens soluble protein (LSP) and used for further analysis.

Crystallin distribution profile by HPLC

The α-crystallin distribution in LSP was analyzed by applying 20 µl (1 mg/ml concentration) of LSP on a TSK-3000 SW column (Tosoh, Tokyo, Japan) using HPLC system (Shimadzu, Japan). The column was equilibrated and the proteins were eluted with 0.05 M sodium phosphate buffer (pH 7.2) containing 0.15 M sodium chloride and 0.05% sodium azide (SPSS buffer) with a flow rate of 1 ml/min. The column was calibrated with a set of known proteins whose molecular weights ranged from 669 kDa (thyroglobulin) to 67 kDa (BSA) prior to analyzing the LSP of cultured lenses.

Isolation of α-crystallin

LSP was applied onto a Sephacryl S-300 HR preparative gel-filtration column (100 cm × 1.5 cm) connected to a FPLC system (AKTA-prime, GE biosciences). The column was equilibrated and crystallins were eluted with TNEN buffer. Fractions corresponding to αHMW (high molecular weight) and αLMW (low molecular weight) crystallins were collected separately and their purity was assessed by SDS-PAGE. Equal quantities (15 µg) of α-crystallin preparations were analyzed on self-made 12% polyacrylamide gels using a discontinuous system using a mini slab-gel apparatus. After electrophoresis, the stacking gels were removed and the separating gels were stained with Coomassie Blue solution. αHMW- and αLMW-crystallins will be referred to as αH- and αL-crystallins throughout the manuscript.

Chaperone-like activity of α-crystallin

The CLA of αL-crystallin was assessed with the βL-crystallin aggregation assay (Reddy et al., 2000, 2002). This assay employs the ability of αL-crystallin to suppress the heat-induced aggregation of βL-crystallin (purified from control rat lenses) at 60°C as

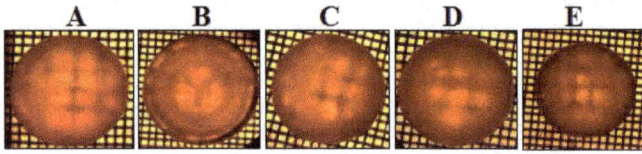

Figure 1. Effect of *E. officinalis* tannoids on lens transparency in organ culture. Representative images from 4 individual experiments are shown. Goat lenses were cultured in modified TC-199 medium in the presence of 30 mM fructose (A), 30 mM glucose (B), 30 mM glucose and 25 µg/ml (C) or 50 µg/ml tannoids (D), and 30 mM fructose +50 µg/ml tannoids (E). The 'Y' shaped suture in the centre and a concentric opaque ring in the cortical region of lens is prominent only in panel B.

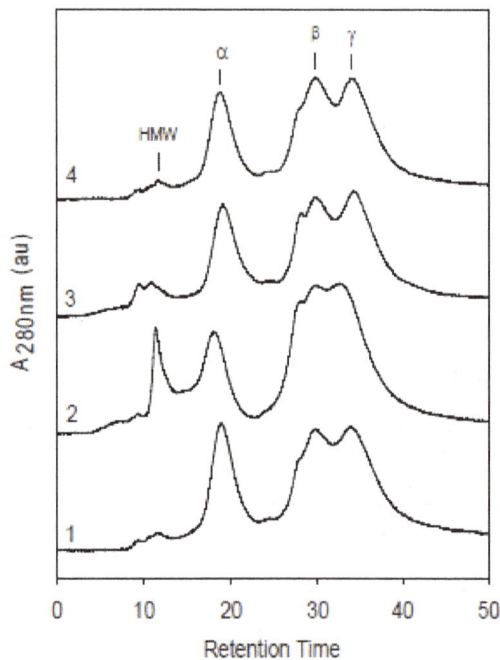

Figure 2. HPLC chromatogram of lens soluble protein. Trace 1 - control lens; Trace 2 - lens incubated with 30 mM glucose; Trace 3 - lens incubated with 30 mM glucose and tannoids 25 µg/ml; Trace 4 - lens incubated with 30 mM glucose and tannoids 50 µg/ml.

monitored by measuring the apparent absorption at 360 nm as a function of time. The relative CLA of αL-crystallin was calculated as a percentage of protection against aggregation using the formula:

% protection = ((Ao − A)/Ao) × 100,

Where Ao and A represent the apparent saturation absorption (after 60 min) in the absence and presence of αL-crystallin.

Circular dichroism (CD) studies

CD spectra were recorded at room temperature using a Jasco J-810 spectropolarimeter. All spectra are the average of five accumulations. Far- and near-UV CD spectra were recorded at

room temperature using cells of 0.01 and 0.02 cm path length, respectively. All spectra were corrected for their respective blanks. The protein concentration used for far- and near-UV measurements was 0.15 and 1.5 mg/ml, respectively.

Fluorescence measurements

Fluorescence measurements were performed using a Jasco spectrofluorometer (FP-6500; Tokyo, Japan). For all measurements, 0.15 mg/ml protein in 20 mM sodium phosphate buffer, pH 7.2 was used. Intrinsic tryptophan fluorescence was recorded by exciting at 280 nm and following the emission between 310 to 390 nm with slit width of 5 nm for both the excitation and the emission filters. The fluorescence of 8-anilino-1-naphthalene-sulfonic acid (ANS) bound to α-crystallin was measured by exciting samples at 390 nm and following the emission between 450 and 600 nm. For this, α-crystallin was incubated with 50 µM ANS for 30 min at room temperature and the fluorescence of the protein-bound dye was measured. The spectra were corrected with appropriate protein and buffer blanks.

Statistical analysis

The Mann-Whitney and Kruskal-Wallis nonparametric tests were performed to analyze the statistical significance of the difference between the distributions of two or multiple independent samples, respectively, using SPSS software (version 11.5).

RESULTS

Goat lenses could be kept viable in culture conditions up to 2 weeks. In this study lenses were incubated in culture for 12 days. Cultured lenses showed no acute protein leakage after 72 h. This suggests that there is no significant damage to the cultured lenses during the dissection and initial incubation. Lenses incubated in 30 mM glucose became opaque and displayed cortical cataracts, with a 'Y' shaped suture in the nucleus of the lens, surrounded by a concentric opaque layer (Figure 1B). Cortical cataracts are one of the predominant lens opacities among diabetic subjects (Leske et al., 1991; Miglior et al., 1994; Rowe et al., 2000; Obrosova et al., 2010). Lenses incubated with high glucose and 25 µg or 50 µg/ml tannoids (Figures 1C and D) showed some slight haziness but no evidence of cataracts compared with untreated lenses (Figure 1B). The morphology of the tannoid-treated lenses was comparable to that of control lenses (Figures 1C, D and A). Incubation of lenses with 30 mM fructose and tannoids 50 µg/ml alone did not cause any change in the morphology (Figure 1E). HPLC analysis of lens soluble protein (LSP) showed that there was an increase in the HMW fraction and a decrease in the α-crystallin fraction in lenses that are maintained in hyperglycemic conditions (Figure 2: Trace 2).

Furthermore, the β and γ fractions of LSP from hyperglycemic lenses were not resolved as distinctly as they were in the control lens (Figure 2: trace 2 vs. 1). Interestingly, the HMW fraction was decreased in lenses incubated with 30 mM glucose and tannoids (Figure 2: Traces 3 and 4) and the effect of tannoids in decreasing

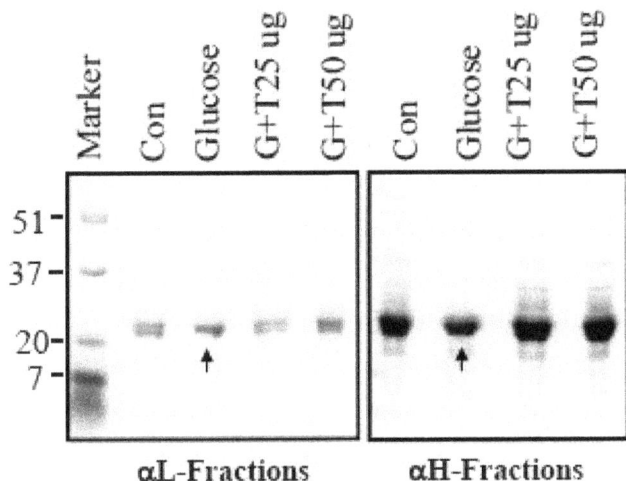

Figure 3. SDS-PAGE profile of αL- and αH-crystallins. LSP from experimental lenses was resolved on a Sephacryl S-300 HR preparative gel-filtration column to obtain αL and αH fractions. The purity of these fractions was analyzed by subjecting them to electrophoresis on 12% SDS-PAGE gels and staining with Coomassie blue (arrow indicates loss of αA subunit).

the HMW fraction was found to be dose dependent. Furthermore, the HPLC profile of the LSP from tannoid-treated samples (50 µg/ml) suggest that the β- and γ-crystallin fractions were resolved separately and the profile was comparable to control lenses (Trace 4 vs. 1).

Concurrently, the increased HMW fraction in lenses grown under hyperglycemic conditions correlated with incidence of cataracts. Furthermore, analysis of the αH- and αL-crystallin fractions on 12% SDS-PAGE revealed that the αA subunit is decreased in the LSP from lenses incubated in hyperglycemic conditions (Figure 3). αL-crystallin isolated from lenses incubated with 30 mM glucose showed a significant loss (approximately 75%) of CLA in suppressing the heat-induced aggregation of βL-crystallin over αL-crystallin isolated from control lens (Figures 4A and B). Interestingly, αL-crystallin isolated from lenses that were exposed to glucose in the presence of tannoids showed improved CLA in a dose dependent manner compared with αL-crystallin from lenses exposed to 30 mM glucose alone (Figure 4). The far-UV CD spectrum for αL-crystallin from control lenses shows negative ellipticity at 217 nm indicating a typical β-sheet structure (Figure 5) (Kumar et al., 2004). The far-UV CD signal for αL-crystallin isolated from lenses incubated with 30 mM glucose showed a loss of negative ellipticity suggesting an altered secondary structure for αL-crystallin isolated from hyperglycemic lenses (Figure 5: Trace 2 vs. 1). However, treatment with tannoids had no modulatory effect on hyperglycemia-mediated secondary structural changes to αL-crystallin. Tryptophan fluorescence of αL-crystallin isolated from lenses incubated with 30 mM glucose was decreased compared with αL-crystallin from control lenses (Figure 6: Trace 2

vs. 1). Loss of tryptophan fluorescence, an indicator of protein tertiary structure changes, suggests a loss of the native conformation of αL-crystallin isolated from hyperglycemic lenses. αL-crystallin isolated from lens incubated with high glucose and tannoids 25 or 50 µg/ml displayed increased tryptophan fluorescence compared with αL-crystallin from lenses treated with glucose alone (Figure 6: Traces 3 and 4 vs. 2). It is generally considered that surface-exposed hydrophobic sites of α-crystallin are involved in the CLA of αL-crystallin (Reddy et al., 2006). Therefore, we measured the surface hydrophobicity in terms of ANS-bound fluorescence.

αL-crystallin from lenses incubated with 30 mM glucose showed less ANS fluorescence than those from control lenses (Figure 7) indicating that there is less accessible surface hydrophobicity in αL-crystallin isolated from lenses incubated in hyperglycemic conditions. However, ANS fluorescence of αL-crystallin isolated from lenses treated with 30 mM glucose and tannoids was higher than that from lenses treated with glucose alone (Figure 7). Decreased accessible surface hydrophobicity of αL-crystallin isolated from lens incubated with hyperglycemic conditions, as observed here, correlates well with its compromised CLA (Figure 4, Trace 3 and Figure 7, Trace 2).

DISCUSSION

Cataracts, which account for nearly half of the blind population, is the most common cause of preventable blindness (Taylor, 1999). Exploring methods to discover anti-cataractogenic agents and pharmacological interventions could help overcome this avoidable blindness. In the present study we adopted a lens organ culture model to study the effect of hyperglycemia on the lens transparency and CLA of α-crystallin. Lens organ culture was used in several studies aimed at understanding the physiology and biochemistry of cataracts (Kamiya and Zigler, 1996; Zigler et al., 1985, 2003). A long-term lens organ culture was employed to monitor lens optical quality (Dovrat and Sivak, 2005) and to determine age-related effects of UV irradiation on the eye lens (Azzam and Dovrat, 2004). Cataract development in this lens culture model shows similarities to experimental cataracts in rodents, during which formation of HMW aggregates was observed (Kumar et al., 2005). Furthermore, incubation of lenses for 12 days with 30 mM glucose resulted in cataract formation that is usually seen in case of diabetic subjects. The LSP fraction of lenses incubated with high glucose has varied significantly from that of control lenses. An increased HMW crystallin fraction, decreased α-crystallin fraction, and altered β- and γ- crystallin profiles were noticed in lenses incubated under high glucose conditions. A similar profile was seen in the LSP from streptozotocin-induced diabetic cataract rat lens (Kumar et al., 2005). It should

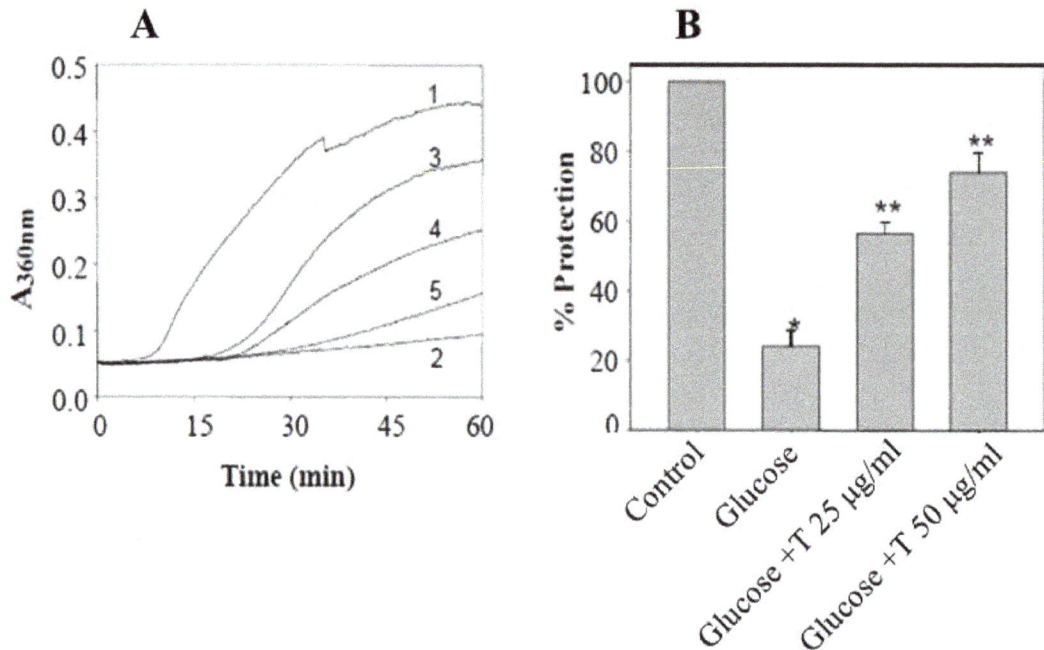

Figure 4. Chaperone-like activity (CLA) of αL-crystallin. A) Representative data of 4 to 5 individual chaperone assays. CLA of αL-crystallin in heat induced aggregation of βL-crystallin (0.2 mg/ml in 50 mM phosphate buffer, pH 7.4) at 60°C. βL-crystallin was incubated in the absence (Trace 1) or presence of 25 µg/ml of αL-crystallin isolated from control lenses (Trace 2), from lenses incubated with 30 mM glucose (Trace 3), from lenses incubated with 30 mM glucose and tannoids 25 µg/ml (Trace 4) or tannoids 50 µg/ml (Trace 5); B) percent protection of heat-induced aggregation of βL-crystallin by αL-crystallin preparations from different treatments. The percent protection by αL-crystallin from control lenses was considered as 100%. Data are expressed as mean ± SE; n = 4. *, p< 0.01, compared with control αL-crystallin, **, p<0.01, compared with α-crystallin from lens incubated with 30 mM glucose.

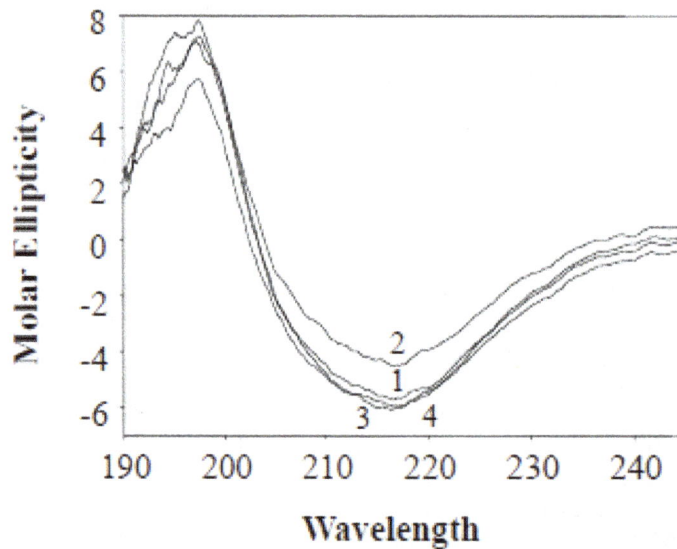

Figure 5. Secondary structure of αL-crystallin. Far-UV CD spectrum of αL-crystallin preparation from a control lens (Trace 1), from a lens incubated with 30 mM glucose (Trace 2), from lenses incubated with 30 mM glucose and tannoids 25 µg/ml (trace 3) or tannoids 50 µg/ml (Trace 4). Data were the average of four assays.

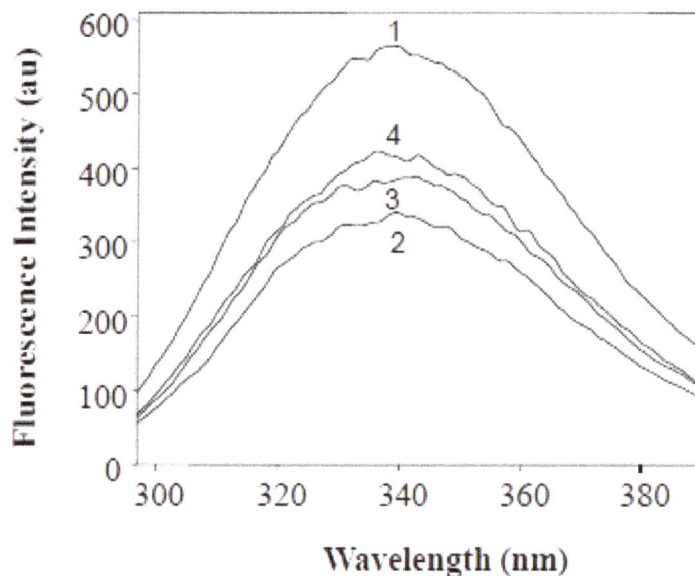

Figure 6. Tertiary structure of αL-crystallin. Tryptophan fluorescence of αL-crystallin from a control lens (Trace 1), from a lens incubated with 30 mM glucose (Trace 2), from-lenses incubated with 30 mM glucose and tannoids 25 µg/ml (Trace 3) or tannoids 50 µg/ml (Trace 4). Data were the average of four assays.

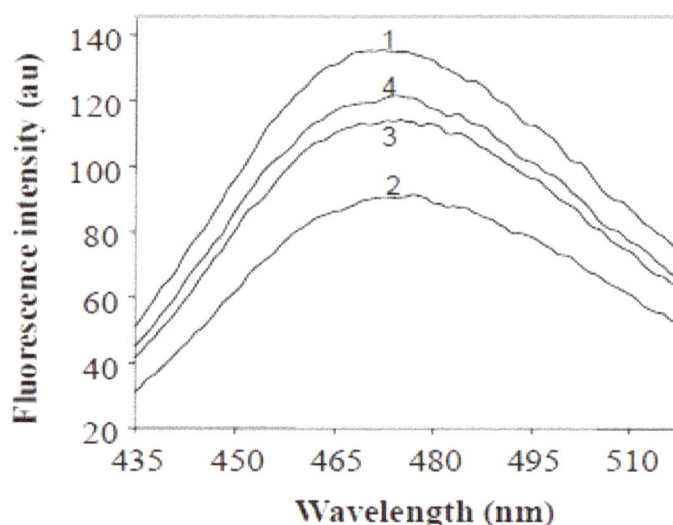

Figure 7. Hydrophobocity of αL-crystallin: ANS bound fluorescence of αL-crystallin from a control lens (Trace 1), from a lens incubated with 30 mM glucose (Trace 2), from-lenses incubated with 30 mM glucose and tannoids 25 µg/ml (Trace 3) or tannoids 50 µg/ml (Trace 4). Data were the average of four assays.

be noted that there is a loss of the αA-subunit in both αH and αL-crystallin preparations from lenses cultured with 30 mM glucose as evident from SDS-polyacrylamide gels. Further, α-crystallin isolated from lenses incubated in hyperglycemic conditions showed decreased CLA. An altered crystallin profile and compromised CLA of α-crystallin from lens incubated with hyperglycemic conditions were comparable to those changes observed with diabetic cataracts in, both, rats and humans (Cherian and Abraham, 1995b; Thampi et al., 2002; Kumar et al., 2005).

A previous study that reported lens opacification due to

lead exposure in an organ culture system also displayed a decrease in α-crystallin CLA (Neal et al., 2010). These data suggest that lens organ culture can be used for screening compounds that modulate α-crystallin CLA in relation to cataractogenesis. Under diabetic conditions, increased activity of aldose reductase leads to the accumulation of sorbitol, an osmolyte, and the resultant oxidative stress is considered as an etiological event that cause diabetic cataracts (Zigler and Hess, 1985; Kador PF 1988; Obrosova et al., 2010). The chaperoning ability of α-crystallin is known to be compromised by oxidative stress (Cherian and Abraham, 1995a; Peluso et al., 2001; Rajan et al., 2006). In addition, non-enzymatic glycation and activation of protein kinase C also contribute to cataractogenesis. Therefore, modulation of α-crystallin's CLA through the inhibition of aldose reductase or prevention of non-enzymatic glycation would provide potential targets to prevent or delay cataract formation (Kumar and Reddy, 2009). Earlier, we reported that the tannoids of *E. officinalis* inhibited aldose reductase in rat eye lenses (Suryanarayana et al., 2004). In addition, feeding tannoids to diabetic rats delayed the incidence of cataracts (Suryanarayana et al., 2007). However, it is unknown whether tannoids modulate the CLA of α-crystallin under hyperglycemic conditions. Interestingly, we demonstrate here that tannoids prevented the formation of HMW aggregates, averted the alteration in αL-crystallin's secondary and tertiary structures, and improved its CLA under hyperglycemic conditions. Therefore, we speculate that the aldose reductase inhibitory potential of tannoids might have minimized the oxidative stress and improved the high glucose-mediated loss of α-crystallin CLA. In summary, the present study suggests that the lens organ culture system can be used to study compounds that modulate α-crystallin CLA during experimental conditions.

From an investigational and potentially therapeutic standpoint, the ability of tannoids to modulate α-crystallin chaperone like activity and prevent cataract formation in a lens organ culture model has immense value. It is our anticipation that modulation of CLA of α-crystallin will provide new insight for tannoids as a novel therapy directed for the treatment of diabetic cataracts.

ACKNOWLEDGEMENTS

This study was supported by grants from Department of Science and Technology and Indian Council of Medical Research, Government of India to GBR. The authors thank Samuel Zigler (NIE and NIH) for suggestions on the lens organ culture and the Indian Herbs Research and Supply Company for providing the tannoid mixture.

REFERENCES

Azzam N, Dovrat A (2004). Long-term lens organ culture system to determine age-related effects of UV irradiation on the eye lens. Exp. Eye Res., 79: 903-911.

Bloemendal H, de Jong W, Jaenicke R, Lubsen NH, Slingsby C, Tardieu A (2004). Ageing and vision: structure, stability and function of lens crystallins. Prog. Biophys. Mol. Biol., 86: 407-485.

Brian G, Taylor H (2001). Cataract blindness--challenges for the 21st century. Bull WHO, 79: 249-256.

Cherian M, Abraham EC (1995a). Decreased molecular chaperone property of alpha-crystallins due to posttranslational modifications. Biochem. Biophys. Res. Commun., 208: 675-679.

Cherian M, Abraham EC (1995b). Diabetes affects alpha-crystallin chaperone function. Biochem. Biophys. Res. Commun., 212: 184-189.

Congdon NG, Friedman DS, Lietman T (2003). Important causes of visual impairment in the world today. JAMA, 290: 2057-2060.

Dovrat A, Sivak JG (2005). Long-term lens organ culture system with a method for monitoring lens optical quality. Photochem. Photobiol., 81: 502-505.

Harding JJ (1991). (ed) Cataract; Biochemistry, Epidemiology and Pharmacology. London: Chapman & Hall.

Horwitz J (2003). Alpha-crystallin. Exp. Eye Res., 76: 145-153.

Kador PF (1988). The role of aldose reductase in the development of diabetic complications. Med. Res. Rev., 8: 325-352.

Kamiya T, Zigler JS Jr (1996). Long-term maintenance of monkey lenses in organ culture: a potential model system for the study of human cataractogenesis. Exp. Eye Res., 63: 425-431.

King H, Aubert RE, Herman WH (1998). Global burden of diabetes, 1995-2025: prevalence, numerical estimates, and projections. Diabetes Care, 21: 1414-1431.

Kumar MS, Reddy PY, Kumar PA, Surolia I, Reddy GB (2004). Effect of dicarbonyl-induced browning on alpha-crystallin chaperone-like activity: physiological significance and caveats of in vitro aggregation assays. Biochem. J., 379: 273-282.

Kumar PA, Reddy GB (2009). Modulation of alpha-crystallin chaperone activity: a target to prevent or delay cataract? IUBMB Life, 61: 485-495.

Kumar PA, Reddy PY, Srinivas PN, Reddy GB (2009). Delay of diabetic cataract in rats by the antiglycating potential of cumin through modulation of alpha-crystallin chaperone activity. J. Nutr. Biochem., 20: 553-562.

Kumar PA, Suryanarayana P, Reddy PY, Reddy GB (2005). Modulation of alpha-crystallin chaperone activity in diabetic rat lens by curcumin. Mol. Vis., 11: 561-568.

Leske MC, Chylack LT Jr, Wu SY (1991). The Lens Opacities Case-Control Study. Risk factors for cataract. Arch Ophthalmol., 109: 244-251.

Miglior S, Marighi PE, Musicco M, Balestreri C, Nicolosi A, Orzalesi N (1994). Risk factors for cortical, nuclear, posterior subcapsular and mixed cataract: a case-control study. Ophthalmic Epidemiol., 1: 93-105.

Moghaddam MS, Kumar PA, Reddy GB, Ghole VS (2005). Effect of Diabecon on sugar-induced lens opacity in organ culture: mechanism of action. J. Ethnopharmacol., 97: 397-403.

Neal R, Lin C, Isom R, Vaishnav K, Zigler JS Jr (2010). Opacification of lenses cultured in the presence of Pb. Mol. Vis. 16: 2137-2145.

Obrosova IG, Chung SS, Kador PF (2010). Diabetic cataracts: mechanisms and management. Diabetes Metab. Res. Rev., 26: 172-180.

Peluso G, Petillo O, Barbarisi A (2001). Carnitine protects the molecular chaperone activity of lens alpha-crystallin and decreases the post-translational protein modifications induced by oxidative stress. FASEB J., 15: 1604-1606.

Rajan S, Horn C, Abraham EC (2006). Effect of oxidation of alphaA- and alphaB-crystallins on their structure, oligomerization and chaperone function. Mol. Cell Biochem., 288: 125-134.

Reddy GB, Das KP, Petrash JM, Surewicz WK (2000). Temperature-dependent chaperone activity and structural properties of human alphaA- and alphaB-crystallins. J. Biol. Chem., 275: 4565-4570.

Reddy GB, Kumar PA, Kumar MS (2006). Chaperone-like activity and hydrophobicity of alpha-crystallin. IUBMB Life, 58: 632-641.

Reddy GB, Reddy PY, Vijayalakshmi A, Kumar MS, Suryanarayana P, Sesikeran B (2002). Effect of long-term dietary manipulation on the aggregation of rat lens crystallins: role of alpha-crystallin chaperone function. Mol. Vis., 8: 298-305.

Rowe NG, Mitchell PG, Cumming RG, Wans JJ (2000). Diabetes, fasting blood glucose and age-related cataract: the Blue Mountains Eye Study. Ophthalmic Epidemiol., 7: 103-114.

Srinivas PN, Reddy PY, Reddy GB (2008). Significance of alpha-crystallin heteropolymer with a 3:1 alphaA/alphaB ratio: chaperone-like activity, structure and hydrophobicity. Biochem. J., 414: 453-460.

Suryanarayana P, Kumar PA, Saraswat M, Petrash JM, Reddy GB (2004). Inhibition of aldose reductase by tannoid principles of Emblica officinalis: implications for the prevention of sugar cataract. Mol. Vis., 10: 148-154.

Suryanarayana P, Saraswat M, Petrash JM, Reddy GB (2007). Emblica officinalis and its enriched tannoids delay streptozotocin-induced diabetic cataract in rats. Mol. Vis., 13: 1291-1297.

Taylor HR (1999). Epidemiology of age-related cataract. Eye (Lond), 13 (Pt 3b): 445-448.

Thampi P, Zarina S, Abraham EC (2002). alpha-Crystallin chaperone function in diabetic rat and human lenses. Mol. Cell Biochem., 229: 113-118.

Tumminia SJ, Qin C, Zigler JS Jr (1994). Russell P. The integrity of mammalian lenses in organ culture. Exp. Eye Res., 58: 367-374.

Ughade SN, Zodpey SP, Khanolkar VA (1998). Risk factors for cataract: a case control study. Indian J. Ophthalmol., 46: 221-227.

Wild S, Roglic G, Green A, Sicree R, King H (2004). Global prevalence of diabetes: estimates for the year 2000 and projections for 2030. Diabetes Care, 27: 1047-1053.

Zigler JS Jr, Hess HH (1985). Cataracts in the Royal College of Surgeons rat: evidence for initiation by lipid peroxidation products. Exp. Eye Res., 41: 67-76.

Zigler JS Jr, Qin C, Kamiya T (2003). Tempol-H inhibits opacification of lenses in organ culture. Free Radic. Biol. Med., 35: 1194-1202.

10

Changes in airway resistance with cumulative numbers of cigarettes smoked

Almaasfeh Sultan

Physics Department College of Science, Al Hussein Bin Talal University, Jordan.

This work was performed with smokers of Virginia tobacco cigarettes without taking into consideration brand or type of cigarettes. An experimental study was conducted for the variation of cigarettes number and the increments in airway resistance for both normal and respiratory disordered subjects using body plethysmograph for testing individuals. Seventeen (17) nonsmokers and ten smokers were studied. The investigation evaluated respiratory function in both smokers and nonsmokers: their airway resistances were recorded, and relationships between height, weight and age were documented. The findings reveal that smoking of higher numbers of self-reported cumulative cigarettes was associated with higher airway resistance and higher total airway resistance. Comparisons of age, height and weight versus airway resistance revealed only slight changes in comparison to those associated with cumulative lifetime cigarette consumption.

Key words: Air way resistance, total air way resistance, plethysmography, normal and disordered respiratory system.

INTRODUCTION

The first few breaths after birth are the most challenging breaths for the neonatal. The neonatal will have to work very hard to overcome the first breaths which are relatively the largest breath the new born conducts at his initial breathing experience. This process is thus done to:1) Inflate and deflate the lungs and chest wall outward (during inspiration) and inward (during expiration), throughout normal breathing; 2) move abdominal structures involved; overcome expected airway resistance (Shapiro et al., 1991; Al Sa'ady, 1997) plus any external resistance of assisting instruments if used during some modes of ventilation (Polese et al., 1991; Al Kadri, 1998);

3) overcome increased impedance may be created by disorders of the respiratory system that may lead to above normal respiratory muscle activity (Beydon et al., 1988; Al Kadri, 1998; Irving and Herman, 2007); 4) meet with demand because of increased ventilation during some circumstances such as exercise (Guyton and Hall, 2009; Al Kadri, 1998; Irving and Herman, 2007).

To fulfill the purpose of the project, instantaneous work and effort were exerted because of friction between molecules of gas itself and between the gas molecules against the tube walls. When flow is laminar or stream-line, Raw varies directly with the viscosity of the gas, the length of the tubes and inversely with the fourth power of the radius of the lumen of tubes (Sykes and Vickers, 1970; Tortora, 2006; Goldman et al., 1976).

$$Raw = \frac{vis\cos ity \times .length}{(radius)^4} \times \frac{8}{\pi}$$

If breathing is conducted normally and quietly, the resistance (R) should remain constant then the equation of the resistance remains:

$$Raw = \frac{\Delta P}{Vdot} \ldots\ldots \frac{cmH_2O}{lit/\sec}$$

This means that Raw = change in driving pressure ΔP in cmH_2O (trans- airway pressure minus atmospheric pressure) per unit change in flow rate $\dot{V}$ (Tortora, 2006).

The aim of the present study was to estimate the variability of airway resistance versus cumulative number of cigarettes. This is because resistance is expected to vary in normal and disordered system and also upon connecting the respiratory system to external circuits or ventilating devices.

MATERIALS AND METHODS

No specific selection was necessary in choosing volunteers, because the aim of the project was to make a clinical comparative study; accordingly, 27 subjects were taken to achieve the aim of the study at Al-Hussein Medical City. Ten of them were good smokers while 17 were nonsmokers. They were of different heights, weights and ages. For the completion of the study, each volunteer breathed (upon appropriate instructions) into the system inlet tube which was provided with mouth piece and gas pillows inside the cage in the glass-room. The subject was asked to sit on chair inside the glass room with the mouth piece linked to his mouth a nose clip on his nose, and the subject was asked to inspire a certain volume of air through the mouth piece. A transducer or sensor was attached in the way to measure the flow rate and pressure and then the body plethysmograph give the value of Raw and Rawt. Values were obtained in a report. Tables 1 and 2 show the information of the smokers and nonsmokers selected

to the study respectively.

Body plethysmography

A plethysmographs was used .The one in Jordan (Al Hussain Medical Center) is of Sensor Medics (Cardiopulmonary care company); 6200 Auto box DI Automated Body Plethysmograph. By making VTG and Raw measurement at rapid rate (panting), very small leaks are acceptable. "Slow" leaks are sometimes introduced to facilitate thermal equilibrium by connecting along tube of small –bore to the atmospheric side of the box pressure transducer or connecting it to a glass bottle within the box. Pressure plethysmograph is suited to measure small volume change (100 ml or less).

The flow box uses a flow transducer in the box wall to measure volume changes into the box. The subject breathes through a pneumotachometer which is connected to the room. Gas in the box is compressed or decompressed to measure the pressure change as gas flows out of the box through the flow opening.

Flow through the wall is integrated and correction is made to record the volume change as the total volume passes through the wall. The volume will be compressed. The flow type of plethysomgraph requires computerization assessment. In addition, maximal-breathing maneuvers such as the VC may be recorded with the subject in the flow box.

Computerized plethysmographs offer the advantage of providing lung volume and airway resistance information immediately after the completion of the manoeuvre. This aids the technologist to select appropriate manoeuvres to get the average.

Most plethysmographs include the necessary hardware to perform physical calibration. Computerized systems allow automated calibration of transducers by means of software – generated correction factors , along with the actual physical calibration . A few manufacturers supply quality control devices such as the isothermal lung analogue. Subjects have ease in performing the required manoeuvres because it is an important feature of the plethysmograph investigations since many subjects may become claustrophobic inside the box. New boxes made of durable plastic transparent and less confining for the subject were also provided by communication system to allow voice contact with the subject.

Here are some of the tests that the subject had to pass with the help of the body plethysmograph to diagnose respiratory system state:

Pulmonary function tests (pft)

Forced vital capacity (FVC)

Forced vital capacity (FVC) is the maximum volume of gas that can be expired as forcefully and rapidly as possible after maximal inspiration.

Forced expiratory volume within the first second (FEV1)

The FEV1% is a statement of FEV for a given interval expressed as a percentage of the subjects actual FVC. It has been reported that a normal subject can expire 50-60% of the FVC in 0.5 s, 75-85% in 1 s, 94% in 2 s, and 97% in 3 s. Slightly, lower ratios may be observed in elderly adults, but in general, subjects without airway disorders obstruction and restriction expire their VC within 4 s. Conversely, subjects with obstructive disease will show a reduced FEV1 in most cases; an FEV1 % lower than 70% is the hallmark of obstructive disease.

Table 1. The values for non- smokers.

Serial number	Sex	Age	Height	Weight	Raw<2.24	Rawt<3.06	Cigarette /day	Year of smoking	Total number of cigarette
1	Male	33	176	96	2.76	4.03	20	2	14600
2	Male	28	175	78	2.56	3.49	10	10	36500
3	Male	32	168	73	1.72	3.12	12	10	43800
4	Male	31	165	75	1.52	2.62	20	10	73000
5	Male	34	165	72	2.57	3.46	20	15	109500
6	Male	39	176	82	3.17	5.16	20	19	138700
7	Male	33	177	82	2.41	5.94	30	13	142350
8	Male	38	167	59	3.49	4.79	40	12	175200
9	Male	32	167	82	4.79	6.05	30	16	175200
10	Male	30	168	103	2.9	3.65	40	12	175200
Sum		330	1704	802	27.89	42.31	242	119	1084050
Average		33	170.4	80.2	2.789	4.231	24.2	11.9	108405
Standard deviation		3.367	4.949	12.381	0.922	1.191	10.475	4.557	62,229.854
Variance		11.333	24.489	153.289	0.850	1.417	109.733	20.767	3,872,554,694.4
f-test		0.021835872	2.7884E-05	8.95741E-09	4.55694E-42	4.54672E-41	4.5569E-42	4.5569E-42	4.54672E-41
t-test		5.46486E-11	4.62884E-16	8.90334E-09	0.000375992	0.000376028	0.000376028	0.00037599	0.000376028

To change units of pressure from cmH_2O to Pascal (Pa) we multiplied by 98.1; to change earway resistance from cmH_2O/Lit/Sec to Pascal/m^3/Sec we multiplied by 98.1x10^3; To change unit of resistance from KPa/Lit/Sec to Pa/m^3/Sec we multiplied 1x10^6.

Subjects with restrictive disease often show a normal or supranormal FEV1 % since their flow rates may be minimally affected, and the FEV1 and FVC are usually reduced in equal proportion. Following this point, both respiratory compliance and resistance of each volunteer were measured by: closing the mouth by a shutter at the end of the tidal inspiration from the spirometer and after one to two s; mouth pressure was measured and then compliance was obtained and releasing the shutter at the inset of flow, both pressure and volume rate of change (i.e. flow) were determined and resistance was calculated (Tortora, 2006; Camroe et al., 1962).

Resistance (Raw)

Airway resistance is the difference in pressure between the mouth (atmospheric) and that in the alveoli, related to gas flow at the mouth.

This pressure difference is created primarily by the friction of gas molecules coming in contact with the conducting airways. Raw is the ratio of alveolar pressure (P_A) to airflow ($\dot{V}$). Its unit is cm H_2O / lit / sec. Gas flow can easily be measured with a pneumotachograph and P_A is measured with a body plethysmograph.

For gas to flow into the lungs (inspiration), P_A must fall below atmospheric pressure; the opposite occurs during expiration. Since the total volume of gas in the lungs and plethysmograph remains constant, the changes in alveolar volume are reflected by reciprocal change in the plethysmograph. Changes in $\dot{V}$ are plotted simultaneously against plethysmographic pressure changes (which is proportional to alveolar volume change) on a storage oscilloscope. The slope of this is $\dot{V}$/Pp Where, $\dot{V}$ is airflow and Pp is the plethysmographic pressure. Immediately after this measurement, an electronic shutter at the mouth piece is closed and changes in plethysmographic pressure are plotted against airway pressure at the mouth, just as is done for measurement of the volume since there is no airflow into or out of the lungs, the mouth pressure approximates P_A.

The slope of this line is P_A/Pp where P_A equal alveolar pressure. This step serves to calibrate changes in P_A to change in Pp for each subject. Raw is then calculated by taking the ratio of these two slopes:

$$Raw = \frac{P_A / P_P}{\dot{V} / P_P} \quad \text{......2.10}$$

Table 2. The values for smokers.

Serial number	Sex	Age	Height	Weight	Raw<2.24	Rawt<3.06
1	Male	34	170	89	1.32	2.8
2	Male	22	170	70	0.62	1.36
3	Male	30	168	74	1.96	2.94
4	Male	28	183	73	1.89	2.79
5	Male	38	176	73	2.79	4.06
6	Male	29	164	68	3.39	4.56
7	Male	34	168	87	2.86	3.66
8	Male	38	166	82	3.61	5.09
9	Male	31	178	78	1.25	2.27
10	Male	31	168	85	0.71	6.85
11	Male	28	176	76	2.96	5.28
12	Male	32	163	80	1.3	2.85
13	Male	22	158	71	1.35	2.21
14	Male	32	165	60	2.62	2.72
15	Male	35	174	85	2.72	4.16
16	Male	35	165	63	0.58	1.25
17	Male	31	170	64	2.07	3.07
	Sum	530	2882	1278	34	57.92
	Average	31.17647	169.5294	75.17647	2	3.407058824
	Standard deviation	4.572134	6.246175	8.705137	0.97638363	1.45801991
	Variance	20.90441	39.01471	75.77941	0.953325	2.125822059
	Sum	1.28E-07	1.16E-09	6.76E-12		0.119003898
	Average	2.51E-15	1.11E-25	4.92E-17		0.00130165

Where, $\dot{V}$ is air-flow, P_A is alveolar pressure and PP is plethysmographic pressure, which is measured with the shutter open and closed.

RESULTS

In all experiments values of Raw , Rawt were obtained for a full procedure period and for each successive resistance test and a comparison was done between variables age, weight and height for nonsmokers (Table 1). Figures 1 and 2 show the Raw and Rawt values versus the three variables, age, height and weight respectively. They illustrate that correlation factors shown on the figures for each of them on the curve associated the trend line equation with R^2 fixed at each curve. The values of $(R = \sqrt{R^2})$ were given as follow for Raw versus the three variables (height, age and weight); R height = 0.32, R age = 0.4 and R weight = 0.053 (Figure 1). Figure 2 shows values of Rawt versus the three variables height, age and weight; R height = 0.19, R age = 0.42 and

R weight = 0.46

Values of t-test and f-test for non-smokers were << 0.05. The figures gave the trend lines of airway resistances that varied with height, weight and age.

Table 2 gives also values and relations between Raw and Rawt versus total number of cigarettes smoked. Figures 3 and 4 illustrate the trend lines with their equations and correlation factors for the three variables (height, weight and age). For Raw, correlation factors are given in Figure 3; R (height) = 0.5, R (age) = 0.43 and R (weight) = 0.21. Rawt for smokers and R values are given on the curves in Figures 3 and 4; R values for weight and age were higher than for nonsmokers' and so on for height; R (weight) = 0.46, R (age) =0.42 and R (height) = 0.19.

Values of t-test and f-test for smokers were << 0.05. Figures 5 and 6 show the values of Raw and Rawt respectively versus cumulative number of cigarettes smoked. They gave the same trend and a noticeable influence of the cumulative number of cigarettes against either Raw or Rawt and the value of R = 0.873 (3).

Values of t-test and f-test were too small << 0.05.

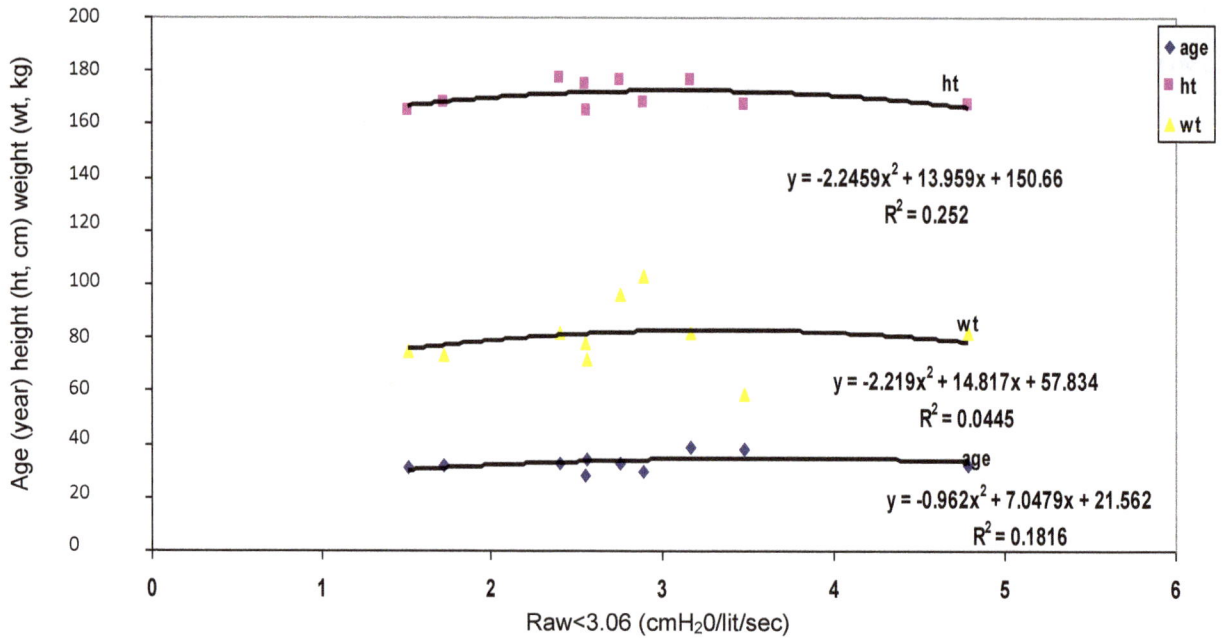

Figure 1. Raw versus age, height and weight.

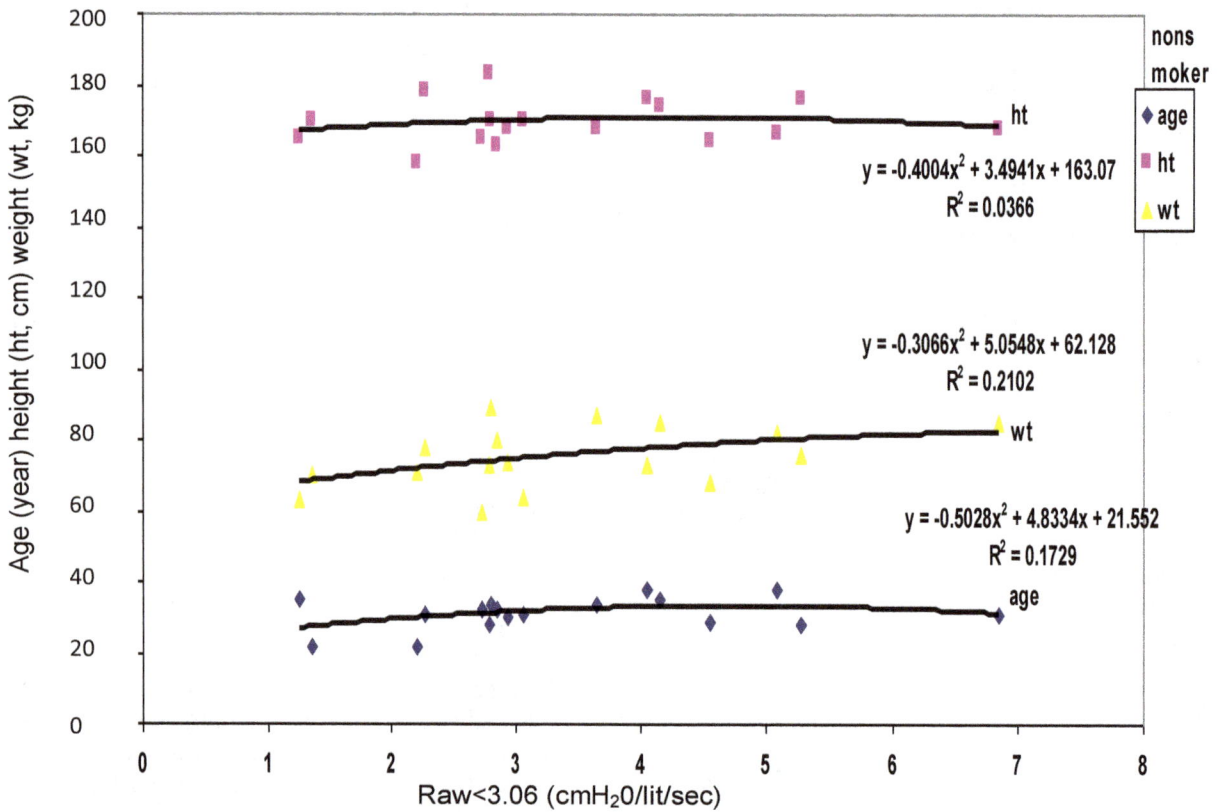

Figure 2. Rawt versus age, height and weight.

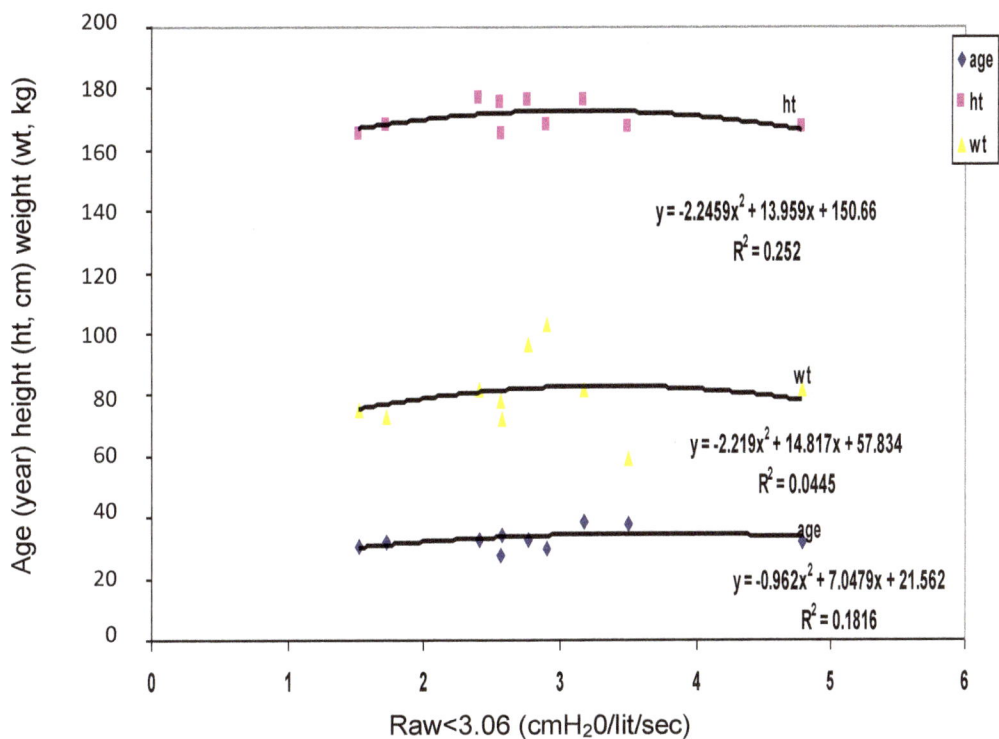

Figure 3. Values of Raw versus height, weight and age in smokers.

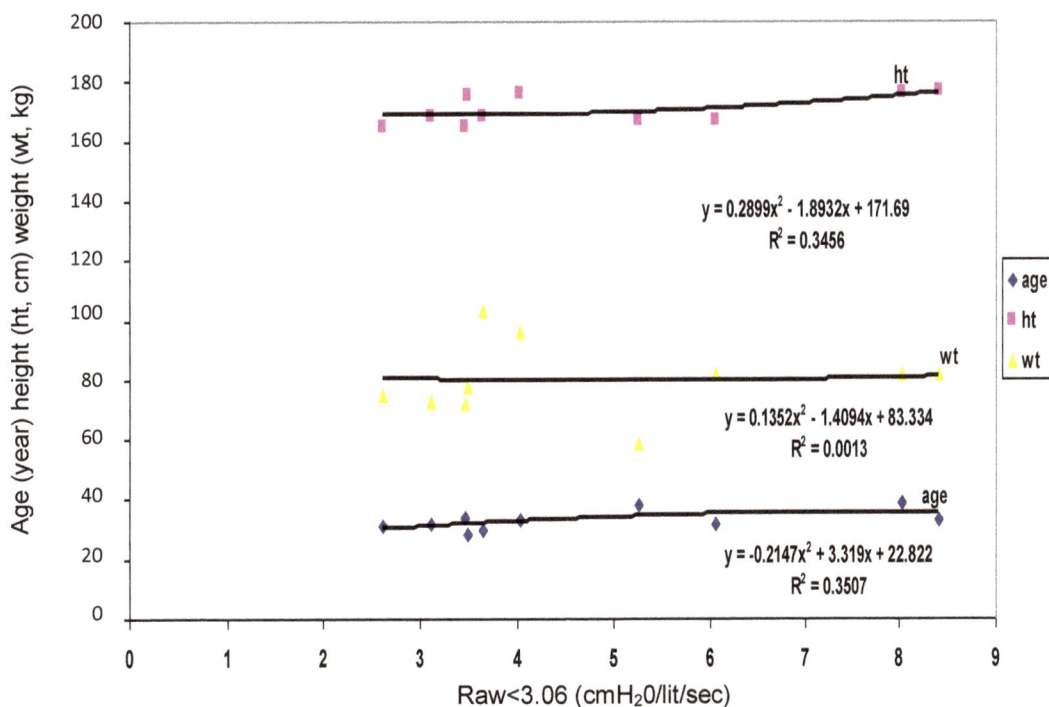

Figure 4. Values of Rawt versus height, weight and age in smokers.

Number of cigarettes versus Rawt

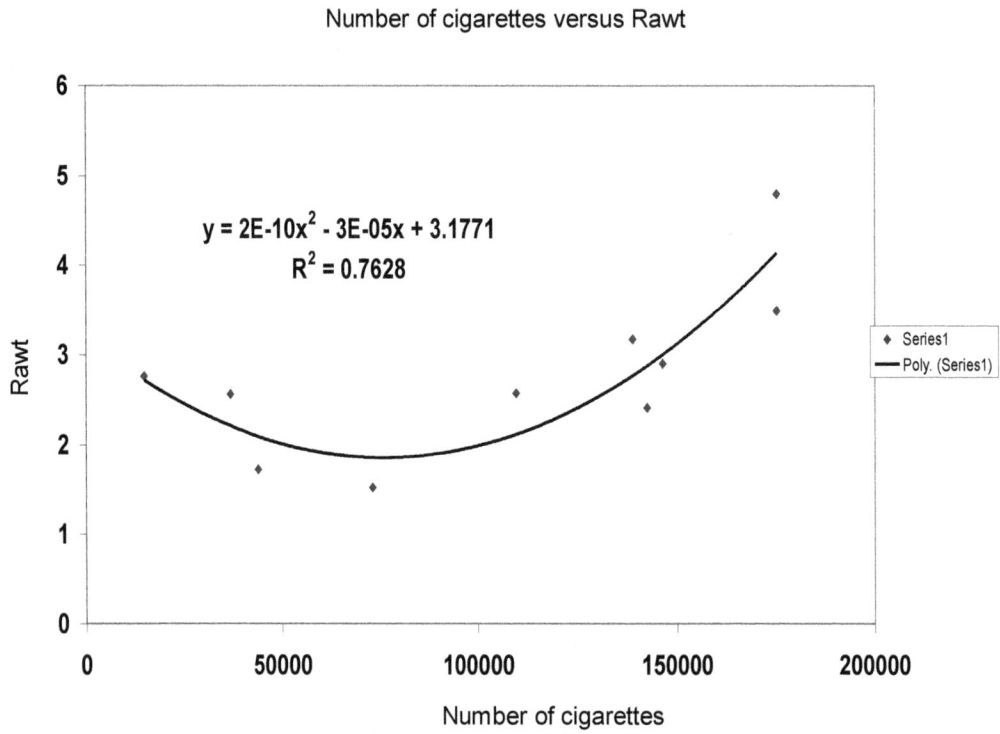

Figure 5. Values of Rawt versus total number of cigarettes smoked.

Number of cigarettes versus Raw

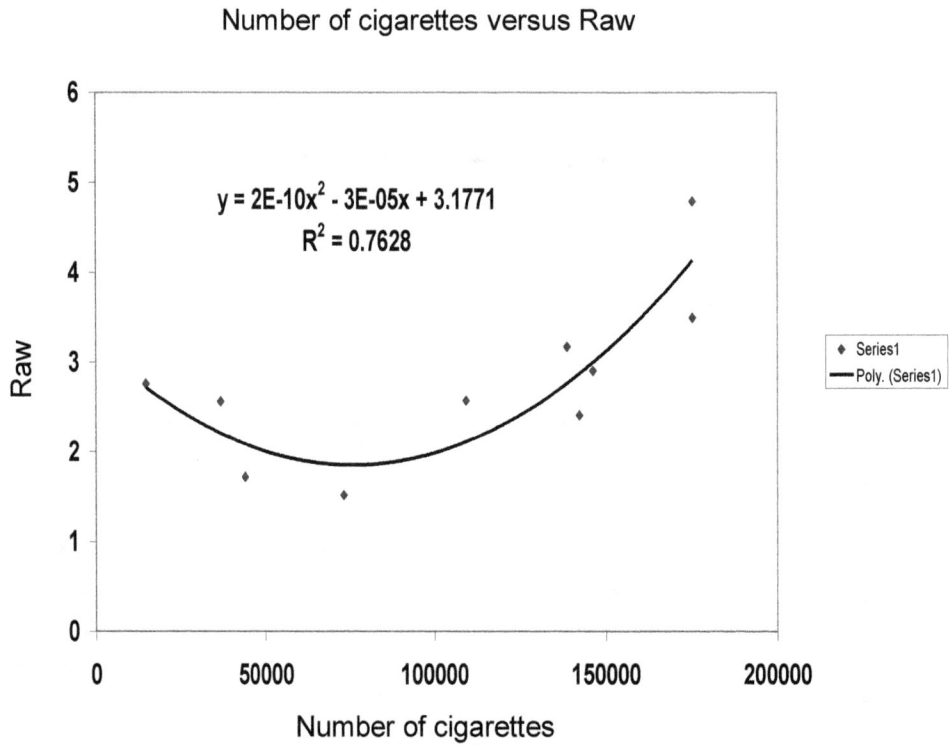

Figure 6. Values of Raw versus total number of cigarettes smoked.

DISCUSSION

The respiratory system is regarded as a continuously variable pressure generator (Mecklenburg et al., 1990; Macklem, 1980; Shapiro et al., 1991; Irving, 2007). Its action would result into maintaining continuous respiretory cycles, thereby facilitating volumes of air to be transported into and out of the lungs (Shapiro et al., 1991; Al Kadri, 1998). The final outcome of this process would be an exchange of gases across the alveolar membrane to provide the body with its needs for oxygen and to expel CO_2 carried into the lungs by the circulation (Cameron et al., 1999; Shapiro et al., 1991; Irving, 2007). Throughout this respiratory process, air with smoking deposits, then work is done and defined as the product of muscle pressure output and volume. Work and efforts are variable with time and both muscle pressure output and also volume are cyclic in their behavior. This description into the status of work is valid whether the respiratory system is normal or disordered but there must be differences in the work done when breathing between disordered status is compared with normal system. The determinants of such variabilities are resistance and compliance if the respiratory muscles are working normally.

In the present study which depends on both of these two parameters, Raw and Rawt were determined by the variation of total number of cigarettes smoked. The application of the study is dependent on its validity which was previously confirmed on both practical and theoretical grounds (Mecklenburg et al., 1992). The study itself covered the determination of the whole of muscle pressure taking into consideration total compliance and resistance of the respiratory system in addition to including any external resistance that may be available in the breathing circuit. The scope of the study spreads to cover the increase in Raw and Rawt values versus total number of cigarettes; or the situation when connecting the subject to external breathing circuits with unknown resistance to breathe through (Mushin et al., 1980; Banner et al., 1994; Sykes, and Vickers 1970; Jason et al., 2006) .

Coats et al. (1994) reported that pulmonary impedance (including elastic and resistive loads) can express the severity and mechanism of lung disease. This is because breathing would be expected to be hindered upon large increase in impedance which would affect pressure generated by the muscles. Coats et al. (1994) found a linear relationship between (Impedance x (min. ventilation)^2) and rate of work of breathing throughout the impedance range used. In the present study, only resistive loading was varied and work was found to increase up to a load (of nearly 1.3269 kP_a/(lit/sec) until work done was nearly diminished.

Respiratory muscle fatigue is induced experimentally by adding high external resistance to breathing (Fitting, 1992a; b; Fitting, 1994; Stephan et al., 2005; Al Sa'ady, 1997), and it is quite possible that in this study, fatigue approached as the resistance reached its maximal values.

In a preface to a book entitled loaded breathing, Pengelly, Rebuck and Campbell wrote:
"Common to many conditions causing breathlessness is the disturbance of mechanism of breathing producing an alteration in the load opposing the respiratory muscle. Clearly therefore an understanding of the way the act of breathing is affected by added mechanical loads is essential to an appreciation of the common clinical problems. Conversely, understanding of the physical processes responsible for homeostasis should be improved by studying patients in which breathing is loaded by disease" (Pengelly et al., 1974; Stephan et al., 2005; Martine et al., 2009).

A great accordance of the present study with the others, increase in smoking leads to increase in tobacco deposits which may lead to restrictive diseases, where values of t-test and f-test were highly significant.

Undoubtedly, the argument must therefore spread into the consideration of the form of generator of work (energy) and on which part of the energy is applied during expiration. The latter point may lead to clarification of whether the respiratory muscles always contribute to the energy stored in the compliance or not. The present method of determining WOB and the variability of external resistances offer expressions of some known forms of WOB in addition to producing a segmentation into its applied portions; investigation into the pressure generator (and thus WOB) will be more attainable in the future.

Conclusion

From the present work, it can be concluded that the respiratory system improves itself when the external resistance is less but at the increase in smoking and the total number of cigarettes, it is expected that the respiratory muscle will suffer disorders until it reaches fatigue, with the high increase in resistance value because of the increase in resistance by smoking. This increase will lead to restrictive diseases (Stephan et al., 2005; Martine et al., October 2009), and if the increase in Raw reaches the three folds, this will create problems with ventilators if used in hospitals at certain cases (Almaasfeh, 1995). Finally, cigarettes are best factor to increase Raw.

RECOMMENDATIONS

It is recommended that further studies should be made

on more subjects and on two variables, (airway compliance and resistance), to prove these variables as the best factors to diagnose, to compare the curve of anybody examined or to deduce a standard figure for certain subjects.

REFERENCES

Al Kadri FM (1998). Changes in respiratory muscle efficiency upon change in airway resistance and compliance. A thesis submitted to the council of the College of Education, Al Mustansiriyeh University.

Al Sa'ady AF (1997). Variability of human respiratory compliance and air way resistance. A thesis submitted to the council of the College of Education of Al Mustansiriyah University.

Almaasfeh Sultan (1995). Indirect Method For Determining Work of Breathing Upon Change in Resistance. Athesis submitted to the council of college of education ALmustansiriyah University.

Banner MJ, Jaeger MJ, Kirby RR (1994). Components of the work of breathing, critical care medicine, (22) no.3 Williams and Wilkins.

Beydon L, Chasse M, Harf A, Le Maire F (1988). Inspiratory work of breathing during spontaneous ventilation using demand valves and continuous flow system. Am, Rev. Respiratory. Dis. 138: 300-304.

Coates AL, Vallinis P, Mallahoo K, Seddon P, and Davis GM (1994). Pulmonary impedance as an index of severity and mechanism of neonatal lung disease, Pediatric pulmonology Wiley-LISS Inc.. 17:41-49.

Fitting JW (1992). Fatigue des muscles respiratoires. (Schweiz Med.Wschr. 122: 302-306).

Fitting JW (1994). Role des muscles respiratoires dans le Sevrage de la ventilation mecanique. (Achweiz med. Wochenschr. 124: 215-220.

Guyton AG, Hall JE (2009). Text book of medical physiology, 13th edition. (W.B. Saunder's Company).

Goldman MD, Grimby G, Mead J (1976). Mechanical work of breathing derived from rib cage an abdominal. V-p- portioning. J. Appl. Physiol. 41(5).

Jason PKirkness,Vidya Krishnan, Susheel P, Patil, Hartmut Schneider (2006). Upper airway obstruction in snoring and upper airway resistance syndrome Prog.RespirRes.Basel,Karger (35) pp. 79 – 89 Randerath Wj, Sanner BM,Somer VK eds Sleep apnea.

Macklem PT (1980). Respiratory muscles: The vital pump. Chest, 78:5 Nov.

Martine Broetima, Nick HT tenHacken, Franke Volbeda, Monique E, Lodewijk MN, Hylkema DS, Postma WT (2009). Air way epithelial changes in smoking but not in ex-smoking asthmatics, PP 1-43 AJRCCM articles in press as doi:10.1164/rccm 2009 06-0828OC.

Mecklenburg JS, Latto IP, Alobaidi TAA, Swai EA, Mapeleson WW (1990). Excessive work of breathing during intermittent mandatory ventilation, Anesth, (58), 1048-1054).

Mecklenburg JS, Alobaidi TAA, Mapeleson WW (1992). A model lung with direct representation of the respiratory muscle activity. BJ. Anaesthesia 68, 603.

Mushin WW, Rendell-Backer L, Thomson, Peter W, Mapleson WW, Hillard EK (1980). Automatic ventilation of the lungs, 3rd edition.(Blackwell Scientific Publications).

Pengelly LD, Rebuck AS, Campbell EJM (1974). Loaded breathing, Churcheill Livingstone

Polese G, Rossi A, Appendini Li, Brandi G, Bates JHT, Brandolese R (1991). Partitioning of respiratory mechanism in mechanically ventilated patients. (J. appl. Physiol. 71 (6), (2425-2433).

Shapiro BA, Kak Marck, Robert MM., Cane, Roy D, Perruzi, William T, Hauptman David (1991). Clinical application of respiratory care, 4th edition, (Mosby Year book).

Stephan J, LaiFook, Yih-Loong Lai (2005). Airway resistance due to alveolar gas compression measured by barometric plethysmography in mice. J. Appl. Physiol. 98: 2204 – 2218.

Sykes MK, Vickers MD (1970). Principle of measurement for anaesthetists, Blackwell Scientific publication, 2nd printing.

Tortora D (2006). Principles of anatomy and physiology 11th Edition, John Wiley and sons.inc.

11

Modeling and proposed mechanism of two radical scavengers through docking to curtail the action of ribonucleotide reductase

Sampath Natarajan* and Rita Mathews

Department of Advanced Technology Fusion, Konkuk University, 1 Hwayang-dong, Gwangjin-gu, Seoul, 143-701, Korea.

Ribonucleotide reductase (RR) is a ubiquitous cytosolic enzyme required for DNA synthesis and repair in all living cells. Therefore, the crucial role of this enzyme in cell division makes it a potential target for designing drugs that inhibit cell growth for cancer therapy. An increased interest in RR as a target for cancer therapy has been documented since the discovery that human RR is regulated by p53 enzyme and that a mutation in p53 leads to several forms of cancer. Cell proliferation stops if normal RR is inhibited. A new strategy to kill the cancer cells would be using specific inhibitors that inhibit the action of RR enzyme. The inhibitor must be a radical scavenger which destroys the tyrosyl radical or an iron metal scavenger (which affects iron center). In this view, modeling studies on human RR-R2 were done to understand its interaction with radical scavengers, flavin (FLA) and phenosafranine (PHE) through docking since they have good reductive property. Radical scavengers are active against RR enzymes at anaerobic condition and their radical scavenging mechanism has been proposed. In aerobic condition RR enzyme will reproduce the radicals and then the radical scavengers fail to act as drug. So, the metal scavengers may be better than the radical scavengers to curtail the action of RR enzyme.

Key words: Ribonucleotide reductase, tyrosyl radical, flavin, phenosafranine, p53 enzyme, human R2.

INTRODUCTION

Ribonucleotide reductase (RR) is a ubiquitous cytosolic enzyme, responsible for converting ribonucleotides into deoxyribonucleotides, the eventual substrates for DNA polymerase (Reichard, 1993) and also repair DNA in all living cells (Eklund et al., 2001). It contains two dissimilar protein components, R1 and R2 (Figure 1). The R1 subunit has homodimeric structure, with a molecular mass of ~170 kDa and has allosteric effector site that controls enzyme activity and substrate specificity (Wright et al., 1990). The R2 subunit is also a homodimer, with a molecular mass ~85 kDa, and forms two dinuclear iron centers that stabilize a tyrosyl free radical which is required for catalytic activity (Wright et al., 1990). The R1 and R2 subunits interact each other at their C-terminal

ends to form an active holoenzyme (Davis et al., 1994). The overall function of RR involves the reduction of the hydroxyl group on the 2- carbon of the ribose moiety of nucleoside diphosphates and triphosphates (NDPs and NTPs). This conversion is achieved by free radicals which are stored in the dinuclear center of the enzymes until they are required for catalysis.

RR enzyme has two important components such as a radical generator (R2 subunit) and a reductase (R1 subunit). The production of radicals by R2 and its storage constitute the first step of the reaction in RR enzymes. Surprisingly the radical generators of all types of RRs are not the same, whereas the reductase component is more or less similar. The RR enzymes isolated so far have been classified into three types (class I, II and III) based on the oxygen dependence and metal cofactors involved in the generation of essential free radicals (Reichard, 1993; Jordan and Reichard, 1998). Mammalian RRs belong to class I reductases, which contain two pairs of

---

*Corresponding author. E-mail: sampath@konkuk.ac.kr or sams76@gmail.com

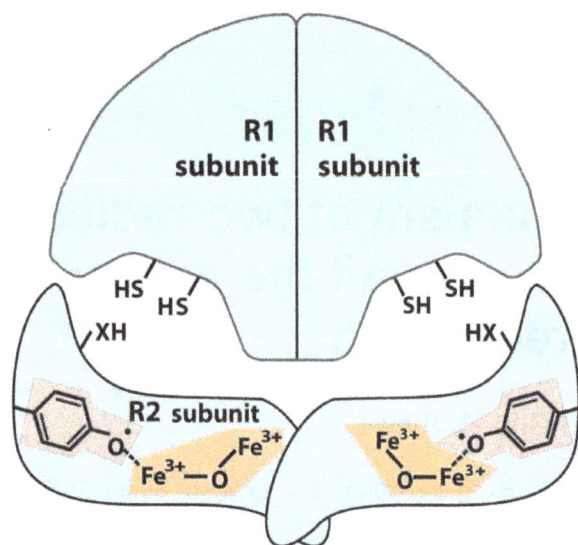

Figure 1. Schematic cartoon diagram of the RR enzyme.

iron and produce the catalytically essential tyrosyl free radical.

Ribonucleotide reductase (RR) enzyme and cancer

An increased interest in RR enzyme as a target for cancer therapy is seen ever since the *Homo sapiens* ribonucleotide reductase (HsRR) of a new type was identified which is regulated by p53 (Lozano and Elledge, 2000; Nakano et al., 2000). The p53 protein actively suppresses the tumor formation, but its mutation causes several forms of cancer. Over 80% of the human tumors are found to contain mutations in p53 (Tanaka et al., 2000). Mammalian RR-R2 is located in the cytoplasm and regulated by the cell cycle. The new R2 gene product called p53R2 is located in the nucleus. The p53 binds to the first intron of p53R2 gene and is important to activate its transcription factor directly (Priya and Shanmughavel, 2009). This gene has been identified in human and mouse cells. The review by Eklund et al. (2001) discusses the possibility of specific inhibitors to RR enzymes. Their argument is based on the fact that cancers often have mutations in the p53 pathway and unable to make p53R2. RR is necessary for DNA replication and thus inhibition of this enzyme will in turn inhibits cell division, whereas normal cells with p53 could survive on the DNA pool supplied by p53R2. A new strategy to kill cancer cells would be to specifically inhibit the RR enzyme, which is essential for cancer cells after DNA damage since they cannot induce the p53R2 due to lack of p53. In contrast, normal cells can repair their DNA damage with the help of the induced p53R2.

Since RR activity is necessary for DNA replication inhibition of this enzyme will inhibit cell division,

(Szekeres et al., 1997). Therefore an understanding of the molecular mechanism of RR is essential in designing new cytostatic drugs. The inhibitor must be a radical scavenger to destroy the tyrosyl radical or an iron metal scavenger (which affects the iron center). The iron or radical site of R2 protein can react with one-electron reductants (e.g., hydroxyurea and hydroxylamine), whereby the tyrosyl radical is converted to a normal tyrosine residue (Sneeden and Loeb, 2004). These drugs are slow and relatively unspecific (Stubbe, 1990). However, other compounds such as flavin (FLA) (Fontecave et al., 1989) and phenosafranine (PHE) (Sahlin et al., 1989) are available to reduce the radical activity.

In view of inhibitor study using radical scavengers, the structure of radical generator, human RR-R2, enzyme was constructed using homology modeling to understand the interactions of drug with the active site residues. Based on the observation and reports, two different drug actions, as radical scavengers (FLA and PHE) or as metal scavengers (thiosemicarbazones) are needed to stop the catalytic activity. Our study has been concluded by finding the most appropriate drug to curtail the catalytic action of RR enzyme.

MATERIALS AND METHODS

Collection of sequences

The complete protein sequences of RR-R2 from *H. sapiens* (P31350) and other RR-R2 sequences were retrieved from NCBI (http://www.uniprot.org/uniprot). The relatedness of sequences deposited in databases was evaluated by Blast (Altschul et al., 1990) implemented via the NCBI server (http://www.ncbi.nim.nih.gov/blast) against the complete training data set. The Blast P (protein query–protein database) comparison was performed using the protein database (PDB).

Sequence alignment

Since RR-R2 radical generators vary from species to species, four well known eukaryotic sequences including mouse RR-R2 (template model), yeast RR-R2, human p53R2 and target human RR-R2 proteins (Figure 2) and one bacterial sequence (*Escherichia coli* RR-R2) were selected for multiple sequence alignment (MSA) using ClustalW program (Chenna et al., 2003). The secondary structure elements of the RR-R2 enzymes are shown on the top of the human enzyme and 100% conserved residues are shown in red colour. The Swiss-prot accession numbers for various species of R2 enzyme sequences are as follows: mouse, P11157 (Thelander and Berg, 1986); human, P31350 (Pavloff et al., 1992); *E. coli*, P69924 (Carlson et al., 1984); human, p53R2 (Q75PY9); and yeast, P09938 (Elledge and Davis, 1987). The P53R2 and *E. coli* RR-R2 sequences have been truncated in the N-termini residues. The dinuclear binding residues are conserved in all sequences including *E. coli*.

Homology modeling of human R2 enzyme

The pair wise alignment between the human RR-R2 and mouse

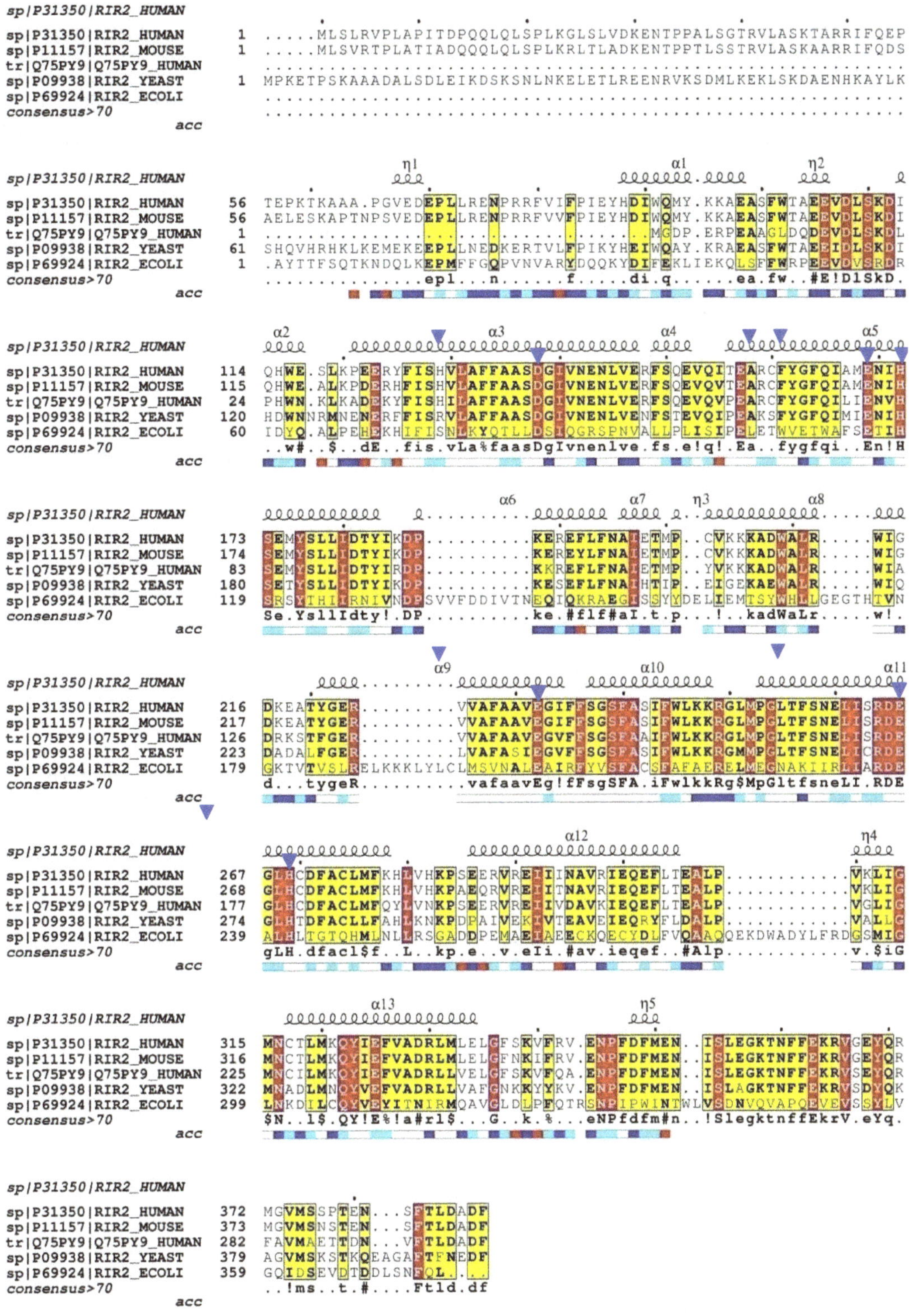

Modeling and proposed mechanism of two radical scavengers through docking to curtail the action of ribonucleotide reductase

97

Figure 2. Multiple sequence alignment of RR-R2 structures from various species. The proteins listed from top to bottom: human RR-R2 from *Homo sapiens* (Uniprot accession number P31350); RR-R2 from mouse (P11157), p53RR-R2 from human (Q75PY9), RR-R2 from yeast (P09938) and RR-R2 from *E. coli* (P69924). Human RR-R2 secondary structure elements are shown in top of the sequence alignment. Fully 100% conserved residues shown in red box with white character and partially more than 80% conserved residues shown in yellow color with black character. All metal binding residues (100% conserved) are shown in blue color triangle box.

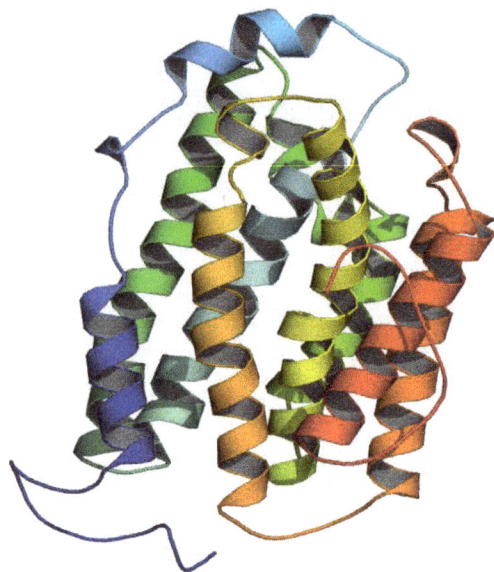

Figure 3. Modeled human RR-R2 enzyme (Swiss-model).

RR-R2 (data not shown) proteins was carried out using ClustalW program and the result showed 95% sequence identity between these two, both are expected to have similar structural reactivity. All active site residues are 100% conserved including the residues Tyr-189 (human RR-R2) and Tyr-177 (mouse RR-R2) which are important for radical generation. So, the crystal structure of mouse RR-R2 (pdb id: 1xsm) was used as a template model to generate the 3D modeled human RR-R2 structure of *H. sapiens* through online server of Swiss-model program (Schwede et al., 2003). The template model structure was directly retrieved from the PDB database (http://www.rcsb.org/pdb/). Visualization and cartoon diagrams were drawn using PyMOL (DeLano, 2002) graphics program. Evaluation and validation of the modeled structure was done using PROCHECK (Laskowski et al., 1993) and WHATIF. These programs generated Ramachandran plots of the amino acid residues in the allowed region and consider the overall G-factors. Out of 389 amino acid residues 288 were taken for constructing the structure. The homology modeled human RR-R2 structure was submitted to Protein Model Database (PMDB; http://mi.caspur.it/PMDB/) and it is assigned the accession code PM0075775.

Molecular docking study

Preparation of radical scavengers (Flavin and phenosafranine)

Tyrosine radical scavengers are very essential to inhibit DNA replication. According to the reports (Eklund et al., 2001; Fontecave et al., 1989), two potential radical scavengers, FLA and PHE, whose structures were built using the program PRODRG2 (Schuettelkopf and Aalten, 2004) and their energy-minimized structure coordinates were used for the docking studies with human RR-R2 protein. In addition, they are known as versatile compounds that can function as electrophiles and nucleophiles. Because of their chemical versatility and potential redox properties, they play a central role in aerobic metabolism through their ability to catalyze two-electron dehydrogenations of numerous substrates and to

participate in one-electron transfer to various metal centers through their free radical states. With this capacity, they frequently form parts of multi redox-center enzymes (Massey, 2000).

In this study the AutoDock program [v.3.0.5] (Morris et al., 1998) was used for docking studies which utilizes a Lamarckian Genetic Algorithm for conformational searching and energy evaluation using a grid-based molecular affinity potential. Water molecules are not used for docking study as they make the analysis complicated. The distance-dependent function of the dielectric constant was used to calculate the energy maps and all other parameters have taken default values. Twenty-five best conformation of the protein ligand complex were retrieved and compounds with highest binding affinity from the best docked complexes were selected. The hydrogen bonding and non-bonded contacts for the complexes were derived using the program HBPLUS (McDonald and Thornton, 1994) and the pictorial representations are drawn using the program LIGPLOT (Wallace et al., 1995).

RESULTS AND DISCUSSION

Structural description of modeled human RR-R2 subunit

The Multiple sequence alignment results are shown in Figure 2 and it suggests that all of the sequences from mouse, human, P53R2 and yeast are highly conserved with appreciable sequence identity except *E. coli*. The dinuclear iron metal binding residues are conserved in all sequences including *E. coli*. The modeled human RR-R2 structure consists of α-helical bundle with 288 residues in the core structure. The dinuclear iron centre is located at the center of helical bundle in R2 subunits in all reported RR-R2 enzymes including template molecule, but the modeled human RR-R2 does not have dinuclear iron center. The overall human RR-R2 structure contains thirteen helices, of which eight long helices form a bundle (Figure 3).

These helices contribute about three quarters of the protein molecule and have an rmsd value of 0.45Å (for Swiss-model) compared to mouse RR-R2 (pdb id: 1xsm). Three of the five shorter helices orient perpendicularly to the long helices in the bundle and the other two are almost parallel to long helices. The long hydrophobic helix is surrounded by six other helices. The structure quality was validated by the Ramachandran map (Figure 4) using PROCHECK program. The torsion angles of 86.2% of residues are occupied within the allowed region and only 1.1% of residues are located in the disallowed region.

Docking study of radical scavengers

Flexible docking was performed for the modeled human RR-R2 enzyme with two well-known radical scavengers, FLA and PHE (Table 1) using the program Auto dock (Morris et al., 1998). Out of twenty-five conformations, one of its best conformations was chosen for the protein-

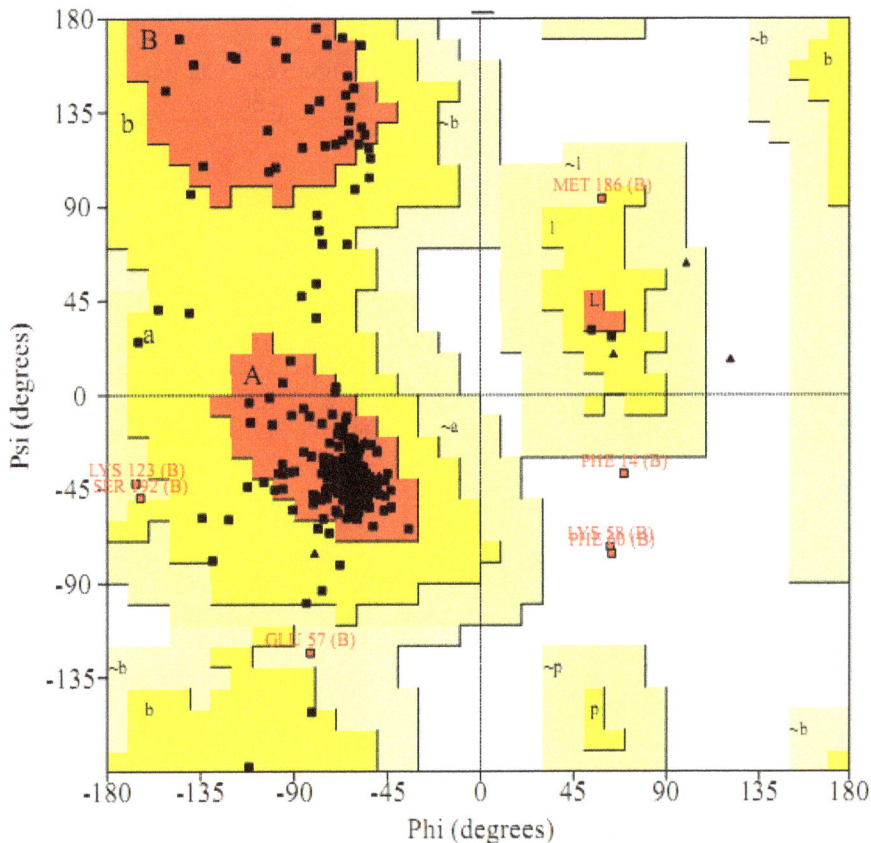

Figure 4. Ramachandran plot for the amino acids of modeled human RR-R2 enzyme occupied in favoured regions (block colour residues). Few residue's conformational angles are occupied in unfavoured region (red color residues with labeled).

Table 1. Details of drugs used for docking study.

S/N	Common name	Molecular formula	Molecular weight (g/mol)	IUPAC name	Structure
1	Flavin (FLA)	$C_{12}H_{10}N_4O_2$	243.23	7,8-dimethylbenzo[g]pteridine-2,4(3H,10H)-dione	
2	Phenosafranine (PHE)	$C_{18}H_{15}N_4$	287.34	3,7-diamino-5-phenylphenazin-5-ium)	

drug interaction study from each ligand. On the basis of docking studies the interaction of radical scavengers with human RR-R2 protein, several energies calculated (docking energy, inter molecular energy, and internal energy) are given in Table 2. The docking energies of the FLA and PHE are noted as −09.37 and −07.69 kcal/mol,

respectively. Highest binding affinity of the drug has been identified based on the lowest docking energy and FLA shows higher affinity than PHE to human RR-R2 protein.

The cartoon and LIGPLOT diagram of FLA and human RR-R2 docking interactions are shown in Figure 5a and b, respectively. Molecule FLA is comfortably occupied at

Table 2. The interaction energy of human RR-R2 and radical scavengers (drug) from molecular docking.

Drug	Binding energy (kcal mol^{-1})	Docked energy (kcal mol^{-1})	Intermolecular energy (kcal mol^{-1})	Internal energy (kcal mol^{-1})
FLA	-08.72	-09.37	-12.22	0.67
PHE	-07.11	-07.69	-9.83	1.38

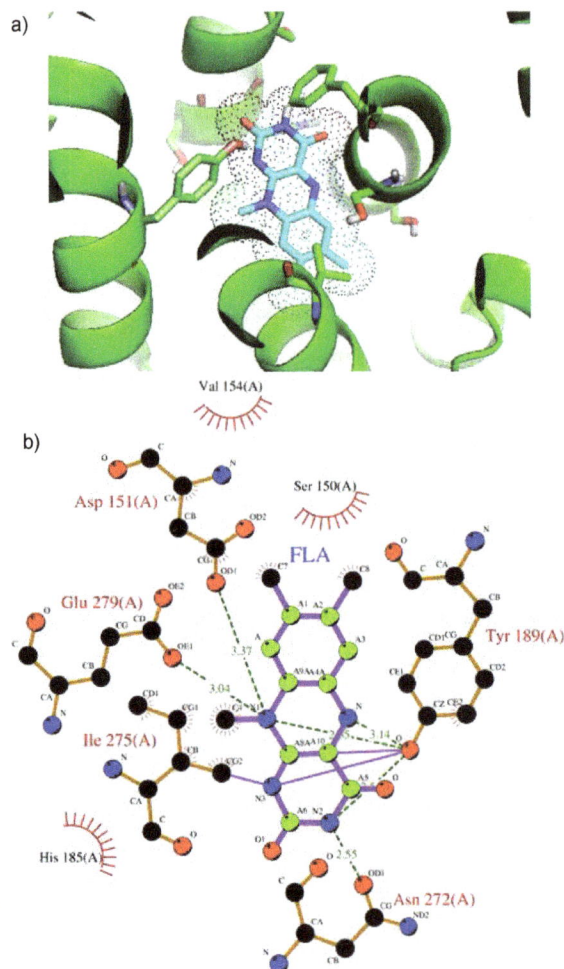

Figure 5. Modeled human RR-R2 enzyme- flavin drug interaction. (a) Cartoon diagram of the protein- drug (Cyan stick) interactions. (b) Ligplot diagram of the human RR-R2-flavin drug interactions. Hydrogen bonds are depicted in green colour dotted lines and hydrophobic interaction residues are shown in red colour.

the active site of human RR-R2 protein via six strong hydrogen bonds using the residues of Tyr-189, Asn-272, Glu-279 and Asp-151 (Table 3). In addition, some hydrophobic non-bonded interactions by the residues of His-185, Val-154 and Ser-150 also help the ligand molecule for its conformational stability. In this model, both the adjacent methyl groups in the inhibitor make

hydrophobic contacts with the Ser-150. The human RR-R2 and PHE complex molecule (Figures 6a and b) is stabilized through four hydrogen bonds rises from two residues Glu-279 and Tyr-189. A few hydrophobic non-bonded interactions (Table 3) by the residues Val-154, Val-244, Ser-150, Glu-245 and Asp-151 are observed.

These hydrophobic forces are attributed due to N, N-dimethyl group in the ligand molecule. In the docking model of both these radical scavengers FLA and PHE, the reduced nitrogen atoms are directly binding to Tyr-189 and it is promising evidence that radical scavengers directly interact with the radical generator residue, Try-189. Based on these evidences radical scavenging reaction mechanism is proposed as follows using these two radical scavengers independently.

Proposed mechanism of flavin (FLA) action as radical scavenger

The FLA drug action as potential radical scavenger is shown in Scheme 1. In anaerobic condition, the oxido oxygen atom of iron (III) cluster scavenge the hydrogen radical from the acidic hydrogen of FLA and form the hydroxyl Fe (III). Another Fe (III) was reduced to Fe (II) and that the reductive step precedes oxygen requiring radical formation (Eliasson et al., 1986; Fontecave et al., 1987). According to this model a reduced R2 should be an intermediate in the reaction. Simultaneously, the unstable FLA radical model (IIa and IIb) is hydrogen bonded (FLA's N-H from uracil moiety) with oxygen atom of oxido-Fe (III) and serves as a driving force for the oxido-Fe (III) bond cleavage. To attain the stable aromatic nature, the IIb FLA radical eliminates one more electron from the piperazine ring and form oxidized FLA (IV). Hydroxyl Fe (III) reacts with second hydrogen radical and form one more reduced Fe (II) and water.

Proposed mechanism of phenosafranine action as radical scavenge

The mechanism of another radical scavenger PHE is shown in Scheme 2. At anaerobic condition, reduced PHE (I) react with oxygen center of dinuclear iron (III) and loses one of its protons and forms neutral PHE (II) and hydroxyl Fe (III). One Fe (III) gets reduced to Fe (II) in the first step. In the next step, dimethyl amine nitrogen

Table 3. Active site residues of human RR-R2 and drug interactions (hydrogen bonds and hydrophobic contacts) involved through the molecular docking.

Drug	Interaction of human RR-R2 and radical scavengers	Distance of hydrogen bonds (Å)	Hydrophobic contact residues
FLA	Tyr-189 (OH...N)	3.14	
	Tyr-189 (OH...N2)	2.50	His-185
	Tyr-189 (OH...N1)	2.35	Val-154
	Asn-272 (OD1...N2)	2.55	Ser-150
	Glu-279 (OE1...N1)	3.04	
	Asp-151 (OD1...N1)	3.37	
PHE	Tyr-189 (OH...N2)	3.13	Ser-150
	Tyr-189 (OH...N4)	2.88	Asp-151
	Tyr-189 (OH...N)	2.24	Val-154
	Glu-279(OE1...N4)	2.88	Val-244
			Glu-245

donates electron to make aromaticity in the molecule and the piperazine moiety expels the hydrogen anion from N-H. This anion H reacts with HO-Fe (II)-R to form water (IV) and gets reduced to Fe (II).

Formation and function of tyrosyl radical

An important observation made by Atkin et al. (1973) is that the incubation of Apo RR-R2 enzyme with Fe (II) and O_2 led to the formation of diferric tyrosyl radical cofactor. The diferrous cluster reacts with O_2 to produce a putative short-lived peroxo intermediate which is reduced to a paramagnetic intermediate dinuclear ferric complex. This peroxo intermediate is then converted into the diferric cluster by oxidation of Tyr residue to the radical form. In presence of oxygen environment, diferric cluster and Tyr radicals are regenerated and then RR-R2 enzyme will be ready for catalytic process to participate in DNA synthesis. So the alternative drugs might be needed to stop the enzyme action which must be better than radical scavengers. Metal chelators will be a good choice to remove the metal ions and reduction of radicals from the enzyme.

Mechanism of Iron chelators

Chelating molecules normally play a vital role to remove or prevent incorporation of iron in the enzymes. When cells lack iron, cell proliferation stops due to specific inhibition of DNA synthesis (Fan et al., 2001; Li et al., 2001). For this reason, iron-chelators have been used as such or in combination with other drugs in anti-proliferative therapy. An early effect due to iron chelators in the cell is that RR activity decreases with the accumulation of

Apo RR-R2 and therefore, DNA synthesis and cell proliferation are curtailed. The drug which belongs to the group of a-(N)-heterocyclic thiosemicarbazones is the most potent inhibitor of mammalian RR-R2, e.g. Triapine (Shao et al., 2006).

In recent year, Triapine's analogs have been tested against human RR enzyme, which showed a strong inhibition. The authors have demonstrated their mechanism as metal chelators having exceptionally strong affinity for iron in aerobic condition (Zhu et al., 2009; Sartorelli et al., 1977). The study reveals that the thiosemicarbazone (TSC) derivatives inhibit the enzyme by destroying the free radical. This is because the radical structure is more exposed in the mammalian reductase (Kjøller et al., 1982). Heterocyclic TSCs are proposed to inhibit RR enzymes by a redox reaction involving reduced iron and oxygen which explain why the iron chelate is the active form of the drug (Moore et al., 1970; Preidecker et al., 1980). The iron chelation process reduces the Fe (III) to Fe (II) which is a necessary process for drug action according to the proposed mechanism.

Thiosemicarbazone (TSC) chelator drugs act directly to destroy the tyrosyl free radicals of RR-R2 enzyme by forming Fe (II)-TSC complex (Thelander and Gräslund, 1983; Finch et al., 2000). According to the above statement, the drug TSCs do not directly chelate the Fe (III) ions in the active site of RR-R2 enzymes, instead they first scavenger the free radicals and reduce the metal ion from Fe (III) to Fe (II). Thereafter, TSCs will chelate the Fe (II) ions and increase the concentration of the apo-RR-R2 level.

Conclusion

The human RR-R2 structure was modeled and docking

(a)

(b)

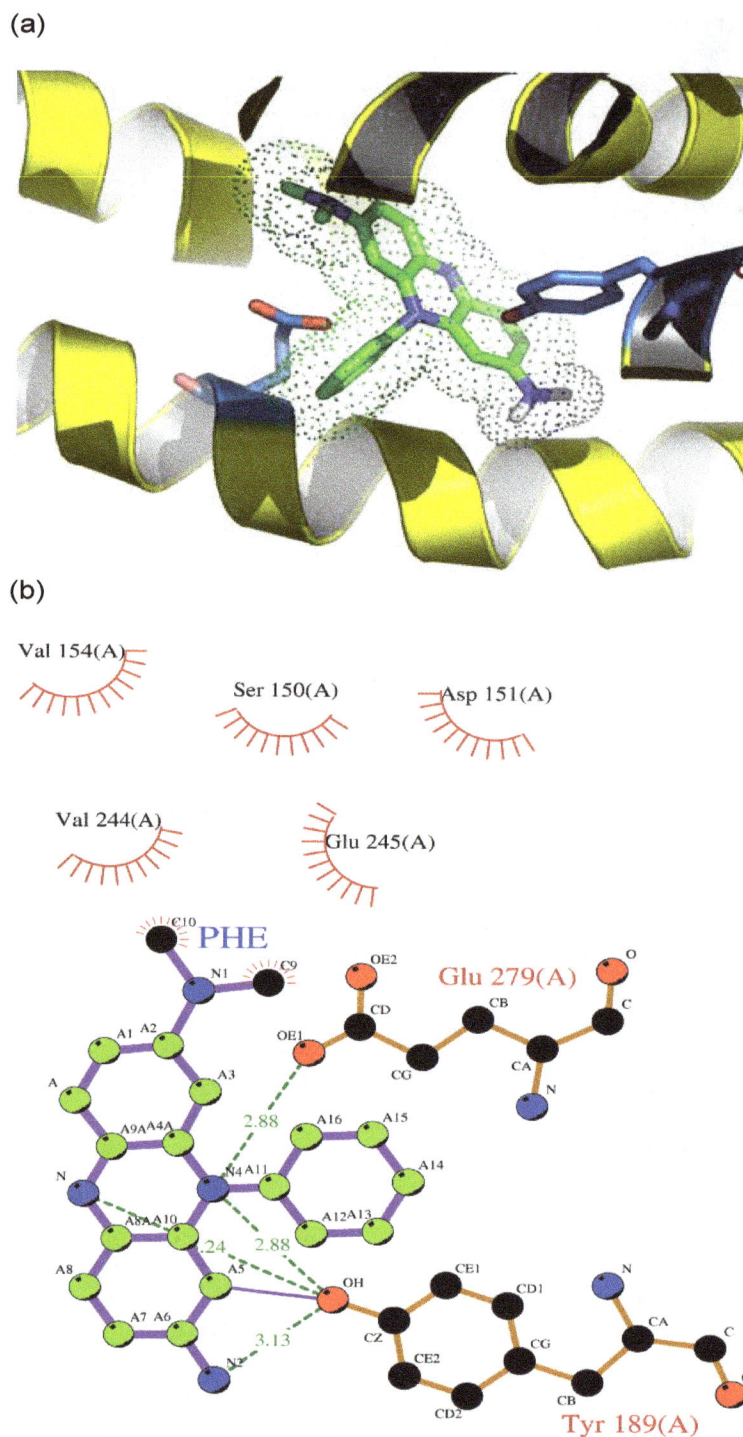

Figure 6. Modeled human RR-R2 enzyme-phenosafranine drug interactions. (a) Cartoon diagram of the protein-drug (green stick) interactions. (b) Ligplot diagram of the human RR-R2-Phenosafranine drug interactions. Hydrogen bonds are depicted in green color dotted lines and hydrophobic interaction residues are shown in red colour.

studies were performed with the radical scavengers, FLA and PHE. Their interactions with human RR-R2 were studied and the plausible scavenging mechanism has been proposed in anaerobic condition. Since attaining

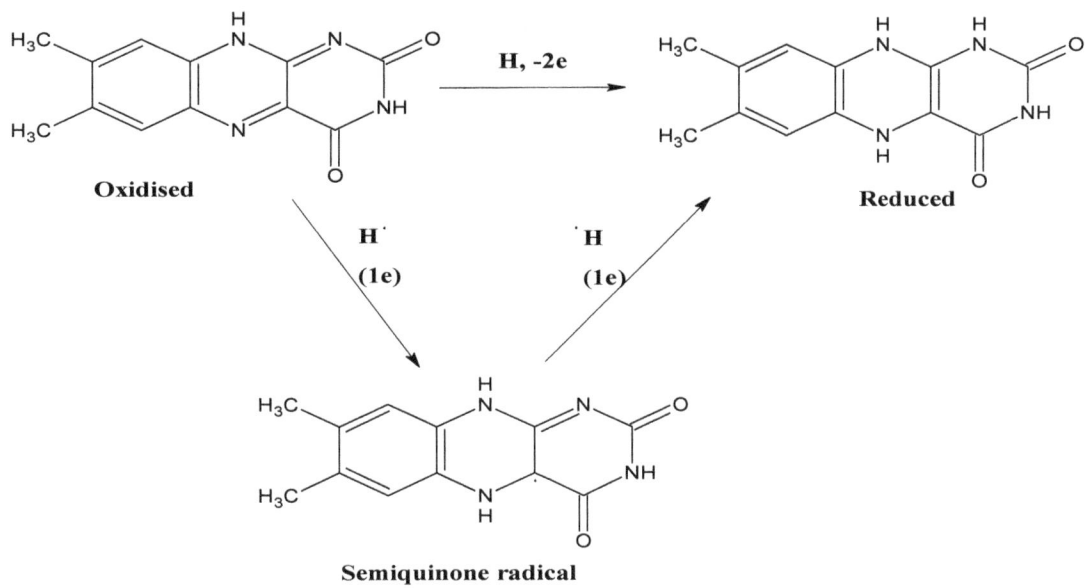

Mechanism

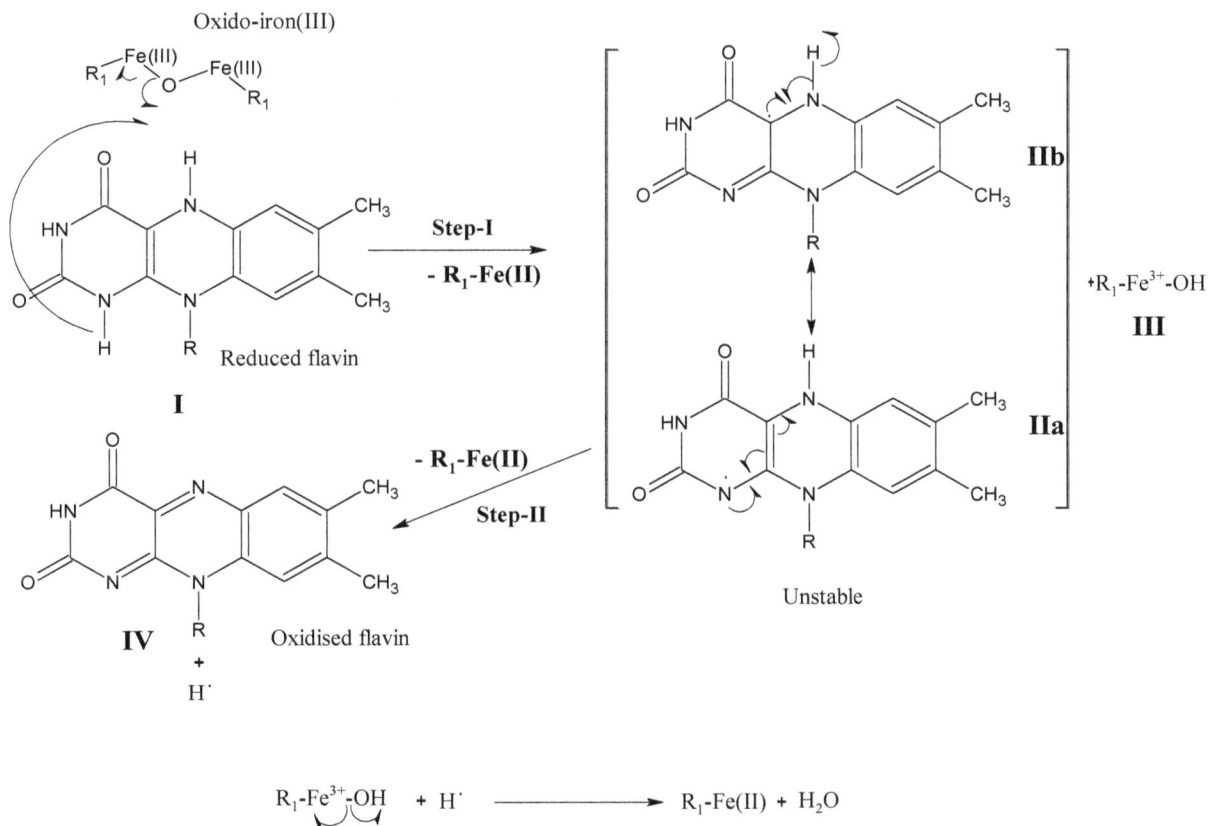

Scheme 1. Possible reaction mechanism of radial scavenger FLA molecule.

anaerobic condition is not possible, using these drugs for cancer therapy becomes difficult. So, alternate drugs are needed to inhibit the activity of the RR enzyme.

According to the various reports, metal scavengers would be more ideal drugs for RR inhibition. Since TSCs have more affinity to Fe (II) and are more effective

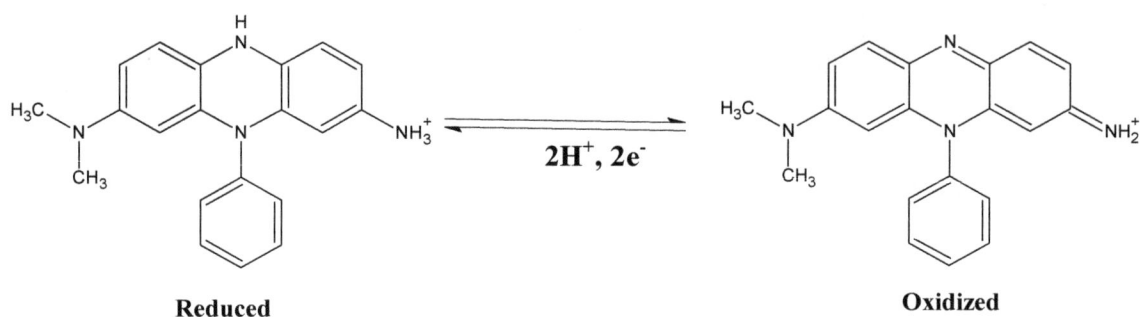

Reduced

2H⁺, 2e⁻

Oxidized

Mechanism:

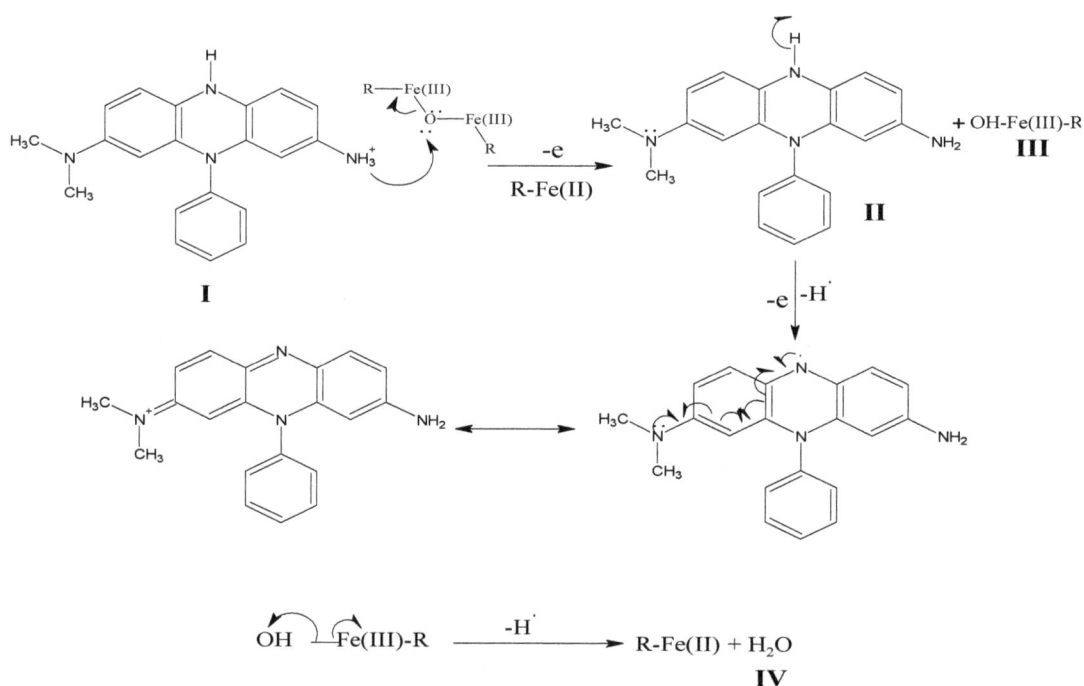

Scheme 2. Possible mechanism of radical scavenger PHE molecule.

against RR enzyme, it may be a more convenient drug to curtail the action of RR and thus prevent cell proliferation by increasing the Apo-RR enzymes. This information about metal scavenger turns to be a reasonable and promising evidence to design selective drugs to inhibit the RR enzyme.

REFERENCE

Altschul SF, Gish W, Miller W, Myers EW, Lipman DJ (1990). Basic local alignment search tool. J. Mol. Biol., 215: 403-410.

Atkin CL, Thelander L, Reichard P, Lang G (1973). Iron and free radical in ribonucleotide reductase. Exchange of iron and Mossbauer spectroscopy of the protein B2 subunit of the *Escherichia coli* enzyme. J. Biol. Chem., 248: 7464-7472.

Carlson J, Fuchs JA, Messing J (1984). Primary structure of the *Escherichia coli* ribonucleoside diphosphate reductase operon.

Proc. Natl. Acad. Sci. USA, 81: 4294-4297.

Chenna R, Sugawara H, Koike T, Lopez R, Gibson TJ, Higgins DG, Thomson JD (2003). Multiple sequence alignment with the Clustal series of programs. Nucleic Acids Res., 31: 3495-3500.

Davis R, Thelander M, Mann GJ, Behravan G, Soucy F, Beaulieu P, Lavallee P, Gräslund A, Thelander L (1994). Purification, characterization, and localization of subunit interaction area of recombinant mouse ribonucleotide reductase R1 subunit. J. Biol. Chem., 269: 23171-23176.

DeLano WL (2002). The PyMOL Molecular Graphics System. DeLano Scientific, San Carlos, CA, USA.

Eklund H, Uhlin U, Färnegårdh M, Logan DT, Nordlund P (2001). Structure and function of the radical enzyme ribonucleotide reductase. Prog. Biophys. Mol. Biol., 77: 177– 268.

Eliasson R, Jörnvall H, Reichard P (1986). Superoxide dismutase participates in the enzymatic formation of the tyrosine radical or ribonucleotide reductase from *Escherichia coli*. Proc. Natl. Acad. Sci. U.S.A. 83: 2373-2377.

Elledge SJ, Davis RW (1987). Identification and isolation of the gene encoding the small subunit of ribonucleotide reductase from

Saccharomyces cerevisiae: DNA damage-inducible gene required for mitotic viability. Mol. Cell. Biol., 7: 2783-2793.

Fan L, Iyer J, Zhu S, Frick KK, Wada RK, Eskenazi AE, Berg PE, Ikegaki N, Kennett RH, Frantz CN (2001). Inhibition of N-myc expression and induction of apoptosis by iron chelation in human neuroblastoma cells. Cancer Res., 61: 1073-1079.

Finch RA, Liu M, Grill SP, Rose WC, Loomis R, Vasquez KM, Cheng Y, Sartorelli AC (2000). Triapine (3-aminopyridine-2-carboxaldehyde-thiosemicarbazone): A potent inhibitor of ribonucleotide reductase activity with broad spectrum antitumor activity Biochem. Pharmacol., 59: 983-991.

Fontecave M, Eliasson R, Reichard P (1987). NAD(P)H:flavin oxidoreductase of *E. coli*: a ferric iron reductase participating in the generation of the free radical of ribonucleotide reductase. J. Biol. Chem., 262: 12325-12331.

Fontecave M, Eliasson R, Reichard P (1989) Enzymatic regulation of the radical content of the small subunit of *Escherichia coli* ribonucleotide reductase involving reduction of its redox centers. J. Bio. Chem., 264: 9164-9170.

Jordan A, Reichard P (1998). Ribonucleotide reductases. Ann. Rev. Biochem., 67: 71-98.

KjØller LI, Sloberg B-M, Thelander L (1982). Characterization of the Active Site of Ribonucleotide Reductase of *Escherichia coli*, Bacteriophage T4 and Mammalian Cells by Inhibition Studies with Hydroxyurea Analogues. Eur. J. Biochem., 125: 75-81.

Laskowski RA, MacArthur MW, Moss DS, Thornton JM (1993). *PROCHECK* - a program to check the stereo chemical quality of protein structures. J. App. Cryst., 26: 283-291.

Li J, Zheng LM, King I, Doyle TW, Chen SH (2001). Syntheses and antitumor activities of potent inhibitors of ribonucleotide reductase: 3-amino-4-methylpyridine-2-carboxaldehyde thiosemicarbazone (3-AMP), 3-amino-pyridine-2-carboxaldehyde-thiosemicarbazone (3-AP) and its water-soluble prodrugs. Curr. Med. Chem., 8: 121-133.

Lozano G, Elledge SJ (2000). p53 sends nucleotides to repair DNA. Nature, 404: 24–25.

Massey V (2000). The Chemical and Biological Versatility of Riboflavin. Biochem. Soc. Trans., 28: 283-296.

McDonald IK, Thornton JM (1994). Satisfying hydrogen bonding potential in proteins. J. Mol. Biol., 238: 777-793.

Moore EC, Zedeck MS, Agrawal KC, Sartorelli AC (1970). Inhibition of ribonucleoside diphosphate reductase by 1-formylisoquinoline thiosemicarbazone and related compounds. Biochemistry, 9: 4492-4498.

Morris GM, Goodsell DS, Halliday RS, Huey R, Hart WE, Belew RK, Olson AJ (1998). Automated Docking Using a Lamarckian Genetic Algorithm and Empirical Binding Free Energy Function. J. Comput. Chem., 19: 1639-1662.

Nakano K, Balint E, Ashcroft M, Vousden KH (2000). A ribonucleotide reductase gene is a transcriptional target of p53 and p73. Oncogene, 19: 4283–4289.

Pavloff N, Rivard D, Masson S, Shen SH, Mes-Masson AM (1992). Sequence analysis of the large and small subunits of human ribonucleotide reductase. DNA Seq., 2: 227-234.

Preidecker PJ, Agrawal KC, Sartorelli AC, Moore EC (1980). Effects of the ferrous chelate of 5-methyl-4-amino-1-formylisoquinoline Ihiosemicarbazone (MAIQ-1) on the kinetics of reduction of CDP by ribonucleotide reductase of the Novikoff tumor. Mol. Pharmacol., 18: 507-512.

Priya PL, Shanmughavel P (2009). A docking model of human ribonucleotide reductase with flavin and phenosafranine. Bioinformation, 4: 123-126.

Reichard P (1993). From RNA to DNA, why so many ribonucleotide reductases? Science, 260: 1773–1777.

Sahlin M, Gräslund A, Peterson L, Ehrenberg A, Sjöberg BM (1989). Reduced forms of the iron-containing small subunit of ribonucleotide reductase from *Escherichia coli*. Biochemistry, 28: 2618–2625.

Sartorelli AC, Agrawal KC, Tsiftsoglou AS, Moore E (1977). Characterization of the biochemical mechanism of the action of α-(N)-heterocyclic carboxaldehyde thiosemicarbazones. Adv. Enzyme Reg., 15: 117-139.

Schuettelkopf AW, van Aalten DMF (2004). PRODRG: A tool for high-throughput crystallography of protein-ligand complexes. ACTA Crystallographica, D60: 1355-1363.

Schwede T, Kopp J, Guex N, Peitsch MC (2003). SWISS-MODEL: An automated protein homology-modeling server. Nucleic Acids Res., 31: 3381-3385.

Shao J, Zhou B, Di Bilio, AJ, Zhu L, Wang T, Qi C, Shih J, Yen Y (2006). A Ferrous-triapine complex mediates formation of reactive oxygen species that inactivate human ribonucleotide reductase. Mol. Cancer. Ther., 5: 586-592.

Sneeden JL, Loeb LA (2004) Mutations in the R2 subunit of ribonucleotide reductase that confer resistance to hydroxyurea. J. Biol. Chem., 279: 40723–40728.

Stubbe J (1990). Ribonucleotide Reductases. Adv. Enzymol. Relat. Areas Mol. Biol. 63: 349-417.

Szekeres T, Fritzer-Szekeres M, Elford HL (1997). The enzyme ribonucleotide reductase: target for antitumor and anti-HIV therapy. Crit. Rev. Clin. Lab. Sci., 34: 503–528.

Tanaka H, Arakawa H, Yamaguchi T, Shiraishi K, Fukuda S, Matsui K, Takei Y, Nakamura Y (2000). A ribonucleotide reductase gene involved in a p53-dependent cell-cycle checkpoint for DNA damage. Nature, 404: 42–49.

Thelander L, Berg P (1986). Isolation and characterization of expressible cDNA clones encoding the M1 and M2 subunits of mouse ribonucleotide reductase. Mol. Cell. Biol., 6: 3433-3442.

Thelander L, Gräslund A (1983). Mechanism of inhibition of mammalian ribonucleotide reductase by the iron chelate of 1-formylisoquinoline thiosemicarbazone. Destruction of the tyrosine free radical of the enzyme in an oxygen-requiring. J. Biol. Chem., 258: 4063-4066.

Wallace AC, Laskowski RA, Thornton JM (1995). LIGPLOT: A program to generate schematic diagrams of protein–ligand interactions. Protein Eng., 8: 127-134.

Wright JA, Chan AK, Choy BK, Hurta RAR, McClarty GA, Tagger AY (1990). Regulation and drug resistance mechanisms of mammalian ribonucleotide reductase, and the significance to DNA synthesis. Biochem. Cell. Biol., 68: 1364–1371.

Zhu L, Zhou B, Chen X, Jiang H, Shao J, Yen Y (2009). Inhibitory mechanisms of heterocyclic carboxaldehyde thiosemicabazones for two forms of human ribonucleotide reductase. Biochem. Pharmacol., 78: 1178–1185.

12

Possible role of 2, 2'- (Diazinodimethylidyne) di - (o-phenylene) dibenzoate, a novel hydrazine as an anti – HIV agent

Rita Ghosh*, Dipanjan Guha, Sudipta Bhowmik and Angshuman Bagchi

Department of Biochemistry and Biophysics, University of Kalyani, Kalyani –741235, W. B., India.

A number of anti-HIV agents act by inhibiting specific steps in the lifecycle of HIV. Inhibition of reverse-transcriptase (RT), a multi-functional enzyme is an important option. The interaction of 2, 2´-(Diazinodimethylidyne)-di-(o-phenylene)-dibenzoate (HZ) with HIV1-RT was studied to explore its possible use in anti-HIV therapy. This novel hydrazine derivative undergoes strong interaction with a number of amino acid residues present in its catalytic domain, some of which are catalytically important like Asp-110, Asp-185 and Asp-186: significant among them is the Asp-186 residue that is highly conserved. Our studies may be useful in the field of structure based drug development in anti HIV therapy.

Key words: HIV 1, drug development, NNRTI, Hydrazine derivative, molecular modeling.

INTRODUCTION

HIV like all retroviruses requires for its infection the integration of double-stranded viral DNA into its host genome. HIV contains two identical single stranded viral RNA molecules (Takahashi et al., 2002). The reverse transcriptase of human immunodeficiency virus (HIV-RT) is responsible for the conversion of the HIV single stranded RNA genome into double stranded DNA (Basu et al., 2008; McBurney et al., 2006). This multi functional enzyme has RNA-dependent DNA polymerase activity, DNA-dependent DNA polymerase activity, strand displacement, strand transfer and RNase H activities (Lanciault and James, 2004). HIV-RT is a heterodimer composed of two subunits, p66 and p51 having molecular weights 66 and 51 kDa, respectively (Lightfoote et al., 1986; Jacobo-Molina and Arnold, 1991).

The polymerase domain is localized at the N-terminus of p66 and the 15 kDa C-terminal domain of p66 contains the RNase H active site (Szilvay et al., 1993). HIV RNase H is a 130 amino acid domain of the viral reverse

transcriptase and is indispensable for the replication of genomic RNA to double stranded DNA. RNase H activity is to degrade the RNA from the RNA-DNA duplex (Tanese et al., 1991), so that the newly formed DNA can act as the template for the DNA dependent DNA polymerase activity in order to form the double stranded viral DNA and the virus can start its lifecycle inside the host cell (Dudding et al., 1991). Biochemical as well as structural analyses show the spatial distance between the two active sites (the polymerase active site and RNase H active site) was about 18 - 19 nucleotides in length when RT is bound to a duplex substrate (Furfine and Reardon, 1991; DeStefano et al., 1991; Gopalakrishnan, 1992; Jacobo-Molina et al., 1993). An effective therapeutic approach against HIV infection is to inhibit the RT activity.

There are two types of inhibitors - the nucleoside/nucleotide RT inhibitors (both called NRTIs for simplicity) and the non nucleoside/nucleotide RT inhibitors (NNRTI) that can act as anti-RT drugs. Some NRTI, 3'-azido-3'-deoxythymidine (AZT, zidovudine) (Fischl et al., 1987), 2',3'dideoxycytidine (ddC, zalcitabine) (Yarchoan et al., 1988), 2',3'- dideoxyinosine (ddI, didanosine) (Lambert et al., 1990; Cooley et al., 1990), 2',3'-didehydro-3'-deoxythymidine (D4T, stavudine) (Riddler et al., 1995) etc. are nucleoside prodrugs that are converted by cellular enzymes to the active deoxynucleotide triphosphates that lack a free 3'- hydroxyl group and lead to a chain

*Corresponding author. E-mail: rghosh_bcbp@klyuniv.ac.in, ritadg2001@yahoo.co.in.

Abbreviations: NRTIs, Nucleoside/nucleotide RT inhibitors; NNRTI, Non Nucleoside/nucleotide RT inhibitors.

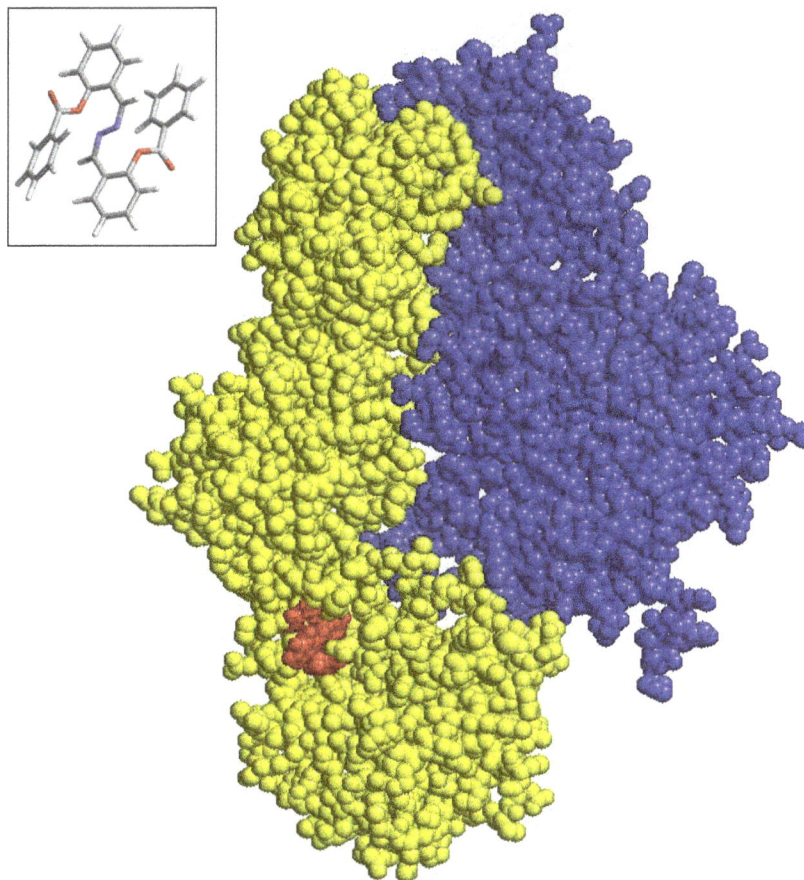

Figure 1. Structure of HIV1-RT complexed with HZ (Red). Subunits p66 and p51 are coloured yellow and blue respectively. Here both protein and ligand are shown in space-fill mode. Structure of HZ is shown in the inset. The coordinate of the HZ derivative was generated using the program Biopolymer of InsightII (MSI/Accelrys) (Honig et al., 1993; Nicholls and Honig, 1991; Sharp et al., 1991). Then energy minimization was done using Conjugate Gradient (CG) algorithm using the program DISCOVER (MSI/Accelrys) with consistent valance force field (cvff) until the derivative reached to 0.001kcal/mol (Dauber-Osguthorpe et al., 1988). The coordinates of enzyme RT were downloaded from the Protein Data Bank (ID: 2I5J). All the water molecules were removed and the protein coordinates were energy minimized using CG algorithm using the program DISCOVER (MSI/Accelrys) with cvff until the derivative reached 0.001kcal/mol. The structure which was energy minimized was used for superimposition and docking experiment. The energy minimised coordinates of hydrazine derivative was superimposed on the coordinates of the ligand, which was already present in 2I5J. The root mean square (r.m.s.) deviation was 0.085Å.The superimposed structure was again energy minimized using CG algorithm using the program DISCOVER (MSI/Accelrys) with cvff until the derivative reached 0.001kcal/mol in order to reduce the bad contacts.

termination and inhibition of viral replication. However, these nucleotide analogues can also get incorporated in the host DNA itself, causing damage to the host. Search for NNRTIs are therefore a favourable option. All known NNRTIs bind to the hydrophobic pocket of RT, but they do not prevent the binding of nucleic acid or nucleoside triphosphate substrates to RT; rather, the NNRTIs block the chemical step of the polymerization reaction (Rittinger et al., 1995; Spence et al., 1995).

Crystallographic studies have shown that the binding of the NNRTIs causes conformational changes by displacement of the β12, β13 and β14 sheet that contains the polymerase primer grip which is important for properly positioning the nucleic acid with respect to the polymerrase active site (Rodgers et al., 1995). Binding of NNRTI in the hydrophobic pocket near RT polymerase active site can also influence its geometry and stops the polymerization reaction that eventually stops the HIV lifecycle (Ding et al., 1998). Examples of such drugs are like, nevirapine, efavirenz and delavirdine that bind near the

Table 1. To find the interactions between the HZ and the HIV-1 RT all the energy calculations were performed using Insight II software on a silicon graphics IndigoII workstation. HZ was also docked using AutoDock in order to get comprehensive result. The residues that are involved in ionic interaction with HZ are shown below. The amino acids showing H bonding and hydrophobic interactions are also indicated in the table.

Software used	Amino acids name with number
Insight II	VAL108, LEU109, ASP110, ASP185, ASP186, LEU187[¶][+], TYR188, LYS223, PHE227[+], LEU228, TRP229, MET230, GLY231, LEU234, GLN 242
AutoDock 4	VAL108, ASP110, TYR181, GLN182, TYR183, ASP186, LEU187, TYR188, PRO217, PHE227, TRP229, LEU234

¶ denotes H-bonding. + denotes hydrophobic interaction.

primer grip region near the polymerase active site and inhibit polymerization (Balzarini, 2004; Smerdon et al., 1994). DHBNH, a hydrazone derivative, also binds to the hydrophobic pocket of RT, but it does not only acts as inhibitor of polymerization but also inhibits RNase H activity, although, its binding is away (<50 Å) from the RNase H activity site (Daniel et al., 2006).

MATERIALS AND METHODS

In the study's attempt to characterize the biological activity of a novel hydrazine derivative 2,2′- (diazinodimethylidyne) di-(o-phenylene) dibenzoate (Chattopadhyay et al., 2008), the authors have tried to investigate from theoretical studies if this molecule has therapeutic potential as HIV-RT polymerase inhibitor (NNRTI). For this purpose, they have generated the 3-D coordinates of the molecule and from docking experi-ments with HIV-1 RT. They have also shown that HZ binds in the hydrophobic pocket near the RT polymerase active site like the other NNRTIs. It interacts with all iden-tified amino acids that are important for catalytic activity of HIV1-RT, particularly those that are conserved. Our results indicate that HZ can act as a NNRTI by blocking the active site by interacting with the primer grip region and therefore has the potential act as anti- HIV drug.

RESULTS AND DISCUSSION

Both the two subunits of HIV1-RT (p66 and p51) consist of fingers (residues 1 - 84 and 120 - 150), palm (residues 85 - 119 and residues 151 - 243), thumb (residues 244 - 322) and connection (residues 323 - 437), but p66 has an extra domain, the RNase H domain (residues 438 - 556). There is an opening between the p66 fingers and thumb in which the non-nucleoside inhibitors can bind (Rodgers et al., 1995).

Comparison with the structures of HIV-RT and non-nucleoside inhibitor-complexed HIV-RT showed that only minor domain rearrangements occur, but there is a significant repositioning of a three-stranded beta-sheet in the p66 subunit which contains the catalytic aspartic acid residues 110, 185 and 186 (which resides in the palm of RT) (Esnouf et al., 1995). The β6-β9-β10 hairpin contains the polymerase active site and the residues that exist

here are 105 - 112, 178 - 183 and 186 - 191 and the β12-β13-β14 hairpin that contains the primer grip consists of the residues 227 - 229, 232 - 235 and 238 – 242 (Jacobo-Molina et al., 1993).

In all known structures of HIV-1 RT with NNRTI complexes, the position of the β12-β13 hairpin or the primer grip is significantly displaced relative to the position in the structure of HIV-1 RT complexed with a double-stranded DNA and in unliganded HIV-1 RT structures. Since the primer grip helps to position the template-primer, this displacement suggests that binding of NNRTIs would affect the relative positions of the primer terminus and the polymerase active site. This could explain biochemical data showing that NNRTI binding to HIV-1 RT reduces efficiency of the chemical step of DNA polymerization, but does not prevent binding of either dNTPs or DNA (Tantillo et al., 1994). The hydrazine derivative (HZ), (Figure 1) complexed with HIV-1 reverse transcriptase is shown in Figure 1. The residues that are present within 4 Å from HZ are shown in Table 1. These amino acids are a part of polymerase active site and primer grip region residues and are within the protein-ligand contact distance (Figure 2) (Diago et al., 2007). Therefore, the ligand has a strong ionic interaction with these residues.

In most cases, it is seen that the ligand-protein interaction site is highly mutative. Where the NNRTI binds the nearby amino acids gets mutated and some steric interactions come into play and disallow the ligand to bind at the active site of the RT polymerase active site (Smerdon et al., 1994). So, a drug designer must have a goal to invent such ligands that have a strong interaction with the conserved site residues, in this case, the ASP186 and TRP229. We found that HZ interacts with both of these two conserved amino acid residues (ASP186 and TRP229). HZ also additionally makes a hydrogen bond with LEU187. HZ also undergoes hydro-phobic interactions with LEU187 and specifically with PHE227.

The hydrophobic side chain of LEU187 comes in contact with the C-skeleton of HZ. The aromatic rings of the HZ remains nearly stacked on to the aromatic ring of PHE227 thereby exposing the Diazino group to undergo H-bonding and other polar interactions with amino acids.

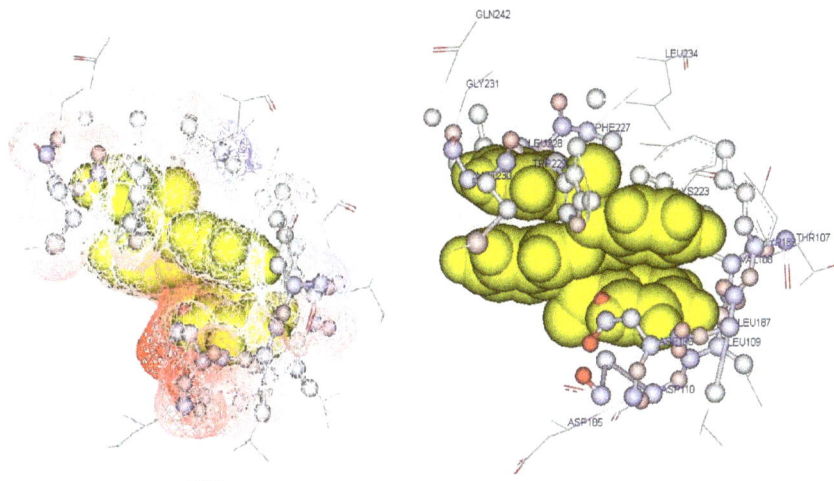

Figure 2. Pattern of charge distribution of residues within 4Å of HZ is shown in the left panel. The right panel shows the names of amino acid residues within 4Å from HZ. This calculation was done using Discovery Studio (2.0) (Honig et al., 1993; Nicholls and Honig, 1991; Sharp et al., 1991). These residues are shown in ball and stick in CPK colours and HZ is in yellow in space fill model. HZ also undergoes hydrophobic interactions with the protein. The hydrophobic interactions including stacking interactions were calculated using Insight II as well as WHAT IF server (Vriend, 1990) and Chimera (Pettersen et al., 2004).

A significant observation is that the conserved residues such as PHE227 and LEU234 (Smerdon et al., 1994) are also in close contact with HZ, implying that this new drug would be less vulnerable to escape mutations. Again from the free energy point of view that is obtained from the AutoDock (Morris et al., 1998) results, we have found that the best-docked free energy was -6.45 Kcal/mol (when the grid volume is 80), which indicates that ligand binds the RT polymerase active site very strongly.

Conclusion

This study has shown that HZ could work as a HIV1-RT inhibitor. The interaction of HZ with HIV1-RT is similar to the interactions of nevirapine, efavirenz, dihydroxy benzoyl naphthyl hydrazone and delavirdine, some well-known drugs in HIV treatment. However, as most of these drugs exhibit considerable side effects it merits search for new drugs. HZ does not only show interactions with all identified amino acids that are important for catalytic activity of HIV1-RT, but also with some other amino acids as well that are present in the catalytic domain. The free energy of binding also ensures that HZ shows a very strong binding with HIV1-RT at polymerase active site. Theoretical studies reveal that HZ has the potential for therapeutic use as HIV1 inhibitor.

ACKNOWLEDGEMENTS

The authors are thankful to Bioinformatics Centre, Bose Institute, Kolkata for extending their computational facility and also to the B.I.F. Centre, University of Kalyani. The authors, D. Guha and S. Bhowmik are provided with fellowships from U.G.C., Govt. of India and University of Kalyani, respectively.

REFERENCES

Balzarini J (2004). Current status of the non-nucleoside reverse transcriptase inhibitors of human immunodeficiency virus type 1, Curr. Top. Med. Chem., 4: 921-944.

Basu VP, Song M, Gao L, Rigby ST, Hanson MN, Bambara RA (2008). Strand transfer events during HIV-1 reverse transcription. Virus Res., 134: 19-38.

Chattopadhyay B, Basu S, Ghosh S, Helliwelld M, Mukherjee M (2008). 2, 2'-(Diazinodimethylidyne) di-(o-phenylene) dibenzoate. Acta Cryst., E64: 0866.

Cooley TP, Kunches LM, Saunders CA, Ritter JK, Perkins CJ, McLaren C, McCaffrey RP, Liebman HA (1990). Once-daily administration of 2', 3'-dideoxyinosine (ddI) in patients with the acquired immunodeficiency syndrome or AIDS-related complex. Results of a phase I trial. N. Engl. J. Med., 322: 1340-1345.

Daniel MH, Stefan GS, Sanjeewa D, Mohammed MH, McCoy SK, Tatiana I, Clark Arthur D Jr., Jennifer LK, John GJ, Patrick KC, Karsten KJ, Ronald ML, Stephen HH, Michael AP, Eddy A (2006). HIV-1 Reverse Transcriptase Structure with RNase H Inhibitor Dihydroxy Benzoyl Naphthyl Hydrazone Bound at a Novel Site. ACS Chem. Biol., 1: 702-712.

Dauber-Osguthorpe P, Roberts VA, Osguthorpe DJ, Wolff J, Genest M, Hagler AT (1988). Structure and energetics of ligand binding toproteins: Escherichia coli dihydrofolate reductase-trimethoprim, a drug-receptor system. Proteins, 4: 31-47.

DeStefano JJ, Buiser RG, Mallaber LM, Myers T, Bambara R, Fay PJ (1991). Polymerization and RNase H activities of the reverse transcriptases from avian myeloblastosis, human immunodeficiency, and Moloney murine leukemia viruses are functionally uncoupled. J. Biol. Chem., 266: 7423-7431.

Diago LA, Morell P, Aguilera L, Moreno E (2007). Setting up a large set of protein-ligand PDB complexes for the development and validation of knowledge-based docking algorithms. BMC Bioinformatics, 8: 310 - 324.

Ding J, Das K, Hsiou Y, Sarafianos SG, Clark AD, Jacobo- Molina A Jr., Tantillo C, Hughes SH, Arnold E (1998). Structure and functional implications of the polymerase active site region in a complex of HIV-1 RT with a double-stranded DNA template– primer and an antibody Fab fragment at 2.8 Å resolution. J. Mol. Biol., 284: 1095-1111.

Dudding LR, Nkabinde NC, Mizrahi V (1991). Analysis of the RNA- and DNA-dependent DNA polymerase activities of point mutants of HIV-1 reverse transcriptase lacking ribonuclease H activity. Biochemistry, 30: 10498-10506.

Esnouf R, Ren J, Ross C, Jones Y, Stammers D, Stuart D (1995). Mechanism of inhibition of HIV-1 reverse transcriptase by non-nucleoside inhibitors. Nat. Struct. Biol., 2: 303-308.

Fischl MA, Richman DD, Grieco MH, Gottlieb MS, Volberding PA, Laskin OL, Leedom JM, Groopman JE, Mildvan D, Schooley RT (1987). The efficacy of azidothymidine (AZT) in the treatment of patients with AIDS and AIDS-related complex: a double-blind, placebo-controlled trial. N. Engl. J. Med., 317: 185-191.

Furfine ES, Reardon JE (1991). Reverse Transcriptase RNase H from the Human Immunodeficiency Virus. Relationships of the DNA polymerase and RNA hydrolysis activities. J. Biol. Chem., 266: 406-412.

Gopalakrishnan V, Peliska JA, Benkovic SJ (1992). Human immunodeficiency virus type 1 reverse transcriptase: Spatial and temporal relationship between the polymerase and RNase H activities. Proc. Natl. Acad. Sci. USA, 89: 10763-10767.

Honig B, Sharp K, Yang AS (1993). Microscopic models of aqueous solutions: Biological and Chemical Applications. J. Phys. Chem., 97: 1101-1109.

Jacobo-Molina A, Arnold E (1991). HIV reverse transcriptase structure-function relationships. Biochemistry, 30: 6351-6356.

Jacobo-Molina A, Ding J, Nanni RG, Clark AD, Lu X, Tantillo C, Williams R L, Kamer G, Ferris AL, Clark P, Hizi A, Hughes SH, Arnold E (1993). Crystal structure of human immunodeficiency virus type 1 reverse transcriptase complexed with double-stranded DNA at 3.0Å resolution shows bent DNA. Proc. Natl. Acad. Sci. USA, 90: 6320-6324.

Lambert JS, Seidlin M, Reichman RC, Plank CS, Laverty M, Morse GD, Knupp C, McLaren C, Pettinelli C, Valentine FT, Dolin R (1990). 2', 3'-Dideoxyinosine (ddl) in patients with the acquired immunodeficiency syndrome or AIDS-related complex: a phase I trial. N. Engl. J. Med., 322: 1333-1340.

Lanciault C, James CJ (2004). Single Unpaired Nucleotides Facilitate HIV-1 Reverse Transcriptase Displacement Synthesis through Duplex RNA, J. Biol. Chem., 279: 32252-32261.

Lightfoote MM, Coligan JE, Folks TM, Fauci AS, Martin MA, Venkatesan S (1986). Structural Characterization of Reverse Transcriptase and Endonuclease Polypeptides of the Acquired Immunodeficiency Syndrome Retrovirus. J. Virol., 60: 771-775.

McBurney SP, Young KR, Nwaigwe CI, Soloff AC, Cole KS, Ross TM (2006). Lentivirus-like particles without reverse transcriptase elicit efficient immune responses. Curr. HIV Res., 4: 475-484.

Morris GM, Goodsell DS, Halliday RS, Huey R, Hart WE, Belew RK, Olson AJ (1998). Automated docking using a Lamarckian genetic algorithm and an empirical binding free energy function. J Comput. Chem., 19: 1639-1662.

Nicholls A, Honig B (1991). A rapid difference algorithm, utilizing successive overorelaxation to solve the Poisson-Boltzmann equation. J. Comp. Chem., 12: 435-445.

Pettersen EF, Goddard TD, Huang CC, Couch GS, Greenblatt DM, Meng EC, Ferrin TE (2004). UCSF Chimera--a visualization system for exploratory research and analysis. J Comput. Chem., 25(13): 1605-1612.

Riddler SA, Anderson RE, Mellors JW (1995). Antiretroviral activity of stavudine (2',3'-didehydro-3'-deoxythymidine, d4T). Antiviral Res., 27: 189-203.

Rittinger K, Divita G, Goody RS (1995). Human immunodeficiency virus reverse transcriptase substrate-induced conformational changes and the mechanism of inhibition by nonnucleoside inhibitors. Proc. Natl. Acad. Sci. USA, 92: 8046-8049.

Rodgers DW, Camblin SJ, Harris BA, Ray S, Culp JS, Hellmig B, Woolf DJ, Debouck C, Harrison SC (1995). The structure of unliganded reverse transcriptase from the human immunodeficiency virus type 1. Proc. Natl. Acad. Sci. USA, 92: 1222-1226.

Sharp B, Nichols A, Friedman R, Honig B (1991). Extracting gydrophobic free energies from experimental data: relationship to protein folding and theoretical models. Biochemistry, 30: 9686-9697.

Smerdon SJ, Jager J, Wang J, Kohlstaedt LA, Chirino AJ, Friedman JM, Rices PA, Steitz TA (1994). Structure of the binding site for nonnucleoside inhibitors of the reverse transcriptase of human immunodeficiency virus type 1. Proc. Nati. Acad. Sci. USA, 91: 3911-3915.

Spence RA, Kati WM, Anderson KS, Johnson KA (1995). Mechanism of inhibition of HIV-1 reverse transcriptase by nonnucleoside inhibitors. Science, 267: 988-993.

Szilvay AM, Nornes S, Kannapiran A, Haukanes BI, Endresen, Helland DE (1993). Characterization of HIV-1 reverse transcriptase with antibodies indicates conformational differences between the RNAse H domains of p66 and p15. Arch Virol., 131: 393-403.

Takahashi H, Sawa H, Hasegawa H, Sata T, Hall WW, Nagashima K, Kurata T (2002). Reconstitution of cleavage of human immunodeficiency virus type-1 (HIV-1) RNAs. Biochem. Biophys. Res. Commun., 293: 1084-1091.

Tanese N, Telesnitsky A, Goff SP (1991). Abortive reverse transcription by mutants of Moloney murine leukemia virus deficient in the reverse transcriptase-associated RNase H function. J. Virol., 65: 4387–4397.

Tantillo C, Ding J, Jacobo-Molina A, Nanni RG, Boyer PL, Hughes SH, Pauwels R, Andries K, Janssen PAJ, Arnold E (1994). Locations of anti-aids drug binding sites and resistance mutations in the three-dimensional structure of HIV-1 reverse transcriptase: implications for mechanisms of drug inhibition and resistance. J. Mol. Biol., 243: 369-387.

Vriend G (1990). WHAT IF: A molecular modeling and drug design program. J. Mol. Graph, 8: 52-56.

Yarchoan R, Thomas RV, Allain JP, McAtee N, Dubinsky R, Mitsuya H, Lawley TJ, Safai B, Myers CE, Perno CF, Klecker RW, Wills RJ, Fischl MA, McNeely MC, Pluda JM, Leuther M, Collins JM, Broder S (1988). Phase I studies of 2',3'-dideoxycytidine in severe human immunodeficiency virus infection as a single agent and alternating with zidovudine (AZT). Lancet, 331: 76-81.

13

Determination of instantaneous arterial blood pressure from bio-impedance signal

Sofiene Mansouri*, Halima Mahjoubi and Ridha Ben Salah

Biophysics Research Unit, Faculty of Medicine, Sousse, Tunisia.

The objective of this study was to determine the instantaneous arterial blood pressure by the peripheral bio-impedance. To achieve this goal, the equation of Ben Salah and Flaud, initially meant to determine the instantaneous aortic pressure according to the signal of thoracic bioimpedance, was successfully applied for the determination of the instantaneous arterial pressure according to the signal of peripheral bio-impedance. The problem was that this equation depends directly on systolic and diastolic pressures which were manually obtained by an electronic sphygmomanometer. Yet the objective was to determine the instantaneous arterial blood pressure without having to measure beforehand diastolic and systolic pressures by the electronic sphygmomanometer. To solve this problem, interpolation function linking respectively the pulse pressure to the impedance variation (with 2,97% error for the determination of the systolic pressure) and the diastolic pressure to the basic impedance were determined (with an error close to zero for the determination of the diastolic pressure). With the proposed method, for each new measurement on a subject we acquire its bioimpedance signal, and then use the interpolation functions to deduce its diastolic and systolic pressures and then, using the Ben Salah equation, in real time, to determine its instantaneous arterial blood pressure. The bio-impedance signal processing, the user interface and the display were managed by LabVIEW.

Key words: Instantaneous arterial pressure, cubic spline interpolation, bioimpedance, electric plethysmography, LabVIEW.

INTRODUCTION

Aortic and arterial pressures are classically determined by invasive methods, based on cardiac catheterization and recently by non invasive methods (Chemla and Lamia, 2009; Siebig et al., 2009; Bogert and Lieshout, 2005). Proposed method in this article is a non-invasive method based on the use of peripheral bioimpedance signal.

Proposed method in this article is a non-invasive method, using the peripheral bioimpedance signal.

The beginning of modern clinical applications of bio-impedance measurements can be attributed to a large part to Nyboer (1970).

The most common application is in the study of the small pulsatile impedance changes associated with heart action. Its goal is to give quantitative and qualitative information about the volume changes in the lung, heart, peripheral arteries, and veins.

The following Nyboer's equation relates the impedance variation (ΔZ) obtained on the thorax or peripheral limbs to the pulsatile blood volume change.

$$\Delta V = \left(\frac{\rho L^2}{Z_0^2}\right)\Delta Z \qquad (1)$$

Where:
ΔV = the pulsatile volume change with resistivity ρ
ρ = the resistivity of the pulsatile volume in Ω-cm (typically the resistivity of blood)
L = the length of the explored section
Z_0 = the impedance measured when the pulsatile volume is at a minimum (diastolic or base impedance)
ΔZ = the magnitude of the pulsatile impedance change or impedance variation.

The idea was to link the variation of pressure to the variation of the bioimpedance. Because when ΔV is maximum in the artery (ΔZ is minimum) and the blood pressure

---

*Corresponding author. E-mail: mansouri_sofienne@yahoo.fr.

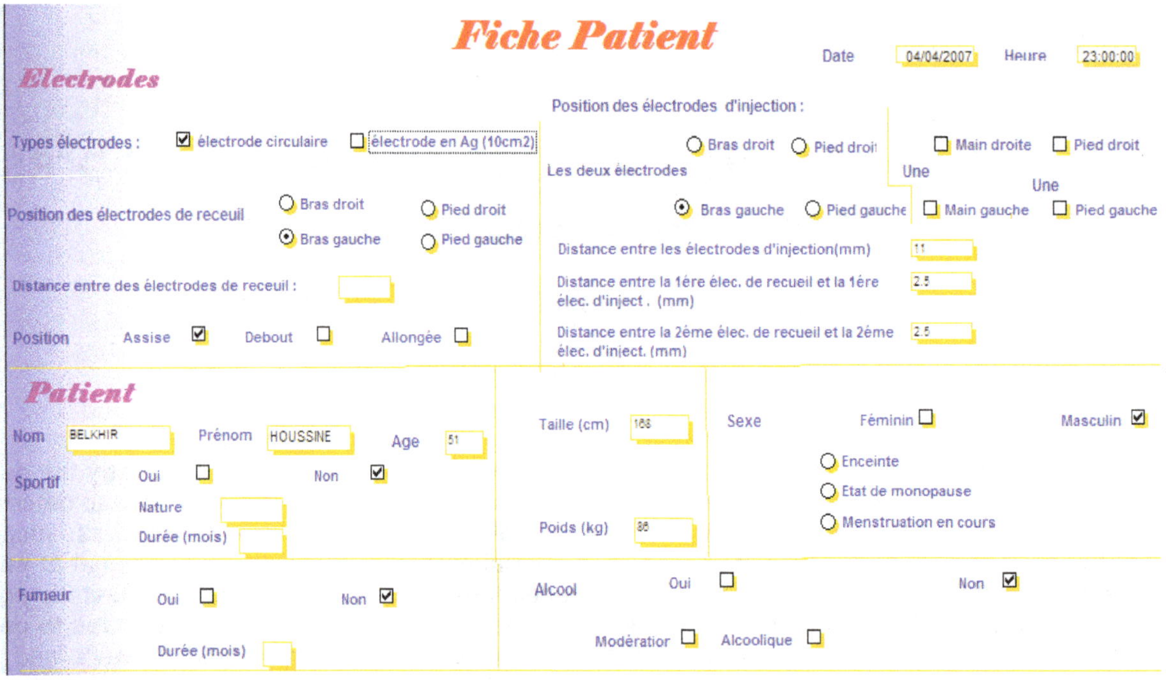

Figure 1. The access base, Z-BIOBASE.

pressure is at its maximum: the systolic pressure. And when ΔV is minimum (ΔZ is maximum), the blood pressure is at its minimum: the diastolic pressure. Knowing that the ΔZ waveform is similar to the aortal blood pressure curve (Grimnes and Martinsen, 2005).

This paper presents a practical work on the implementation of a new technique for non invasive determination of instantaneous arterial blood pressure directly from the peripheral bioimpedance signal.

MATERIALS AND METHODS

Materials

We used plethysmograph (Siemens, Direktrheagraph 933) in this study. The instrument has an injection module, generating a square pulse with a 30 KHz frequency and 1 mA intensity. In output, the device delivers an analogic signal representing the variation of the impedance of the explored section. Disposable electrodes (those used for the electrocardiograph) were used for the injection of the square pulse and the collection of bioimpedance signal. An electronic sphygmomanometer model, MP3 FUZZY was used.

The acquisition of the bioimpedance signal on PC was made easier by the use of a National Instrument data acquisition device, the NI USB 6009. This card, placed between the plethysmograph and the PC, assures the digitalization of the bioimpedance signal and its transfer on PC for further processing. The results of the different measurements were stored in an access database, the Z-BIOBASE (Figure 1).

The determination of the coefficients of interpolation was achieved using the version R2007a of Matlab. The bioimpedance signal processing, the display as well as the user interface are managed by the LabVIEW Professional Development System; version: 8.6.

Methods

The objective of this study was to determine the continuous non-invasive arterial pressure by the peripheral bioimpedance.

In a previous research, a theoretical relationship (2) between thoracic electrical bioimpedance signal and instantaneous aortic pressures was established. This study was based on the pressure variation according to radius variation during cardiovascular activity (Salah, 1988, 1986; Flaud, 1979).

$$P(t) = P_{Dia} + \left(P_{Sys} - P_{Dia}\right)\frac{Z(t)}{Z_{max}} \tag{2}$$

Where:
P (t): Instantaneous aortic pressure
P_{Sys}: systolic pressure in mmHg
P_{Dia}: diastolic blood pressure in mmHg
Z (t): The instantaneous bioimpedance signal
Z_{max}: The maximum of the bioimpedance signal curve

The P_{Sys} and P_{Dia} were determined by an electronic sphygmomanometer.

Validation for the peripheral use

The validity of this equation was demonstrated for the determination of the aortic pressure according to the thoracic bioimpedance signal. It is necessary to validate it for the arterial pressure and for the peripheral bioimpedance signal.

To do that, we conducted simultaneous measurements, on the left arm, of bioimpedance signal, of its maximum, Z_{max}, we also measured the systolic and the diastolic pressure with an electronic sphygmomanometer. We replaced all in the equation (2). We drew the instantaneous pressure and we found that always the maximum and the minimum of the curve represent the systolic and the diastolic pressures. Thus this equation was then validated to be used for the computing of the continuous arterial pressure from the peripheral bioimpedance signal.

Automatic computing of the instantaneous pressure

Disadvantage of this method is that to get the instantaneous arterial pressure we are dependent on the sphygmomanometer. This does not enable us to determine automatically and in real time instantaneous pressure from only the bioimpedance signal.

In order to avoid this, the idea is to try to link empirically, by interpolation, the diastolic pressure (P_{Dia}) to base's impedance (Z_0) and the systolic pressure (P_{Sys}) to the maximum signal of bioimpedance (Z_{max}).

When we build the interpolation function between (P_{Sys}) and (Z_{max}), and the interpolation function between (P_{Dia}) and (Z_0), each new acquisition of a bioimpedance signal allows us to have Z (t), the (Z_0) and (Z_{max}). Having (Z_0), from the function interpolation we determine the correspondent (P_{Dia}), and having (Z_{max}), we determine the correspondent (P_{Sys}). We replace then, in equation (2), (P_{Sys}), (P_{Dia}), (Z_{max}) and Z (t) by their values and we then compute automatically the instantaneous arterial pressure.

Measurements

In order to build the interpolation function, we carried out a series of simultaneous measurements of the bioimpedance signals, to extract Z_0 and Z_{max} from it, as well as measurements by an electronic sphygmomanometer of the systolic and diastolic's pressures.

These measures concerned three hundred subjects divided into three groups, a healthy one, a sportsmen group (the national team of weightlifting) and a group of patients from Internal Medicine Department of the University Hospital (La Rabta - Tunis, Tunisia).

Cubic spline interpolation

From the different measures we kept only one hundred and forty five, the most reliable ones, which we divided into two groups: A 'learning group' and a 'test group'. A 'learning group' is made up of one hundred people who were used to build the four vectors needed for the interpolations. A vector representing the P_{Dia}, a second one corresponds to the P_{Sys}, a third one for Z_0 and a fourth one to Z_{max}. And a 'test group' made up of forty five people (15 healthy subjects, patients and athletes each) were used to check the effectiveness of the interpolations.

Four interpolation methods were tested: the linear, quadratic, cubic and cubic spline interpolation. Cubic spline gave the best interpolation, but it present some weaknesses in particular with athletes and patients. Notice that the estimation of the diastolic blood pressure shows that it is always good. However, this is not always the case with the systolic pressure. Improvements were required for this reason.

To better estimate the systolic pressure, we do not consider any more P_{Sys} in function of Z_{max}, but we test one time the pulse pressure ΔP in function of the difference between Z_0 and Z_{max} (ΔZ) and in another time, the mean pressure in function of the mean impedance. We use cubic splines to perform the interpolation.

Interpolation of pulse pressure and impedance variation

For the estimation of systolic pressure from the pulse pressure ΔP, deduced by spline interpolation from ΔZ, we proceeded as follows.

From the P_{Sys} and P_{Dia} vectors, we create the ΔP vector, where $\Delta P = (P_{Sys} - P_{Dia})$. And from the Z_0 and Z_{max} vectors, we create the ΔZ vector, where $\Delta Z = (Z_0 - Z_{max})$. We will have a cubic spline interpolation between the two vectors. When a new measure is performed, from Z_0 and Z_{max} obtained directly from the bioimpedance signal, we compute ΔZ. Then we use the interpolation function to determine ΔP. Then having P_{Dia} and ΔP we deduce the systolic pressure by: $P_{Sys} = \Delta P + P_{Dia}$. And now having P_{Dia} and P_{Sys} we can use the equation (2) to obtain the instantaneous arterial blood pressure.

Interpolation of mean pressure and mean impedance

For the estimation of systolic pressure from the mean pressure we use two methods as follows:
Using the mean pressure well known formula, from the P_{Sys} and P_{Dia} vectors, we created the mean pressure vector:

Where
$$P_{mean} = \frac{(P_{Sys} + 2P_{Dia})}{3}$$

And from the Z_0 and Z_{max} vectors we created the mean impedance vector:

$$Z_{mean} = \frac{(Z_{max} + 2Z_0)}{3}$$

We have done a cubic spline interpolation between the two vectors. Whenever a new measure is performed, from the bioimpedance signal we use Z_0 and Z_{max} to compute Z_{mean}. Then we use the interpolation function to determine P_{mean}. Then from P_{Dia} and P_{mean} we deduce the systolic pressure by $P_{Sys} = 3 \cdot P_{mean} - 2 \cdot P_{Dia}$.
And now having P_{Dia} and P_{Sys} we can use the equation (2) to get the instantaneous arterial blood pressure.

Using mean pressure of Chemla (Chemla et al., 2004; Chemla and Lamia, 2009), we followed the above steps but we use the new equation of P_{mean} and Z_{mean}:

$$P_{mean} = \sqrt{P_{Dia} P_{Sys}} \quad \text{and} \quad Z_{mean} = \sqrt{Z_0 Z_{max}}$$

In the next section, we will compare these three methods.

We can summarize the retained method as follows:

Estimating the diastolic pressure P_{Dia} from the base impedance by using the cubic spline interpolation. Estimating ΔP from ΔZ by cubic spline and then deducing the systolic pressure P_{Sys}. Determination of the instantaneous blood pressure from equation (2).

RESULTS

The first task was the study of the relevance of using peripheral bioimpedance to estimate the instantaneous arterial pressure with the same equation expressing the dependence of the thoracic bioimpedance to the instantaneous aortic pressure. This has been verified successfully with three hundred subjects divided into 3

Table 1. The interpolation error of the determination of the pressures by the four methods (results obtained with the one hundred measures of the 'lea group').

	Pressures obtained by linear interpolation (%)	Pressures obtained by quadratic interpolation (%)	Pressures obtained by cubic interpolation (%)	Pressures obtained by cu spline interpolation (%)
Systolic error	2.17	1.87	1.53	0.92
Diastolic error	9.1	8.62	6.49	0.18

With systolic error = P_{Sys} (obtained by the interpolation method) − P_{Sys} (obtained by the electronic sphygmomanometer). Diastolic error = P_{Dia} (obtained b interpolation method) − P_{Dia} (obtained by the electronic sphygmomanometer).

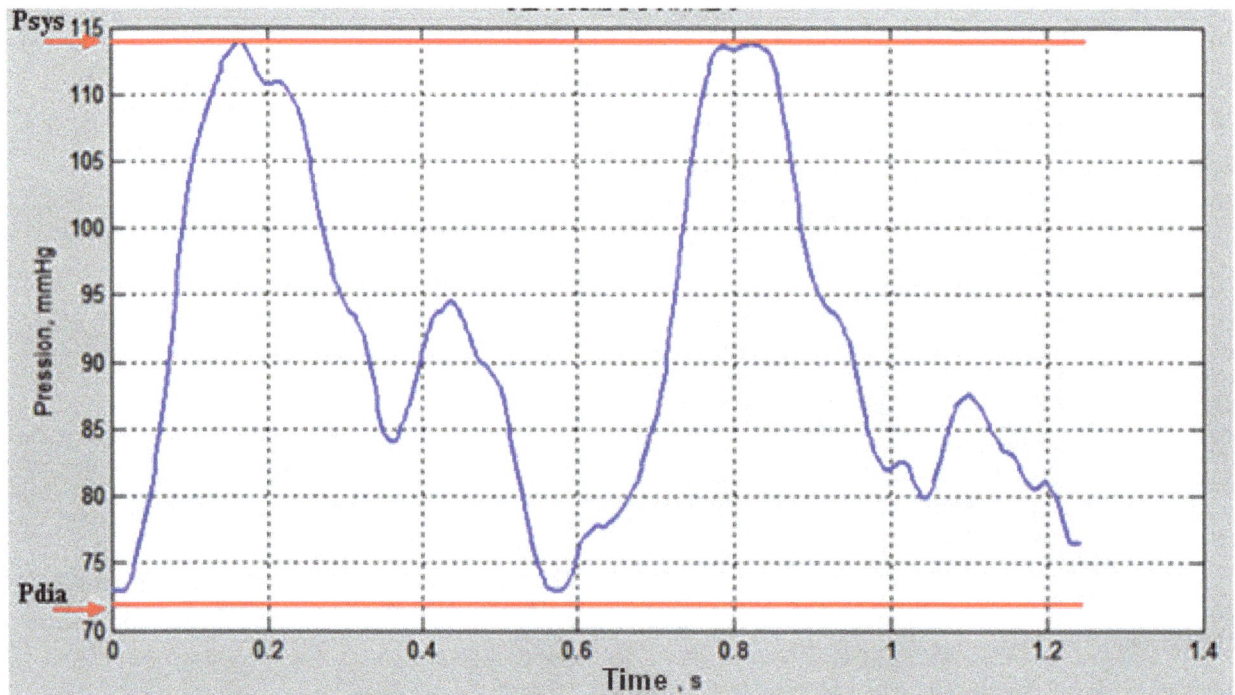

Figure 2. Curve of the instantaneous arterial pressure obtained by the cubic spline.

categories: a healthy group (people between 18 and 25 years, one hundred and eighty subjects) patients (sixty subjects) and athletes (sixty subjects).

For each of these subjects, we measured with an electronic sphygmomanometer, the systolic and diastolic pressures obtained on the left arm, then we replaced the values in the equation (2). Then we have drawn the instantaneous blood pressure curve and from which we determined the minimum that represents the diastolic pressure and the maximum that represents the systolic pressure. We compared these pressures with the values obtained by the sphygmomanometer and we found that the results are very close. We deduced that we can use the equation (2) for peripheral studies.

From the different measures we retained the most precise ones, one hundred and forty five divided into two groups: A 'learning group' used to build the vectors needed

for the interpolations. And a 'test group' used to verify the performance of the interpolations.

We had chosen that the selection criteria between the various methods are the minimum difference between the systolic pressure (respectively diastolic pressure) obtained by an electronic sphygmomanometer and the systolic pressure obtained from the bioimpedance signal. First of all we compared the performance of the interpolation between linear, quadratic, cubic and cubic spline interpolation.

The comparison between the various methods allows us to choose the cubic spline which gave the best approximations (Table 1 and Figure 2).

Although the estimate of the diastolic pressure is quasi correct, the determination of the systolic pressure presented sometimes some weaknesses particularly with athletes and patients (Figure 3).

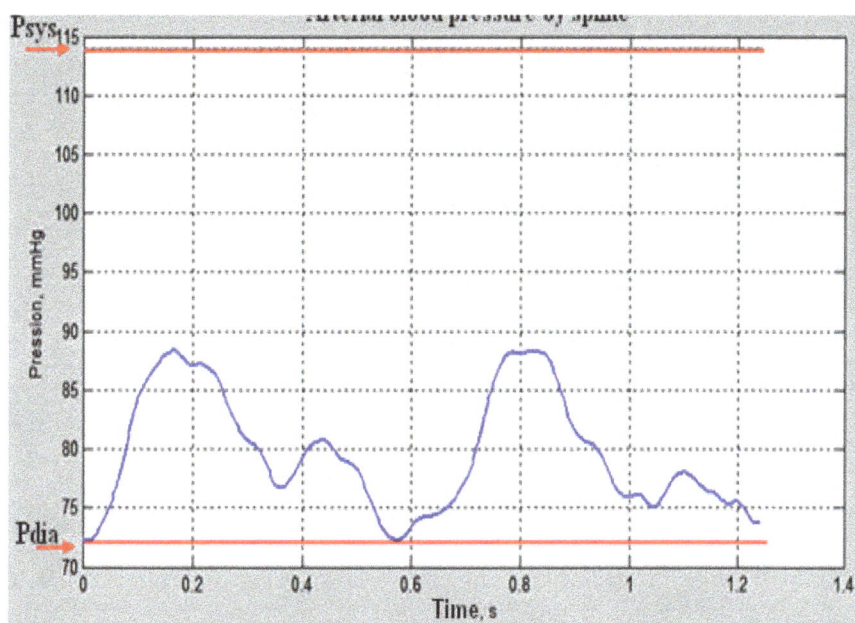

Figure 3. Limits of the cubic spline interpolation (P_{DIA} correct but the P_{SYS} is false).

Table 2. The interpolation error of the determination of the pressures by the four methods (results obtained with the forty five measures of the 'test group').

	Pressures obtained by cubic spline with (Pmax, Zmax)	Pressures obtained by cubic spline with standard (Pmean, Zmean)	Pressures obtained by Cubic spline with Chemla's (Pmean, Zmean)	Pressures obtained by cubic spline with pulse pressure (ΔP, ΔZ)
Systolic error	21.95%	4.1%	42.31%	2,97%
Diastolic error	≈ 0%	≈ 0%	≈ 0%	≈ 0%

To improve the method, we have chosen cubic spline for the direct determination of the diastolic pressure from the basic impedance.

For the systolic pressure we have tested three methods:

- The estimation of systolic pressure from the pulse pressure ΔP, deduced by spline interpolation from the bioimpedance variation (ΔZ).
- The estimation of systolic pressure from the mean pressure (two methods, the standard's one, and the mean pressure of Chemla).
- To estimate the diastolic pressure (P_{Dia}) from Z_0, the cubic spline provided the best results (Table 1).

To estimate the systolic pressure (P_{Sys}) the most precise results were given by the cubic spline interpolation between pulse pressure and bioimpedance variation. Table 2 shows the comparison between the different methods.

The Figures 4, 5 and 6 illustrate for a given subject the obtained curves of instantaneous blood pressure by the three methods.

Then we successfully tried to determine in real time the instantaneous arterial pressure, the systolic and diastolic pressures and several other cardiac parameters by using the following methodology.

The plethysmograph provides an analogical bioimpedance signal which was digitized and transferred to the computer by the NI 6009. The processing, the display and the user interface were managed by LabVIEW.

DISCUSSION

The extension of Equation 2 to the use of the peripheral bioimpedance and the arterial pressure is justifiable because the model of Nyboer described in the introduction section applies perfectly in our case (the left

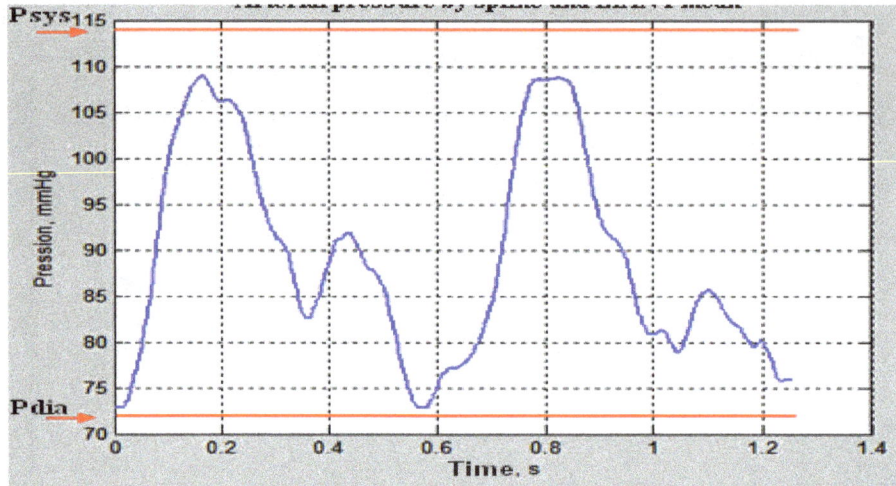

Figure 4. Curve of the blood pressure obtained by the standard mean pressure method.

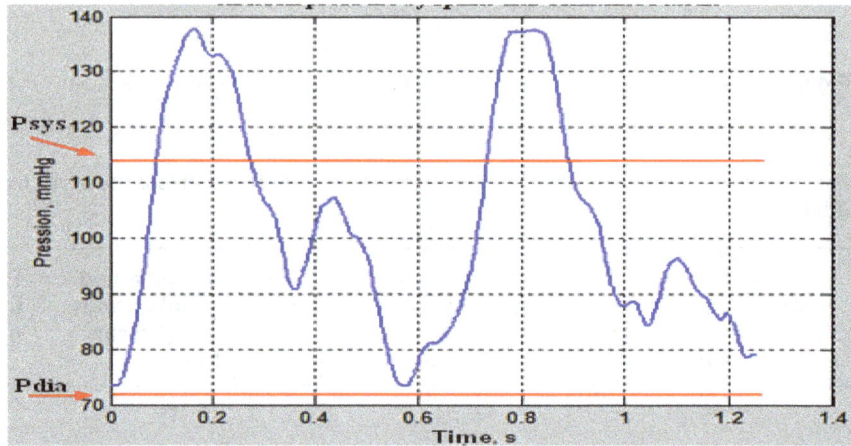

Figure 5. Curve of the instantaneous arterial pressure obtained by Chemla's method.

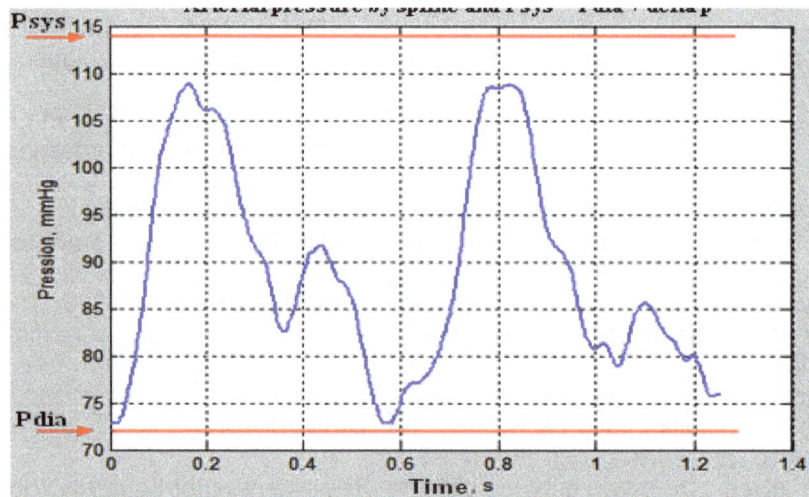

Figure 6. Curve of the blood pressure obtained by the interpolation between the pulse pressure (ΔP) and the bioimpedance variance (ΔZ).

arm) especially if the explored section is not large. That is the distance (L) between the two electrodes of collection must be lower than 5 cm, typically L= 4 cm.

The result of the comparison between the four methods of interpolation was foreseeable, since the cubic splines are largely recognized as being the best functions of interpolation.

Spline interpolation is preferred over polynomial interpolation because the interpolation error can be made small even when using low degree polynomials for the spline. Thus, spline interpolation avoids the problem of Runge's phenomenon which occurs when using high degree polynomials (Runge, 1901).

In the cardiac cycle when the pulsatile volume is at its minimum, which corresponds to the diastolic pressure, all of the conducting tissues and fluids are represented by the minimum volume. This volume can be a heterogeneous mixture of all the non time-varying tissues such as fat, bone, muscle, etc. in the region under measurement. The only information needed about this volume is its impedance Z_0. That is why when we tried to interpolate the diastolic pressure and the impedance base by a spline cubic interpolation, the results found were precise.

For the systolic pressure we thought in the beginning that the use of the standard formula of the average blood pressure was going to offer the best results to us, however the tests carried out showed the opposite.

The interpolation of the pulse pressure according to the variation of the impedance was more precise, because during the systole the pressure increases with the increase in blood volume whereas the impedance decreases. Thus, the dependence is in term of pulse volume and thus of the pulse pressure and the variation of the impedance and not in terms of mean volume mean pressure and mean impedance.

Conclusion

The monitoring of the instantaneous arterial pressure in a noninvasive way remains a precious objective for a best

and quick exploration of the cardiac hemodynamic parameters.

We tried through this paper to carry out this objective by using the bioimpedance method.

The described method allows with precision, in real time, in an inexpensive and noninvasive way for the determination of continuous arterial pressure directly from peripheral bioimpedance signal.

REFERENCES

Ben Salah R (1988). 'Pléthysmographie électrique thoracique localisée. Application à la détermination des paramètres cardiovasculaires et au diagnostic des cardiopathies.' Thèse d'état, Tunis – Tunisia. pp. 56-63.
Ben Salah R (1986). 'Diagnostic of valvular diseases by cepstral simulation of aortic plethysmogram'. 2nd European Simulation Congress (E.S.C). Antwerp, Belgium, Proceeding E.S.C P.pp.724-728.
Chemla D, Vincent C, Yves L, Marc H, Gérald S, Philippe H (2004). New formula for predicting mean pulmonary artery pressure using systolic pulmonary artery pressure. Chest, 126:1313-1317
Chemla D, Lamia B (2009). 'Noninvasive hemodynamic monitoring in the ICU: Recent advances in arterial tonometry', Reanimation. 18(3):197-200.
Flaud P (1979). 'Influence des propriétés rhéologiques non linéaires sur la dynamique des écoulements dans un tuyau déformable.' Thèse d'Etat, PARIS VI.
Grimnes S, and Ø.G. Martinsen (2005) 'Bioimpedance and Bioelectricity Basics' Academic Press .275 pp.
Bogert LWJ and van Lieshout JJ (2005). 'Non-invasive pulsatile arterial pressure and stroke volume changes from the human finger'. Exp. Physiol., 90:437-446.
Nyboer J (1970). Electrical Impedance Plethysmography. 2nd ed. Charles C. Thomas, Springfield. IL. pp 96-97
Runge Carl (1901). "Über empirische Funktionen und die Interpolation zwischen äquidistanten Ordinaten". Zeitschrift für Mathematik und Physik, 46: 224–243.
Siebig S, Rockmann F, Sabel K, Zuber-Jerger I , Dierkes C, Brünnler T, Wrede CE (2009). 'Continuous Non-Invasive Arterial Pressure Technique Improves Patient Monitoring during Interventional Endoscopy'. Int. J. Med. Sci., 6:37-42.

14

Ultra-low doses of melafen affect the energy of mitochondria

I. V. Zhigacheva[1], E. B. Burlakova[1], I. P. Generozova[2], A. G. Shugaev[2] and S. G. Fattahov[3]

[1]Russian Academy of Sciences, N. M. Emanuel Institute of Biochemical Physics, ul. Kosygina 4, 119334 Moscow, Russia.
[1]Russian Academy of Sciences, K. A. Timiryazev Institute of Plant Physiology ul. Botanicheskaya 35, 127276, Moscow, Russia.
[3]Russian Academy of Sciences, A. E. Arbuzov Institute of Organic and Physical Chemistry, ul. Akademika Arbuzova 8, 420083 Kazan' Research Center, Kazan', Tatarstan, Russia.

Addition of the organophosphorous plant growth regulator - melafen to the mitochondria incubation medium resulted in modification of energy thereof. The modification was dose dependent one. The melafen concentrations 2×10^{-12}, 2×10^{-18} and 2×10^{-21} M raised the maximum rates for oxidation of NAD^+-dependent substrates, elevated the efficiency of oxidative phosphorylation and activated electron transport in the terminal step of mitochondrial respiratory chain. Melafen stimulated electron transport during oxidation of succinate by rat liver mitochondria, but had no effect on the rate of this substrate oxidation by sugar beet root mitochondria, which was an evidence for adaptive properties of the preparation. Water stress resulted in decreasing the maximum rates of oxidation of NAD^+-dependent substrate and decreasing the electron transport rates at the end of the respiratory chain by 30%. A pretreatment of pea seeds with a 10^{-7}% solution of melafen led to elimination of differences in energetic of mitochondria of sprouts growing under standard conditions and under conditions of low moisture. By stimulating the activity of NAD^+-dependent dehydrogenases and activation electron transport in the cytochrome oxidase part of respiratory chain, melafen stimulated energy metabolism in the cells and these effects determine the adaptive properties thereof.

Key words: Cytochrome oxidase, mitochondria, oxidative phosphorylation, respiratory chain.

INTRODUCTION

Many natural and synthetic biologically active substances (BAS) exhibit their activity in a range of low (10^{-10} - 10^{-4}M) and ultra-low concentrations (10^{-20} - 10^{-11}M). The level of biological organization, at which the effect of ultra-low doses (ULD) is observed is very diverse: from macromolecules, cells, organs and tissues to animal and plant organisms and even populations (Zinkevich et al., 2002; Belov et al., 2002; Terekhova et al., 2002; Ashmarin et al., 2005). It does not follow from the aforesaid that the effect was observed at ultra-low doses of any one of biologically active substances on any one of biological objects. The observable effect at the substance concentrations 10^{-13} - 10^{-17} M and lower can not be attributed to any definite structure or a level of biological organization (Burlakova et al., 2003).

The effects of ultra-low doses of biologically active substances have common characteristics that do not depend on the substance nature. These characteristics manifest themselves most visibly in studies on dose dependences. In some cases, the dependence is bimodal: the effect increases at ultra-low doses of a preparation, then, the effect decreases, as the dose is increased and is succeeded by a "dead zone" and increases again. Sometimes, the dose dependence has a stage of "a change of sign". For example, if an inhibiting activity was observed in the region of ultra-low doses, it changed for a stimulating one as the concentration was increased, and

*Corresponding author. E. mail: zhigacheva@mail.ru.

then again an inhibiting effect was observed. There are known cases when the effect did not depend on a dose within a wide concentration range. In particular, in one of the work, wherein the effect of a herbicide of the class of hydro peroxides on a plant cell culture was studied, it was discovered that the preparation has an equal effect at doses that are six orders different (10^{-13} - 10^{-7} M) and the effect is absent in the range of intermediate concentrations (Burlakova, 2003).

The nature of the dependences may be accounted for by the fact that BAS ultra-low doses have common targets. These targets may be cell and sub cellular membranes that play a key role in the cell metabolism. A change in physicochemical properties of the latter caused by various factors including that of BAS results in changing the activity of membrane-associated enzymes (Palmina, 2009).

Although there are a great number of works on effects of BAS ultra-low doses, the mechanism thereof is not well studied yet. The aim of this work is to study the biological effects of plant growth regulators at ultra-low doses.

Natural phytohormones and their synthetic derivatives play a key role in the regulation of plant metabolism in all steps of their ontogenesis (Shevelukha, 1992; Kulaeva, 2002). At present, many biologically active substances able to stimulate plant growth were identified or synthesized (Korol et al., 2001; Thomas et al., 1993; Morohashi, 1984). Application of these compounds enables possibility to prevent drowing of grain crops, accelerate germination and maturation of plants, increase their yields and improve the quality of agricultural products by increasing their resistance to pathogenic microorganisms and parasites (Khan N.A., Samiulallah, 2001; Korol' et al., 2001; Kirillova et al., 2003; Romanov, 2009).

However, the use of natural plant growth regulators (phytohormones) in agricultural practice is often coupled with difficulties, e.g., high cost of final products, fast loss of their useful properties under the action of environmental factors, etc. Therefore, a search for and the synthesis of substances able to stimulate plant growth even when used at low and very low concentrations are currently underway. Such compounds concern the melamine salt of bis (ox methyl) phosphoric acid (melafen) with the structural formula:

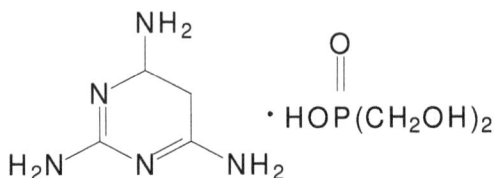

for an example. At very low concentrations, melafen increases the general productivity of some agricultural crops (Fattakhov et al., 2002; Kostin et al., 2006). At 3 × 10^{-9} M, it increases the photosynthesis rate and accelerates

respiration of plant cells by 15% (Fattakhov et al., 2004).

The pre-sowing treatment of seeds of cereals and leguminous and solanaceous cultures with melafen results in a 5 - 25% gain in the energy of germination, an increase in the productivity and vegetative mass of plants and improvement of quality and nutritive value of the product (Fattakhov et al., 2004).

Since the activation of synthetic processes requires considerable energy expenditures, it is safe to suppose that melafen regulates vital processes of plant cells by influencing the energy metabolism therein and primarily, the energy of mitochondria. The respiratory chains of mitochondria of plants and animals are organized similarly; the main differences are in the CN-resistant electron transfer and the structure of the NADH-dehydrogenase region of the respiratory chain (Palmer, 1976). Therefore, the test subjects used were rat liver mitochondria and mitochondria of plants: a storage parenchyma of sugar beet (*Beta vulgaris* L.) and mitochondria of pea sprouts (*Pisum sativum*).

Since most of plant growth regulators possess the antistress activity, melafen may also possess such activity. Indeed, the action of stress factors promotes the generation of reactive oxygen species (ROS) in the electron transport chains of mitochondria (Baraboi, 1991; Kurganova, 2001; Cadenas et al, 1977). In the situation when the generation of reactive oxygen species increases and the antioxidant system can not cope with the increasing ROS pool, the lipid per oxidation (LPO) processes are activated in membranes (Kulinsky, 1999; Grabelnych, 2005). The antistress preparations must obviously reduce the level of LPO products in biological membranes; the effect is achieved by various pathways including restructuring the mitochondria energy. We used a model of pea seeds germination under conditions of insufficient watering - water stress. Melafen, like many other physiologically active substances, may exhibit a dose dependence of the effect on metabolic processes and the activity of the preparation may be expected for low and ultra-low doses. Based on the above assumptions, the aim of this work was to study the concentration dependences of the effect of the preparation on the energy of mitochondria and the influence of a water stress and pressuring treatment of pea seeds with melafen on the energy of mitochondria isolated from pea sprouts.

MATERIALS AND METHODS

The experiments were carried out on mitochondria from sugar beet root, sprouts of peas and rat liver.

Germination of pea seeds

Pea seeds were germinated in a control group, rinsed with water and soap and then with a 0.01% KMnO4 solution and were soaked in water for 60 min in the experimental group, the seeds were in water for 60 min. In the experimental group, the seeds were soaked in a 10^{-7}% melafen solution for 30 min and then in water for 30 min. In a day, half of the control seeds and half of the seeds treated with melafen were carried over onto a dry filter paper in open corvettes. In two days of the "drought", the seeds were carried over into closed corvettes with a periodically damped filter paper, wherein the seeds were left for another 5 days. On the fifth day, we calculated the number of sprouting seeds, measured the length of hypocotyls and isolated mitochondria.

Isolation of mitochondria

Mitochondria were isolated from rat liver (Zhigacheva et al., 1995) by differential centrifugation. The isolation medium comprised of 0.25 M sucrose, 5 mM MOPS, pH 7.4. The primary centrifugation lasted for 10 min at 600 g; the secondary, for 10 min at 10000 g. The precipitate was suspended in 0.5 ml of the isolation medium.

Mitochondria from sugar beet root storage parenchyma (Shugaev et al., 1982) and sprouts of peas (Popov, 2003) were isolated by differential centrifugation. Epicotyls having a length of 3 to 6 cm (20 to 25 g) or 30 g of sugar beet root storage were poured with the isolation medium and disintegrated in a blender. The isolation medium comprising 0.4 M sucrose, 5 mM EDTA, 20 mM KH_2PO_4 (pH 8.0), 10 mM KCl, 2 mM 1, 4 -Dithio-dl-theritol, and 0.1% BSA (free of fatty acids) was placed in a cup of a glass homogenizer and homogenized manually for 1 min. The tissue/medium ratio was 1:2. The first centrifugation was performed at при 25000 g for 5 min. The precipitate was resuspended in 8 ml of the rinsing medium and centrifuged at 3000 g for 3 мин. The rinsing medium comprised of 0.4 M sucrose, 20 mM KH_2PO_4 (pH 7.4), 5 mM EDTA, 10 mM KCl and 0.2% BSA (free of fatty acids). Mitochondria were precipitated by centrifugation at 11000 g for 10 min. The precipitate was resuspended in 2 to3 ml of a medium comprising 0.4 M sucrose, 20 mM KH_2PO_4 (pH 7.4), 0.1% BSA (free of fatty acids) and then mitochondria were precipitated again by centrifugation at 11000 g for 10 min. Protein was determined by the biuret method.

Rate of mitochondria respiration

The rate of mitochondria respiration was measured with the aid of Clarke oxygen electrodes and LP-7 polarograph (Czechia). Sugar beet root and pea sprout mitochondria were incubated in a medium containing 0.4 M sucrose, 20 mM HEPES-Tris buffer (pH 7.2), 5 mM KH_2PO_4, 4 mM $MgCl_2$ and 0.1% BSA. The incubation medium for study of the electron transport in respiratory chain of rat liver mitochondria contained 0.25 M sucrose, 10 mM Tris-HCl, 2 mM $MgSO_4$, 2 mM KH2PO4 and 10 mM KCl, pH 7.5 (28°C).

Reagents

sucrose("Sigma-Aldrich" USA), FCCP (carbonylcyanide-p-trifluorometoxyphenylhydrozone) ("Fluka", Germany)., rotenone ("Sigma-Aldrich" USA), antimycin A ("Sigma-Aldrich" USA), ascorbate("Sigma-Aldrich" USA), malate ("Sigma-Aldrich" USA), glutamate("Sigma-Aldrich" USA), succinate ("Sigma-Aldrich" USA), BSA (Bovine serum albumine) (Fraction V, free fatty acids) ("Sigma" USA), N,N,N',N'- tetramethylphenyldiamide (TMPD) ("Sigma" USA); HEPES (4-(2-Hydroxyethyl)piperazine-1-ethanesulfonic acid) (BioChemika Ultra, for molecular biology) ("Fluka", Germany), MOPS (3-(N-Morpholino)propanesulfonic acid) ("Fluka", Germany), Tris (hydroxymethyl)aminjmethan)(" MP Biomedicalis, LLC", Germany); KCl (Potassium chloride purees) ("Fluka", Germany), 1,4 –dithio-dl-theritol ("Fluka", Germany).

RESULTS

The study of the effect of melafen on the energy of mitochondria was commenced with a study of the effect on the electron transport maximum rate in a respiratory chain of mitochondria of rat liver and sugar beet root storage parenchyma. The electron transport rates were studied in the presence of a FCCP (carbonylcyanide-p-trifluorometoxyphenylhydrazone) - protonophore reducing a pH gradient and providing thereby for maximum rates of oxidation of substrates. We studied also the rates of oxidation of substrates in the presence of ADP (State 3), introduction of which to the mitochondria incubation medium results in activation of ATP synthesis and, as a consequence, in a decrease in $\Delta\mu_H^+$ and in enhancement of the electron transport rate.

The addition of melafen to the incubation medium for the mitochondria of rat liver or sugar-beet storage roots changed the mitochondria energy. The changes were dose-dependent ones. At the concentrations 2×10^{-5} and 2×10^{-14} M, the preparation reduced rates of oxidation of NAD^+-dependent substrates by liver mitochondria by 50 and 12%, respectively and by sugar beet root mitochondria by 30% (Figure 1). The most efficient were the concentrations 2×10^{-12} M and $2 \times 10^{-18} - 2 \times 10^{-21}$. Melafen used in these concentrations increased the rates of oxidation of NAD^+-dependent substrates in the respiratory chain of liver mitochondria by 35 - 43% in the presence of ADP and by 52% in the presence of FCCP (carbonylcyanide-p-trifluorometoxyphenylhydrazone).

The respiratory control rate (RCR) increased in 1.2 - 1.4 times (Table 1). In the respiratory chain of sugar beet root mitochondria, the rates of oxidation of NAD^+-dependent substrates increased by 13 - 23% in the presence of ADP and the respiratory control rate increased in 1.3 times (2×10^{-18} and 2×10^{-21}) (Table 2).

It should be noted that in all cases the increase or decrease (2×10^{-5}; 2×10^{-14} M) in the RCR in the oxidation of NAD^+-dependent substrates was due to an increase or decrease in the rates of oxidation of substrates by liver mitochondria in the presence of ADP (in State 3).

The differences between plant and animal mitochondria were observed when the oxidation substrate used was succinate. The effect of the preparation on the oxidation of succinate by rat liver mitochondria was similar to that on the rates of oxidation of NAD^+-dependent substrates (Table 3). However, all concentrations studied had no effect on the maximum rates and the respiratory control rate in oxidation of succinate by sugar beet root mitochondria, which is an evidence for the adaptive character of the melafen effect (Table 4).

Mitochondria of storage organs are characterized by relatively low rates of oxidation of NAD^+-dependent substrates (Shugaev, 1985) Melafen enhances the activity of NADH-dehydrogenases and, evidently, activates the

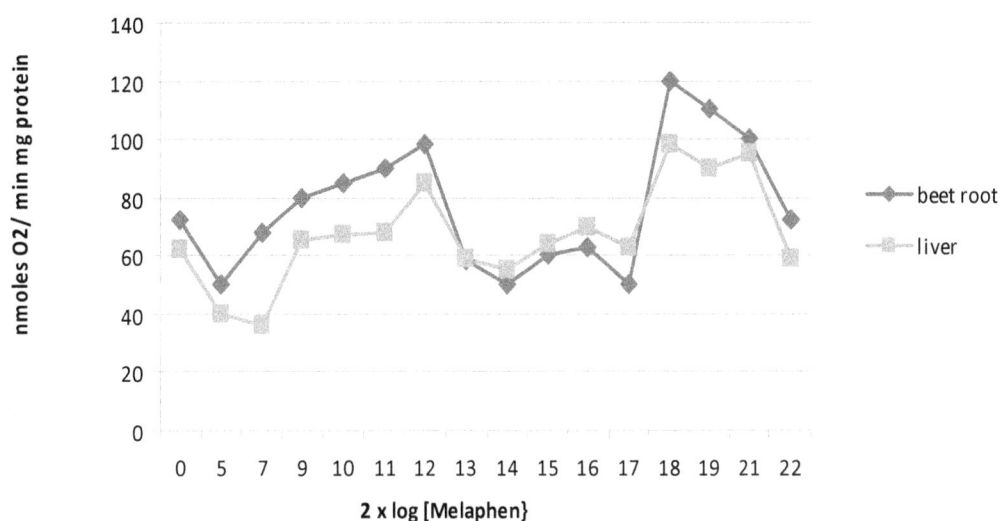

Figure 1. Effects of melafen on respiration rate of mitochondria in the presence of FCCP. Ordinate: rates of glutamate + malate oxidation in the presence of FCCP in n atoms O2/ mg protein min.. Abscissa: 2 × l g concentration of melafen. Ordinate in n atoms/ mg protein min.

Table 1. Effect of various melafen concentrations on the efficiency of oxidative phosphorylation in the respiratory chain of rat liver mitochondria in oxidation of NAD^+-dependent substrates. The respiration rate was measured in n atomsl O_2/min x mg protein.

Melafen, M	State 2	State 3	State 4	RCR	FCCP
-	16.0 ± 1.3 (8)	50.4 ± 2.4 (8)	20.5 ±1.5 (8)	2.45 ± 0.1 (8)	58.9 ± 1.6(8)
2×10^{-5}	14.0 ± 1.2 (5)	40.0 ± 2.5 (5)	21.0 ±1.0 (5)	1.90 ± 0.1(5)	51.0 ± 4.2 (5)
2×10^{-12}	16.5 ± 0.2 (6)	58.2 ± 1.6 (6)	19.7 ± 0.9 (6)	2.95 ± 0.10	62.9 ± 0.6 (6)
2×10^{-18}	22.1 ± 3.2 (5)	72.5 ± 4.3 (5)	21.0 ± 2.6 (5)	3.45 ± 0.2 (5)	90.0 ± 5.2 (5)
2×10^{-19}	17.5 ± 3.0 (5)	62.4 ± 2.1 (5)	24.0 ± 2.4 (5)	2.6 ± 0.2 (5)	55.0 ± 2.7 (5)
2×10^{-21}	22.5 ± 1.4 (6)	68.4 ± 4.6 (6)	22.8 ± 3.4 (6)	3.00 ± 0.2 (6)	90.0.0 ± 3.2 (6)

Incubation medium: Contain 0.25 M sucrose, 10 мM tris-HCl, 2 mM KH_2PO_4, 5 mM $MgSO_4$, 10 mM KCl, pH 7.5. Other additives: 200 µM ADP, 10^{-6}M FCCP, 4 mM glutamate, 1 mM malate.

Table 2. Effect of various melafen concentrations on the efficiency of oxidative phosphorylation in the respiratory chain of beet root mitochondria in oxidation of NAD^+-dependent substrates. The respiration rate was measured in n atomsl O_2/min x mg protein

Melafen, M	State 2	State 3	State 4	RCR	FCCP	KCN
-	15.8 ± 1.0 (6)	66.4 ± 3.9 (6)	28.8 ± 2.5 (6)	2.3 ± 0.1 (6)	72.0 ± 4.6 (6)	6.3 ± 0.4 (6)
2×10^{-5}	12.7 ± 2.4 (5)	50.1 ± 3.0 (5)	27.6 ± 2.0 (5)	1.81 ± 0.2 (5)	50.0 ± 3.8 (5)	5.9 ± 0.6(5)
2×10^{-12}	17.5 ± 2.0 (5)	75.5 ± 3.0 (5)	26.1 ± 1.0 (5)	2.90 + 0.1(5)	95.0 + 3.2 (5)	8.2 + 0.2 (5)
2×10^{-14}	12.0 ± 2.4 (6)	60.0 ± 3.0 (6)	33.5 ± 2.0 (6)	1.79 ± 0.2 (6)	48.6 ± 4.0 (6)	5.7 ± 0.4 (6)
2×10^{-18}	20.0 ± 3.2 (5)	80.0 ± 2.4 (5)	27.1 ± 1.6 (5)	2.95 ± 0.3 (5)	120.0 ± 4.0 (5)	6.0 ± 0.2 (5)
2×10^{-21}	20.5 ± 2.4 (5)	81.9 ± 4.6 (5)	27.8 ± 2.2 (5)	2.95 ± 0.2 (5)	100.0.0 ± 40 (5)	5.2 ± 0.5

Incubation medium: 0.4 M sucrose, 20 mM HEPES-tris-buffer (pH 7.2), 5 mM KH_2PO_4, 4 mM $MgCl_2$, 0.1% BSA.,: 5 mM malate, 10 mM glutamate. Other additives: 125 µM ADP, 0, 5 µM FCCP (carbonylcyanide-p-trifluorometoxyphenylhydrazone).

energy processes in cell and provides thereby for a high energy of seeds germination.

The activation of energy processes in cell is provided also by the effect of the preparation on the electron transfer

Table 3. Effect of melafen on the kinetics of consumption of oxygen by rat liver mitochondria in oxidation of succinate. The respiration rate was measured in n atomsl O_2/min x mg protein.

Melafen, M	State 2	State 3	State 4	RCR	FCCP
-	42.3 ± 3.8 (6)	113.2 ± 2.5 (6)	45.7 ± 3.1(6)	2.48 ± 0.10 (6)	129.0± 7.2 (6)
2×10^{-12}	19.0 ± 1.4 (5)	151.8 ± 1.5 (5)	46.0 ± 0.2(5)	3.30 ± 0.09 (5)	64.0 ± 1.0 (5)
2×10^{-18}	38.7 ± 4.1(5)	158.7 ± 1.4 (5)	46.0 ± 1.7(5)	3.45 ± 0.10 (5)	130.0± 2.9 (5)
2×10^{-21}	41.8 ± 3.0 (5)	136.2 ± 3.4 (5)	45.4 ±2.5 (5)	3.00 ± 0.10 (5)	123.0± 4.0(5)

Incubation medium: Contain 0.25 M sucrose, 10 мM tris-HCl, 2 mM KH_2PO_4, 5 mM $MgSO_4$, 10 mM KCl, pH 7.5. Other additives: 200 μM ADP, 10^{-6}M FCCP, 5 mM succinate.

Table 4. Effect of melafen on the kinetics of consumption of oxygen by sugar beet root mitochondria in oxidation of succinate. The respiration rate was measured in n atomsl O_2/min x mg protein.

Melafen, M	State 2	State 3	State 4	RCR	FCCP
-	42.3 ± 3.8(6)	113.2 ± 2.5 (6)	45.7 ± 3.1 (6)	2.48 ± 0.3 (6)	129.0 ± 7.2(6)
2×10^{-5}	42.0 + 2.4(5)	120.0 + 2.4 (5)	53.0 + 1.6 (5)	2.26 + 0.3 (5)	139.0 + 4.0 (5)
2×10^{-12}	45.0 + 3.2 (5)	120.0 + 2.4 (5)	53.0 + 1.6 (5)	2,26 + 0.3 (5)	123.0 + 4.0 (5)
2×10^{-18}	38.7 ± 4.1 (5)	115.0 ± 1.4 (5)	46.0 ± 1.7 (5)	2.50 ± 0.3 (5)	130.0 ± 2.9 (5)
2×10^{-21}	41.8 ± 3.0 (5)	111.7 ± 3.4(5)	45.4 ± 2.5 (5)	2.46 ± 0.1 (5)	123.0 ± 4.0 (5)

Incubation medium: 0.4 M, sucrose 20 mM HEPES-tris-buffer (pH 7.2), 5 mM KH_2PO_4, 4 mM $MgCl_2$, 0.1% BSA, 10 mM succinate. Other additives: 125 μM ADP, 0, 5 μM FCCP (carbonylcyanide-p-trifluorometoxyphenylhydrazone).

transfer rate at the terminal cytochrome oxidase region of the respiratory chain of both plant and animal mitochondria. The dependence of the electron transport rate on the concentration of the preparation was also dose-dependent one. At the concentration 2×10^{-5}; 2×10^{-14} and 2×10^{-17}M, melafen decreased the rates of oxidation of ascorbate by sugar beet root storage mitochondria in the presence of TMPD (N,N,N',N'-tetramethylphenyldiamine) and by rat liver mitochondria (2×10^{-5}; 2×10^{-17}M) by 13 and 40 - 20%, respectively. Other effect was observed on addition of 2×10^{-12}M and 2×10^{-18} - 2×10^{-21}M melafen to incubation medium. The rates of oxidation of ascorbate in the presence of TMPD by liver mitochondria increased from 82.4 ± 1.2 to 105.1 ± 1.4 nmol O_2/mg.protein.min. At the same concentrations, melafen promoted the electron transport at the cytochrome oxidase site of the respiratory chain of sugar beet root mitochondria. The rates of oxidation increased by 12 - 15%. The increase was not caused by the activation of the alternative CN-resistant oxidase (AO) of mitochondria, since the electron transport was completely suppressed by cyanide. Evidently, "a change of sign" of the effect of the preparation on the electron transport rate at the end site of the respiratory chain determines the dose dependence of oxidation rates of NAD^+-dependent substrates. The observable increase in the activity of cytochrome oxidase and oxidation rates of NAD^+-dependent substrates in the presence of melafen promotes, evidently, the energy metanolism in cell, which

results in increasing the heat generation by plant cells and activation of synthetic processes described in the literature (Fattakhov et al., 2004.). Evidently, the regulation of metabolic processes in plant cell by melafen is effected due to modification of physicochemical properties of biological membranes and, consequently, of the activity of enzymes bound thereto.

We verified the supposition in respect of adaptive properties of melafen in terms of a model of mitochondria of pea sprouts growth under conditions of water stress. Low moisture led to decreasing the maximum rates of oxidation of NAD^+-dependent substrates by pea sprout mitochondria. The rate of oxidation of glutamate + malate in the presence of FCCP decreased from 105.0 ± 2.1 to 75.0 ± 3.4 n atoms of O_2/ mg protein.min. Hence, the respiratory control rate decreased from 2.27 ± 0.1 to 1.7 ± 0.2. The pretreatment of seeds with melafen prevented from drought-induced change in the efficiency of oxidative phosphorylation. In addition, the pretreatment resulted in reduction of the rates of oxidation of NAD^+-dependent substrates in the presence of ADP or FCCP to the control values. The decrease in the maximum rates of oxidation of NAD^+-dependent substrates under low-moisture conditions may be caused by a decrease in the electron transfer rate at the end of the respiratory chain. From Table 5, the rates of oxidation of ascorbate in the presence of N, N, N', N'- tetramethylphenylenediamine (TMPD) by mitochondria isolated from drought-conditioned pea sprouts were less than the control values

Table 5. Effects of melafen on the rate of electron transport in the respiratory chain of sugar beet root mitochondria in the presence of ascorbate and TMPD. The respiration rate was measured in n atomsl O_2/min x mg protein.

Group	State 2	TMPD		
		200 µM	200 µM	400 µM
Standard conditions of germination	8.0 ± 0.3 (5)	250.0 ± 10 (5)	336.0 ± 11.0 (5)	500.0 ± 32 (5)
Water stress	5.0 ± 0.2 (8)	175.0 ± 25 (8)	240.0 ± 21 (8)	356.0 ± 32 (8)
Melafen treatment + water stress	7.0 ± 0.4 (6)	242.0 ± 15 (6)	340.0 ± 17 (6)	498.0 ± 36 (6)

Incubation medium: 0.4 M sucrose, 20 mM HEPES-Tris buffer, 5 mM KH2PO4, 2 mM MgCl$_2$, 5 мM EDTA, 10 mM ascorbate, 60 µM rotenone, 5 µM antimycin A, 0,5 µM FCCP, pH 7.4.

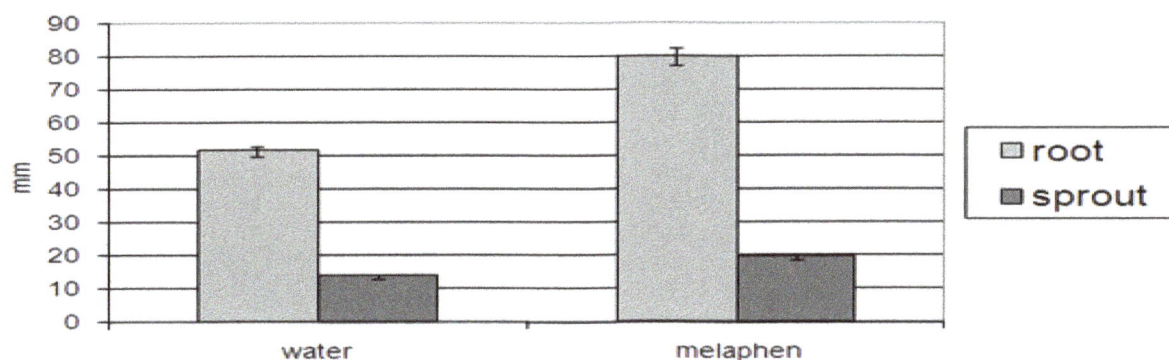

Figure 2. Growth of roots of etiolated pea germs under low - moisture conditions. Control - standard sprouting conditions; drought - sprouting of seeds under low- moisture conditions. Ordinate - length of hypocotyls, mM. The number of pea seeds in each group was 100 (n = 100); p < 0.05.

almost by 30%. In this case, the pretreatment with melafen is prevented from change at this site of the respiratory chain. It is safe to suppose that shifts in the physiological parameters were caused by changes in the metabolic activity of mitochondria: The pretreatment of pea seeds with melafen stimulated the growth of sprouts (by 18 - 24%) both for the control and for plants growing in drought; however, germination of treated and untreated seeds differed significantly. Under drought conditions, the germination of seeds in control decreased by 46%; that of melafen-treated seeds remained almost unchanged. Moreover, melafen at 1.5 times stimulated the growth of sprout roots under drought conditions, which is of great importance for adaptation (Figure 2). Melafen increases the maximum oxidation rates of NAD$^+$-dependent substrates and the efficiency of oxidative phosphorylation and thus promotes the activation of energy processes in cell and provides for a high energy of seeds germination.

DISCUSSION

The adaptive effect of the compound may be attributed to activation of the electron transfer at the end of the mitochondrial respiratory chain, which results in inhibiting the formation of reactive oxygen species (ROS) at the site of complex III (Cadenas, 1977; Sviryaeva and Ruge, 2006). Even ultra-low concentrations of melafen affect the functional state of membranes.

It should be noted that the discrete character of the concentration dependences of the effect of ultra-low doses of melafen obtained in our experiments (Tables 1 - 3 and Figure 1) is qualitatively consistent with published data on the effect of ultra-low doses of biologically active substances on living systems of various degrees of complexity (Burlakova et al., 2003; Myagkova et al., 2003; Zhernovkov, 2006). There exist a great number of hypotheses as to the mechanisms of the effect of ultra-low concentrations of BAS in the literature. According to one of the hypotheses, BAS, in a range of physiological concentrations of up to 10^{-9} M, are incorporated into membranes and interact with lipids. At concentrations of 10^{-9} - 10^{-17} M, BAS interact with membrane ligands and receptors located at particular membrane sites (rafts). Thus, the receptor signal may be amplified by a factor of 10^6 - 10^{10} due to highly efficient systems of transmission and amplification of signals (Palmina, 2009). As to the region of "virtual concentrations" (10^{-17} - 10^{-25} M), the authors

authors interpret the observable effects in terms of the concept about the effect of ultra-low doses of physical factors and chemical substances on the structure of water (Burlakova et al., 2003).

In our opinion, the results obtained for melafen may be interpreted in terms of the physicochemical behavior of highly diluted solutions of the preparation. Konovalov et al. (2008) showed that melafen in a concentration of 10^{-20} - 10^{-4} mol/l forms nanoassociates of the size of about 100 - 200 nm with participation of water. The concentration dependences obtained by the authors for a size and electrokinetic potential (ζ-potential) of nanoassociates formed in aqueous solutions of melafen at low and ultra-low concentrations (Ryzhkina et al., 2009), are comparable with the biological effects of the preparation as described in this work and published previously (Zhigacheva et al., 2008), which is the evidence for the role of melafen nanoassociates in biological effects. The most plausible evidence for the fact that ultra-low concentrations of melafen that affect the energy of mitochondria are discrete in character is the formation of associates of a different polarity and probably, of different level of complexity depending on the melafen concentration in a solution (Konovalov et al., 2008).

REFERENCES

Ashmarin IP, Korazeeva EP, Lelekova TV (2005). Efficiency of ultra-low doses of endogenous bioregulators and immunoactive compounds. J. Microbiol. Epidemiol. Immunobiol., 3:109-116.

Baraboi VA (1991). Stress mechanisms and lipid peroxydation. Uspekhi Sovremennoi Biologii (Advances in Modern Biology) (Rus) 11(6):923-932.

Belov VV, Mal'tseva EL, Palmina NP (2002). Effects of a broad range of alpha-tocopherol concentrations on structural characteristics of membrane lipid bilayers in mouse liver endoplasmic reticulum in vitro // Proceedings of the Third International Symposium "Mechanisms of action of ultra-low doses" (3-6.12.2002), Moscow. pp. 7

Burlakova EB, Konradov AA, Mal'tseva EL (2003). Effect of ultra-low doses of biologically active substances and low-intensity physical factors // Khimicheskaya Fizika (Chemical Physics). 22:21-40.

Cadenas E, Boveris A, Ragan CI, Stoppani ADM (1977). Production of superoxide radicals and hydrogen peroxide by NADH-ubiquinone reductase and ubiquinol-cytochrome c reductase from beef-heart mitochondria. Arch. Biochem. Biophys., 180(2): 248-257.

Khan NA, Samiulallah (2001).Phytohormones and crop productivity under different enviroments. Scientific Publish (India) Hardcover pp.233

Fattakhov SG. Reznik VS, Konovalov AI (2002). Melamine salt of bis (ox methyl)phosphoric acid (MELAPHEN) as a new generation of regulator of plant grows // Proceedings of the 13th International Conference on Chemistry of Phosphorus Compounds. Saint-Petersburg. pp. 80.

Fattakhov SG, Loseva NI, Konovalov AI, Reznik VS, Alyab'ev YuA, Tribunskikh VI (2004). Melafen Effects on Grows and Energetic Processes in Plant Cell. Dokl. Akad. Nauk 394:127-129.

Grabelnych OI (2005). The energetic function of plant mitochondria under stress. J. Stress Biol. Biochem. 1:37-54.

Konovalov AI, Rizhkina IS, Murtazina LI, Timashova AP, Shagiddulin RR, Chernova AV, Avakumova LV, Fattakhov SG (2008). Ultra-molecul system on the basis of dehydrate of melamine salt of bis(hydroxymethyl) phosphonic acid and the surface-active

substances. Communication I .Composition and self-association of melafen in water and chloroform. Izvestiya Akademii Nauk (Rus), (Proc. Russian Acad. Sci. Chem. S., 6: 1207-1213.

Kirillova IG, Evsyunina AE, Puzina TI, Korableva NP (2003). Effect of Ambiol and 2-Chlorethylphosphonic acid on the content of phytohormones in potato leaves and tubers. Applied Biochem. Microbiol., 39(2):210-214.

Kostin VI, Kostin OV, Isaichev VA (2006). Study of Melafen applications in agriculture. Condition of researches and prospects of application of a regulator of growth of plants "Melafen" in agriculture and biotechnologies, Kazan'. pp 27-37.

Korol 'VV, Kirillova IG, Puzina TI (2001). Changes in the Hormonal Balance and Physiological Processes in Potato Plants under Grows Regulator and Microelement Treatments. Vestn. Bashkirsk. (Rus). Univ., 2:84-86.

Kulaeva ON (2002). Recent advances and outlooks in cytokinin research Fiziologiya Rastenii (Plant Physiol.) (Rus.). 49(4):455-458.

Kulinsky VI (1999). Reactive oxygen species and oxidative modification of molecules: benefits, determents and protection. Sorosovskiy obrazovatelny zhurnal (Sorosovsky educational Journal). 2:2-7.

Kurganova LN (2001). Lipid peroxidation as a component of fast response to stress. Sorosovskiy obrazovatelny zhurnal (Sorosovsky Educ. J.,) 6:76-78.

Morohashi Y (1984). Effect of benzyladenine on the development of mitochondrial activities in Imbred Black Gram Cotyledons. Plant Physiol., 116: 235-298.

Myagkova MA, Abramenko TV, Panchenko ON, Epshtein OI (2003). Ultra-low doses of antibodies to delta--dream peptide: effect in enzyme-linked assays. Bulleten' Eksperimentalnoi Biologii i Meditsiny (Bull. Exp. Biol. Med.,) (Rus.). Suppl,. 1:10-12.

Palmer IM (1976). The Organization and Regulation of Electron Transport in Plant Mitochondria. Annu. Rev. Plant Physiol., 27:133-157.

Palmina NP (2009). Mechanism of Effect of ultra-low doses. Khim. I Zhizn. (Chemistry and Life) (Rus) 2:10-13.

Popov VN, Ruuge EK, Starkov AA (2003). Influence ingibitors of electron transport on formation of active forms of oxygen at oxidation of succinate by mitochondria. Biohimiya. Biochemistry. 68 (7): 910-916.

Romanov GA (2009). How cytokines affect a cell. Fiziol. rastenii. Plant Physiol. (Rus). 56(2):295-319.

Ryzhkina IS, Murtazina LI, Kiseleva Yu V, Konovalov AI (2009). Properties of supramolecular nanoassociates formed in aqueous solutions of low and ultra-low concentrations of biologically active substances. Dokl. Ross. Akad. Nauk. 428(4):487-491.

Shugaev AG, Vyskrebentseva EI (1985). Effects of rotenone and exogenous NAD on malate oxidation by mitochondria isolated from sugar beet roots at different steps of plant ontogenesis Fiziologiya Rastenii (Plant Physiology) (Rus.). 32:259-267.

Shevelukha VS (1992). Plant growth and its regulation in ontogenesis. Moscow. Kolos Press, 594pp

Shugaev AG, Vyskrebentseva EI (1982). Isolation of intact mitochondria from sugar beet roots. Fiziologiya Rastenii. Plant Physiology. (Rus). 29:799-803.

Sviryaeva IV, Ruge EK (2006). Generation of oxygen free radicals in heart mitochondria: Effect of hypoxia reoxigenation, Biofizika (Biophysics) (Rus)51(3): 478-484.

Terekhova SF, Grechenko TN (2002). Regulation of neuronal functional state by ultra-low doses of biologically active substances Proceedings of the Third International Symposium "Mechanisms of action of ultra-low doses" (3--6.12.2002). Moscow. pp. 35.

Thomas JC, Bohnert HJ (1993) Salt stress precipitation and plant grows regulators in the halophyte Mesembryan themum crystallinum. Plant Physiol. 103:1299-1304.

Zhernovkov VE, Bogdanova NG, Palmina NP (2006). Structural changes in endoplasmic reticulum membranes induced by ultra-low doses of thyreoliberin in vitro./ Biologicheskie Membrany (Biological Membranes) (Rus.). 23(1):52-59.

Zhigacheva VI, Burlakova EB, Shugaev AG, Generozova IP, Fattakhov SG, Konovalov AI (2008). The organophosphorus plant grows regulator Melafen: resistant of plant and animal cells to stress factors.

Biologicheskie Membrany (Biol. Membranes) (Rus.) 25 (3) :183-189.

Zhigacheva IV, Kaplan E Ya, Pakhomov VYu, Rozantseva TV, Khristianovich DS, Burlakova EB (1995). The drug «Anphen» and the energy status of the liver. Dokl. Akad. Nauk (Rus.), 340:547-550.

Zinkevich EP, Ganshin VM (2002). Possible mechanisms of low dose sensitivity in olfactory reception of vertebrates // Proceedings of the Third International Symposium "Mechanisms of action of ultra-low doses" (3--6.12. 2002), Moscow. pp10.

15

Influence of gibberellic acid and arbuscular mycorrhizae inoculation on carbon metabolism, growth, and diterpene accumulation in *Taxus wallichiana* Zuccarini var. mairei

A. Misra*, N. K. Srivastava, A. K. Srivastava and S. K. Chattopadhyay

Central Institute of Medicinal and Aromatic Plants, P. O. CIMAP, Lucknow-226015, India.

Changes in growth parameters and $^{14}CO_2$ and [U-^{14}C]-sucrose incorporation into the primary metabolic pools and diterpene 10 DAB compound were investigated in leaves and stems of *Taxus wallichiana* Zucarnii treated with gibberellic acid (GA) and inoculated with arbuscular mycorrhizae (AM). Compared to the control, GA(1000 ppm) and AM (1 kg/ha each) with AM-GA combined treatments, induced significant phenotypic changes and a decrease in chlorophyll content, CO_2 exchange rate and stomatal conductance. Treatment with AM-GA led to increased total incorporation of CO_2 into the leaves whereas total incorporation from ^{14}C sucrose was decreased. When $^{14}CO_2$ was fed, the incorporation into the ethanol soluble fraction, sugars, organic acids, and essential oil was significantly higher in AM-GA treated leaves than in the control. However, [U-^{14}C] -sucrose feeding led to decreased label incorporation in the ethanol-soluble fraction, sugars, organic acids, and diterpenes compared to the control. When $^{14}CO_2$ was fed to AM-GA treated leaves, label incorporation in ethanol-insoluble fraction, sugars, and oils was significantly higher than in the control. In contrast, when [U-^{14}C]-sucrose was fed the incorporation in the ethanol soluble fraction, sugars, organic acids, and oil was significantly lower than in the control. Hence the hormone treatment induces a differential utilization of precursors for oil biosynthesis and accumulation and differences in partitioning of label between leaf and stem. GA and GA-VAM influence the partitioning of primary photosynthetic metabolites and thus modify plant growth and 10-DAB compound accumulation.

Key words: Amino acids, chlorophyll, CO_2- and C-sucrose incorporation, organic acids, primary photosynthetic metabolites, stem, stomatal conductance, sugars, transpiration rate.

INTRODUCTION

Taxus wallichiana Zuccarini var. mairei is one of the main sources of diterpene 10 DAB Compound taxoids, which are used widely in, pharmaceutical industry (Suffnes, 1995). Diterpene biosynthesis in Himalayan Yew (*T. wallichiana* Zucarnii) including other monoterpenes bearing plant is strongly influenced by several intrinsic and extrinsic factors (Lawrence, 1986; Bernard et al., 1990) including temperature (Clark and Menary, 1980a),

photoperiod (Burbott and Loomis, 1967), photosynthetic photon flux density (Clark and Menary, 1980b), nutrition (Srivastava and Luthra, 1994, Srivastava et al., 1997),genotype (Srivastava and Luthra, 1991), ontogeny (Srivastava and Luthra, 1991b), and osmotic stress (Charles et al., 1990). Diterpenes especially Taxoids, composed mainly of Paclitaxol (Taxol[R]) which is mainly occurring naturally as highly taxane diterpene amides are synthesized through the mevolanate-isoprenoid pathway in the epidermal oil glands which are carbon-heterotrophic and hence depend on the adjoining mesophyll cells for precursors (McGarvey and Croteau, 1995). However, these diterpenes may not only be accumulated but also biosynthesized in leaf mesophyll cells (Gershenzon et al., 1989). Among precursors, CO_2 and sucrose are preferred for monoterpenes and diterpene

---

*Corresponding author. E-mail: amisracimap@yahoo.co.in.

Abbreviations: Chl, chlorophyll; *E,* transpiration rate; *g_s,* stomatal conductance; /"N, net photosynthetic rate.

biosynthesis (Gershenzon and Croteau, 1991, 1993). Diterpene and monoterpene biogenesis is also linked to the contents of primary metabolites (Srivastava and Luthra, 1991a), and a positive but insignificant association has been shown with net photosynthetic rate, P_N (Srivastava et al., 1990). Thus the secondary metabolic pathway is closely associated and dependent on the primary metabolic pathway. Taxoids are produced in the leaf needles. Most paclitaxol taxoids produced via semisynthetic conversion of 10 - DAB (10-deacetylbaccatin III). Paclitaxol is a naturally occurring taxane diterpene amide that has been proven effective in treating various types of cancer such as ovarian carcinoma, metastatic breast cancer, non small cel cancers, adinocarcinoma and squmous cell carcinoma of the oesophagus (Morita et al., 2005).

Growth hormones play a dominant role in the regulation of growth and development by affecting sink-source relationship (Marschner, 1986). El-Keltawi and Croteau (1986a, b) reported the influence of phosphon-D, cycocel, ethephon, and daminozide on the constituents of essential monoterpene oil(s) of M. piperita. Farooqi and Sharma (1988) reported influence of growth retardants on growth and essential oil accumulation in M. arvensis whereas Srivastava and Sharma (1991) reported the influence of triacontanol on photosynthetic characteristics and oil accumulation in M. arvensis. Most of the growth hormone studies on monoterpene bearing plants attribute the effects to the influence on enzymes of biosynthetic pathways and on plant and growth characters such as herb yield and leaf/stem ratio. However, it is not clear what changes occur in the photosynthetic C-metabolism of the hormone treated plant and translocation of assimilates to the Diterpene accumulation. While studying the influence of growth hormones on yield and growth, we observed significant and persistent effect of GA and GA-AM on plant phenotype. T. wallichiana Zucarnii (Himalayan Yew) is a very slow growing high altitude forest tree. Hence, this hormone treatment alone and in combination with AM study on Himalayan Yew has been taken for rapid growth and carbon metabolism on the photosynthesis and photosynthatates effect on the physiology of the plant.

Further, in the present study we report the influence of GA and GA-AM on the photosynthetic efficiency and diterpene 10-DAB paclitaxol accumulation studies during the incorporation of $^{14}CO_2$ and [U-^{14}C]-sucrose into primary photosynthetic metabolites, sugars, amino acids, and organic acids, and simultaneously into the taxol bearing compounds of T. wallichiana plants seedlings brought from high altitude CIMAP field station Purara, Almora treated plants. Changes in P_N, chlorophyll (Chi) content, and stomatal conductance (g_s) were also determined.

MATERIALS AND METHODS

Uniform suckers of M. spicata cv. MSS-5 obtained from the farm

nursery of the Institute were treated with GA and GA-VAM (1 kg nr^3 each) by dipping in respective solution for 24 h. Later the treated suckers were planted in 10 000 cm^3 earthen pots maintained in a glasshouse at ambient temperature (30 - 35 °C) and irradiance (800 - 1000) |amol rtr^{2}s^{-1}, measured by a LiCOR light meter model 188 B). Values of growth characters, essential oil, and tracer feeding were recorded 100 d after the treatment.

Chl (a + b) content was measured on the third brances of the needle like leaf. A known mass of leaf tissue was extracted with 80% acetone and the absorbance was recorded by a Milton Roy spectrophotometer Spectronic 21 D using the method of Arnon (1949). P_N, initial transpiration rate (£), and g_s of the third leaf were measured in a closed system using a portable computerized photosynthesis model Li-6000 {LiCOR, Lincoln, USA) as described in Srivastava and Luthra (1991a). For determining the extraction of taxoids from the control plants (untreated) or after feeding of $^{14}CO_2$ or [U-^{14}C]-sucrose, a known mass of shoot (leaf + stem) material was finely chopped and subjected to percolate in pure methanol at room temperatre as described by earlier (Chattopadhyay and Sharma, 1995). The radioactivity in ether samples was determined in a scintillation counter (LKB Rack Beta 1215) using a PPO-POPOP-toluene cocktail (Srivastava and Luthra 1991a).

The tracer studies were carried out with $^{14}CO_2$ and [U-^{14}C]-sucrose that were fed to the freshly excised shoots of treated and control plants and the amounts of label incorporated into 10-DAB and simultaneously into the pool of primary photosynthetic metabolites (sugars and sugar phosphates, amino acids, and organic acids) were determined. Before the labelling studies, the shoots were cut under water and tested to ensure that they were able to take up water properly. For $^{14}CO_2$ studies, 12 unbranched main shoots (of GA-AM treated, GA treated, and control plants) having 6 leaf pairs were placed in vials with the cut ends dipped in half strength Hoagland and Arnon (1938) solution. The vials were then placed in a sealed plexiglass chamber (20 000 cm^3 capacity) around a central vial containing Na$_2$$^{14}CO_3$ solution (3.7 MBq, 2.8 GBq mmol-^{1}) obtained from the isotope division of Bhabha Atomic Research Centre (BARC), Trombay, India. $^{14}CO_2$ was generated by injecting 4 N H$_2$SO$_4$ into carbonate solution through a PVC tube and uniformly distributed with the help of a small electric fan. The leaves were allowed to assimilate $^{14}CO_2$ for 1 h in sunlight (800 - 1000 |amol m$^{"2}$ s$^{"1}$). At the end of 1 h, a saturated solution of KOH was run into the central vial and left for 15 min to absorb excess $^{14}CO_2$. The chamber was then opened for the remaining incorporation period of 6 h (Srivastava and Luthra 1991a). For feeding experiments with [U-^{14}C] -sucrose, unbranched shoots having 6 leaf pairs each were placed in vials containing an aqueous solution of 1 |amol [U-^{14}C]-sucrose (185 kBq) obtained from the Isotope Division of BARC, Trombay, India (specific activity 21.5 GBq mmol"). After the uptake of labelled material, the vials were kept filled with half strength Hoagland solution, and the samples were harvested after 6 h.

After exposure to $^{14}CO_2$ or [U-^{14}C]-sucrose feeding, plants were separated into leaf and stem, finely chopped, and divided into two parts:

1.) A known weight of leaf, stem and root tissues were processed for determining the incorporation of current photosynthetic metabolite in total diterpenes. The radioactivity in alkaloid fraction was determined using PPO-POPOP-Toluene cocktail in a liquid scintillation counter. The unit of expression was Bq/g.dry wt. of tissue (leaf, stem and roots).

2.) A known weight of leaf, stem and root tissues were immediately fixed into boiling ethanol so that the current metabolic status was maintained .The plant material was ground in ethanol, filtered, filtrate evaporated and diluted in a known volume of aqueous phase; termed as ES fraction. This aqueous phase was further extracted with chloroform and this CS fraction contained pigments

(A)   **(B)**   **(C)**

Figure 1. Changes in plant characters *of Taxus wallichiana* due to hormone and hormone-AM treatment. Left (A): Control:, middle: (B): AM-GA, and right: (C): GA.

and some of the terpenoid pathway derived end metabolites. The remaining plant material termed, as EIS fraction was further hydrolyzed by enzyme diastase in 0.05 M acetate buffer (pH 5.2) at 50°C (Srivastava et al. 1990). The label in ^{14}C in ES and in EIS fraction was determined in Bray's scintillation fluid and in CS fraction in PPO-POPOP-Toluene cocktail in a liquid scintillation counter. The unit of expression was Bq/g.fresh wt. of tissue (leaf, stem and roots). Total ^{14}C incorporated was expressed as sum of values of ES+EIS+CS fraction. The ES fraction was further separated into metabolic pool consisting of neutral (sugar+sugar phosphates) acidic (organic acids) and basic (amino acids) fractions by separation through Amberlite ion exchange column chromatography. The ^{14}C content in eluates after column chromatgraphy was determined in Bray's scintillation fluid in a liquid scintillation counter (Srivastava and Luthra 1991b), for determining the incorporation of label into taxoids 10-DAB a known mass of shoot material (leaf + stem) was processed as described earlier (Chattopadhyay and Sharma, 1995). Further, for determining incorporation of label into primary photosynthetic metabolites a known mass of tracer-fed leaves and stem was extracted immediately in boiling 80% ethanol. The stem sample did not include the basal portion which had been immersed in labelled sucrose. The ethanol soluble material was separated into neutral (sugars and sugar phosphates), basic (amino acids), and acidic (organic acids) fractions by *Amberlite* ion exchange column chromatography. Ethanol-insoluble material was hydrolyzed by diastase in 0.05 M acetate buffer (pH 5.2) at 50°C. The radioactivity in hydrolyzed alcohol insoluble material and in eluates after ion exchange separation was measured using Bray's scintillator (Srivastava and Luthra 1994). Total ^{14}C incorporated was calculated as the sum of the total label incorporated in ethanol-soluble and -insoluble fraction and expressed on fresh mass basis.

All measurements were taken in triplicate and the results are given as means ± SE. Values were statistically analysed for significance by paired /-test.

RESULTS AND DISCUSSION

Treatments with GA and GA-AM significantly increased

plant phenotype which was evident even at 100 d of growth (Figure 1). Normally the plant metabolizes the externally applied hormones and even if there are some phenotypic differences, these are temporary and the plant reverts soon to its normal phenotype, but in the present case the hormone effects were evident much longer. This was accompanied by marked changes in physiological characteristics.

The GA treated plants had significantly lower contents of ChL *(a+b)*, Chi *a,* increased P_N, g_s, *E,* and plant height as compared to control (Table 1). Thus the overall growth was in increased. There was a difference in utilization pattern of $^{14}CO_2$ and [U-^{14}C] -sucrose. When $^{14}CO_2$ was fed, the total $^{14}CO_2$ fixed in leaves in GA treatment was significantly higher than in the control. Also the ethanol-insoluble fraction, the sugars, organic acids, and essential oil had a significantly higher ^{14}C-incorporation in GA treated leaves than in the control (Table 2). Thus the GA-AM treated plants allocated more photosynthetic metabolites towards diterpene taxoids than the control plants. Partitioning of photosynthetic metabolites between leaf and stem is an important factor in yield determination (Srivastava and Luthra 1991a). In stems, ^{14}C incorporation in ethanol soluble fraction, sugars, and organic acids was significantly higher in the GA-AM treated plants than in the control. Thus, ethanol soluble compounds remained untranslocated in the stem (Table 2). Overall, if we will go for phenotypic changes the plants in Figure 1 showed the more increase in height then the control one. Generally, the increase in height is only3 inches in a year where as this height increase is only in 4 months.

When [U-^{14}C]-sucrose was fed to GA treated leaves, the total ^{14}C incorporation was significantly higher than in the control. Incorporation into ethanol soluble fraction was significantly higher than that measured in the

Table 1. Changes in growth and yield characters of *T. wallichiana* treated with etherel and gibberellic acid (GA). Chi = chlorophyll; $P^\wedge$ = net photosynthetic rate.

Characters	GA	Control	GA-AM
Chl *a* [g kg-'(FM)]	2.19 ± 0.06*	2.72 ± 0.21	1.79 ± 0.03*
Chl b[gkg-'(FM)]	0.74 ± 0.08*	1.15 ± 0.08	0.55 ± 0.03*
Chl *(a+b)* [g kg-'(FM)]	2..93 ± 0.05 *	3.87 ± 0.27	1.34 ± 0.01*
P_N [ng(CO_2) m-^{2} s"^{1}]	164 ± 7*	129 ± 4	134 ± 12*
Initial transpiration rate [mmol nr^2 s"^{1}]	476 ± 30*	691 ± 90	446 ± 20*
Stomatal conductance [mmol m-^{2} s-']	229 ± 11*	429 ± 10	271 ± 20*
Plant height [cm]	95 ± 0.21**	61 ± 0.10	81.85 ± 0.05**

*/** Mean values significant at 5/1% level of significance by pair r-test; NS – non significant.

Table 2. Changes in incorporation pattern of $^{14}CO_2$ into primary photosynthetic metabolic pool and diterpene: taxoids in leaves and stems *of T. wallichiana* treated with GA-AM and gibberellic acid (GA).

	Fractions	GA-AM	Control	GA
	Ethanol-soluble fraction	238 ±16[NS]	282 ± 27	2691 ± 5[NS]
	Ethanol-insoluble fraction	1749 ±13**	1256 ±53	1818 ± 49*
	Total incorporation	1917 ±43**	1301 ± 57	4395 ±41*
Leaves	Sugar	108 ±6**	49 ± 2	75 ±1*
	Amino acids	317 ±2[NS]	171 ± 72	747 ± 83[NS]
	Organic acids	127 ±5**	77 ± 2	111 ± 2[NS]
	Taxoids 10-DAB	1.41 ± 0.02**	0.42 ± 0.01	1.31 ± 0.02*
	Ethanol-soluble fraction	147 ± 6*	79 ± 2	382 ± 23*
	Ethanol-insoluble fraction	1301 ± 71[NS]	401 ±07	2791 ± 47"
Stem	Total incorporation	1577 ± 71[NS]	523 ±19	3298 ±301**
	Sugar	127 ±25*	49 ±3	189 ±27[NS]
	Amino acids	901 ± 103[NS]	148 ± 33	1491 ± 329[NS]
	Organic acids	154 ± 7*	39 ± 3	186 ± 17*

All values in 10^3 dps kg'"(FM). */** Mean values significant at 5/1% level of significance by pair test; NS – non significant.

insoluble fraction. However, the label in sugars, organic acids, and essential oil fraction was significantly lower than in the control (Table 3). When these fractions were analyzed in the stem, the ethanol-insoluble fraction had significantly higher label whereas the ethanol-soluble fraction had significantly lower amounts of labeled sugars, amino acids, and organic acids than the controls (Table 3). Thus, the amount of compounds derived from added [U-^{14}C]-sucrose was higher in leaves and was significantly lower in stems in GA treated plants. Hence the capacity to utilize end products of photosynthetically fixed $^{14}CO_2$ and the externally applied sucrose was entirely different. Ontogenic changes exist for distribution of photosynthetically fixed $^{14}CO_2$ in peppermint leaves. The incorporation of $^{14}CO_2$ into sugars was maximum followed by organic acids, amino acids, and essential oil at all stages of leaf development. The incorporation into sugars and amino acids declined as the leaf matured whereas the incorporation into monoterpenes and

organic acids increased with leaf expansion and then decreased (Srivastava and Luthra 1991b). In onions, the older was the plant the more of C-assimilate left the source leaf (Khan, 1981).

The GA-AM treated plants had significantly lower contents of ChL pigments, P_N, *E,* and g_s, however the plant height was significantly higher than in the control (Table 1). GA-AM treatment resulted in both higher total fixation of $^{14}CO_2$ and ^{14}C incorporation in ethanol-insoluble fraction and sugars of leaves. Significantly higher amounts of photosynthetic metabolites were translocated towards essential oils because the label was significantly higher in essential oil (Table 2). Amino acid and organic acid contents were not significantly affected over control. Similarly, the stem of GA-AM treated plants showed significantly higher total incorporation, contents of ethanol-soluble and -insoluble fraction, whereas the contents of organic acids, amino acids, and sugars were not significantly different than in the control (Table 2).

Table 3. Changes in incorporation pattern of [U- C]-sucrose into primary photosynthetic metabolites and in diterpene: taxoids in *T. wallichiana* treated with GA-AM and gibberellic acid (GA).

	Fractions	GA-AM	Control	GA
Leaves	Ethanol-soluble fraction	6113 ± 27"	3571 ± 731	1429 ± 323*
	Ethanol-insoluble fraction	2258 ± 148[NS]	3192 ± 321	4321 ± 1108[NS]
	Total incorporation	9714 ± 139"	5839 ± 1407	2142 ± 1408"
	Sugar	694 ± 92"	2831 ± 220	1955 ± 87[NS]
	Amino acids	549 ± 12[NS]	76 ± 4	148 ± 3*
	Organic acids	53 ± 7*	137 ± 3	128 ± 4*
	Taxoids 10-DAB	1.03 ± 0.04**	1.19 ± 0.07	9451 ± 170**
Stem	Ethanol-soluble fraction	10549 ± 1489*	12201 ± 1158	14629 ± 325[*]
	Ethanol-insoluble fraction	3754 ± 133*	1261 ± 299	2512 ± 377[NS]
	Total incorporation	14534 ± 1613[NS]	14521 ± 1234	18251 ± 212[NS]
	Sugar	823 ± 47	1128*49	1191 ± 11[NS]
	Amino acids	61 ± 1*	72 ± 4	98 ± 3[NS]
	Organic acids	88 ± 2*	116 ± 3 *	98 ± 3*

All values in 10^3 dps kg'''(FM). */** Mean values significant at 5/1% level of significance by pair test; NS – non significant.

Thus overall incorporation of $^{14}CO_2$ into metabolites and their higher subsequent translocation to oil biosynthetic pathway were higher in GA-AM treated plants. Here, the GA-AM treated plants showed the increase in height and the increase in number of branches in comparison with the control (Figure 1).

As far as the utilization pattern of [U-^{14}C]-sucrose is concerned, GA-AM treatment resulted in leaves in significantly higher total incorporation, incorporation in ethanol-soluble fraction, amino acids, and essential oil, whereas ethanol-insoluble fraction and sugar contents were not significantly influenced (Table 3). In contrast, the contents of all these metabolites in stem were significantly not affected (Table 3).

Application of GA significantly increases growth in terms of height and physiological parameters which negatively affects herb yield. Hormone application in general does not bring a simultaneous increase in growth and diterpenes. In Japanese mint, chlormequat chloride increased monoterpene content but inhibited growth whereas ethephon decreased growth but had no significant effect on monoterpene content (Farooqi and Sharm, 1988). Hormones such as phosphon-D and daminozide influence enzymes and interconversion in monoterpene biosynthesis (El-Keltawi and Croteau 1996a, b) and endogenous content of other hormones. However, it is not known how the carbon fixation capacity is affected by hormone application. Despite the decrease in herb yield, both hormone-treated plants contained higher amounts of the $^{14}CO_2$ fixation products. This probably results in greater translocation of photosynthetic metabolites to the oil biosynthetic pathway. The higher contents do not necessarily mean higher $^{14}CO_2$ efficiency; it could also mean that the fixed $^{14}CO_2$ is not utilized by the plant growth process whereas in control plants it is utilized and its content is lower.

The fed sucrose was poorly utilized for oil biosynthesis and simultaneously the content of photosynthetic metabolites was also low. Thus the utilization of $^{14}CO_2$ and sucrose for oil biosynthesis was different. The changes in growth could also be due to differences in partitioning of available assimilates between leaf and stem. The monoterpene/diterpene biosynthesis is an integration of several metabolic pathways which require linking of several steps such as continuous production of precursors, their transport and translocation to the active site of synthesis, and finally the diterpene taxoids 10-DAB. This sequence of steps depends on normal functioning of associated metabolic pathways. Any disruption in normal metabolic pathways affects the sequence of steps in oil biosynthesis. Thus a plant may alter/adopt its metabolic pathway in response to particular effect, such as nutrient imbalance especially the P and Zn, hormone application, *etc.* Under GA and GA-AM treatment there is higher accumulation of photosynthetic metabolites, nevertheless, the decrease in herb yield and growth may be due to energy deficiency, membrane effects, or other control mechanisms which need to be investigated. Further the increase in height due to the application of GA, GA-AM visualized the rapid elongation in the slow growing Himalayan Yew.

ACKNOWLEDGEMENT

The authors thank the Director for providing necessary facilities and encouragement during this study.

REFERENCES

Arnon DI (1949). Copper enzymes in isolated chloroplasts: Polyphenoloxidase in *Beta vulgaris*. Plant Physiol. 24: 1 -15.

Bernard V, Nathaile B, Christine B (1990). Effect of day length on monoterpene conversion in leaves Of *Mentha piperita*. Phytochemistry 29: 749 -755.

Burbott AJ, Loomis WD (1967). Effects of light and temperature on the monoterpene of peppermint. Plant Physiol. 42: 20 -28.

Charles DJ, Joly RJ, Simon JE (1990). Effect of osmotic stress on the essential oil content and composition of peppermint. Phytochemistry 29: 2837-2840.

Chattopadhyay SK, Sharma RP (1995). A taane from the Himalayan Yew, Taxus wallichiana. Photochemistry 39: 935 - 936.

Clark RJ, Menary RC (1980a). Environmental effects on peppermint *{Menthapiperita* L.). I. Effect of day length, photon flux density, night temperature and day temperature on the yield and composition of peppermint oil. Aust. J. Plant Physiol. 7: 685 -692.

Clark RJ, Menary RC (1980b). Environmental effects on peppermint *(Menthapiperita* L). II. Effects of temperature on photosynthesis, photorespiration and dark respiration in peppermint with reference to oil composition. Aust. J. Plant Physiol. 7: 693 - 697.

El-Keltawi NE, Croteau R (1986a). Influence of ethephon and daminozide on growth and essential oil content of peppermint and sage. Phytochemistry 25: 285 -288.

El-Keltawi NE, Croteau R (1986b). Influence of phosphon-D and cycocel on growth and essential oil content of sage and peppermint. Phytochemistry 25: 1603 -1606.

Farooqi AHA, Sharma S (1988). Effect of growth retardants on growth and essential oil content in Japanese mint. Plant Growth Regul. 9: 65 -71.

Gershenzon J, Croteau R (1991). Regulation of monoterpene biosynthesis in higher plant. In: Towers, G.H.N., Stafford, H.A. (ed.): Biochemistry of Mevalonic Acid Pathway to Terpenoids. Plenum Press, New York pp. 99 -159.

Gershenzon J, Croteau R (1993). Terpenoid biosynthesis: The basic pathway and formation of monoterpenes, sesquiterpemes and diterpenes. In: Moore, T.S., Jr. (ed.): Lipid Metabolism in Plants. CRC Press, London pp. 339 - 388.

Gershenzon J, Maffei M, Croteau R (1989). Biochemical and histochemical localization of monoterpene biosynthesis in the glandular trichomes of spearmint *(Mentha spicata)*. Plant Physiol. 89: 1351-1357.

Hoagland DR, Arnon DI (1938). The water culture method for growing plants without soil. Circ. Calif, Agr. Exp. Stat. 347: 1-32.

Khan AA (1981). Effect of leaf position and plant age on the translocation of ^{14}C-assimilates in onion. Cambridge J. Agr. Sci. 96: 451- 455.

Lawrence BM (1986). Essential oil production. A discussion of influencing factors. In: Parliment. T.H., Croteau, R. (ed.): Biogeneration of Aroma. Amer. Chem. Soc. New York. pp. 363-369.

Marschner H (1986). Growth. In: Marschner, H. (ed.): Mineral Nutrition of Higher Plants. Academic Press, New York pp. 269-340.

McGarvey D, Croteau R (1995).Terpenoid metabolism. Plant Cell 7: 1015-1026.

Srivastava NK, Luthra R (1991a). Interspecific variation in mints for photosynthetic efficiency, and ^{14}Cprimary metabolic pool in relation to essential oil accumulation. J. Plant Physiol. 138: 650-654.

Srivastava NK, Luthra R (1991b). Distribution of photosynthetically fixed ^{14}CC>2 into essential oil in relation to primary metabolites in developing peppermint *(Menthapiperita)* leaves. Plant Sci. 76: 153 -157.

Srivastava NK, Luthra R (1994). Relationships between photosynthetic carbons metabolism and essential oil biogenesis in peppermint under Mn-stress. J. Exp. Bot. 45: 1127-1132.

Srivastava NK, Luthra R, Naqvi A (1990). Relationship of photosynthetic carbon assimilation to essential oil accumulation in developing leaves of Japanese mint. Photosynthetica 24: 406 - 411.

Srivastava NK, Misra A, Sharma S (1997). Effect of Zn deficiency on net photosynthetic rate, ^{14}C partitioning, and oil accumulation in leaves of peppermint. Photosynthetica 33: 71-79.

Srivastava NK, Sharma S (1991). Effect of tricontanol on photosynthetic characteristics and Essential oil accumulation in Japanese mint. *Mentha arvensis* L. Photosynthetica 25: 55 -60.

Suffnes M (1995). Taxol (ed.): Science and Application, CRC Press. Boca Raton, FL. pp.111-116.

BPES analyses of a new diffusion-advection equation for fluid flow in blood vessels under different bio-physico-geometrical conditions

M. Dada[1], O. B. Awojoyogbe[1], K. Boubaker[2]* and O. S. Ojambati[1]

[1]Department of Physics, Federal University of Technology, Minna, Niger-State, Nigeria.
[2]Unité de Physique de Dispositifs à Semiconducteurs -UPDS- Faculté des Sciences de Tunis, Campus Universitaire 2092 Tunis, Tunisia.

In human physiological and pathological flow systems, it is not possible to rule out diffusion in all advective processes because perfusion goes hand in hand with diffusion processes. It is the perfusion throughout the capillary bed and then the diffusion of fluids throughout the tissue that is the subject of most magnetic resonance functional imaging procedures. It is observed from literature that basic theory of perfusion is mostly based on experimental observation which makes it entirely computational with quite a lot of data fitting. Therefore, it is quite rigorous and has many phenomena that seem not to have a common background. It is very important to attempt developing a theory that would take most issues (if not all) into consideration under a common phenomenon. In this study, based on the Bloch NMR flow equations along with the Boubaker polynomials expansion scheme (*BPES*), we describe analytically the dynamics of perfusion processes by an equation which combines both diffusive and advective properties.

Key words: Bloch NMR flow equations, diffusion-advection equation, blood vessels, BPES scheme.

INTRODUCTION

Distribution of oxygen to every corner of the body is accomplished by the cardiovascular system, with the help of the most important fluid in the body: the Blood, the stream of life. Life depends so much on blood such that its importance cannot be over emphasized. It has been investigated that any obstacle to the normal flow of blood causes a malfunctioning in the body system that leads to cardiovascular related diseases.

Functional magnetic resonance imaging (Martinez et al., 2002; Valfouskaya and Adler, 2005; Segnorile et al., 2006) consists of several different imaging methods that are used to visualize and, in some cases, quantify blood and fluid movement beyond the general vascular system. It is the perfusion through out the capillary bed and the

diffusion of fluid throughout the tissue that is the subject of most magnetic resonance functional procedures (Sprawls, 2000).

Most perfusion processes within the human body are always changing from time to time with regards to a lot of body conditions. These processes take place in tube - like vessels (for example, the blood vessel) which are always under some sort of pressure and since they are elastic in nature, we need to characterize the flow velocity and the diffusion coefficient from point to point. If we take for example, the case of a sudden rush of blood to a part of the body tissue, the blood vessel carrying blood to the part of the tissue would suddenly become larger because of increased pressure and at that point, the flow velocity and diffusion coefficient changes. If the cause of the sudden demand for more blood is removed, the vessel goes back to its normal shape. Therefore, it would be very crucial to account for the velocity and diffusion (Awojoyogbe, 2004) coefficient at all points for an accurate description of the process under investigation.

---
*Corresponding author. E-mail: mmbb11112000@yahoo.fr.

Hence, based on the Bloch NMR flow equations (Zoppou and Knight, 1997; Awojoyogbe et al., 2010; Awojoyogbe, 2007; Awojoyogbe, 2003; Awojoyogbe, 2002; Awojoyogbe, 2008), we must use the diffusion-advection equation with spatially varying diffusion coefficients as proposed in this study.

MATHEMATICAL METHOD

In this study, a mathematical (analytical) technique in the form of a plane wave is applied to transform the time dependent Bloch NMR flow equation to diffusion-advection equation for the qualitative analysis of nuclear magnetization. We consider the perfusion (or transport) of any specific blood component as one dimensional since blood flow within the vessels is directional and, even in bifurcations, flow has a resultant direction of fluid flow. Therefore, for any NMR sensitive substance of interest, the perfusion process is given by the NMR advection - diffusion equation derived from the Bloch NMR flow equations (Awojoyogbe, 2004)

$$v^2 \frac{\partial^2 M_y}{\partial x^2} + 2v \frac{\partial^2 M_y}{\partial x \partial t} + v T_o \frac{\partial M_y}{\partial x} + T_o \frac{\partial M_y}{\partial t} + \frac{\partial^2 M_y}{\partial t^2} + \Omega M_y = F_o \gamma B_1(x,t) \quad (1a)$$

Where;

$$\Omega = T_g + \gamma^2 B_1^2(x,t); \quad F_o = \frac{M_o}{T_1}; \quad T_g = \frac{1}{T_1 T_2} \text{ and } T_0 = \frac{1}{T_1} + \frac{1}{T_2},$$

γ is the gyromagnetic ratio, D is the diffusion coefficient, v is the fluid velocity, T_1 is the spin lattice relaxation time, T_2 is the spin relaxation time, M_o is the equilibrium magnetization, $B_1(x,t)$ is the applied magnetic field and M_y is the transverse magnetization.

Solutions to Equation (1a) have been discussed by a number of analytical methods (Awojoyogbe, 2008; Oyodum et al., 2009), and for the present purpose it is sufficient to design the NMR system in such a way that the transverse magnetization M_y, takes the form of a plane wave,

$$M_y(x,t) = A e^{mx+nt} \quad (1b)$$

Subject to the following theoretical conditions:

$$\frac{1}{T_1 T_2} >> \gamma^2 B_1^2(x,t) \quad (2)$$

$$n = -2vm \pm \frac{\sqrt{4(v^2 m^2 - T_g)}}{2} \quad (3)$$

Where m and n are dependent on the NMR flow parameters and B_1 is independent of x and t. based on equations (1b, 2 and 3), we can write equation (1a) in the form of diffusion-advection equation for the nuclear magnetization.

$$\frac{\partial}{\partial x}(v(x) M_y) + \frac{\partial M_y}{\partial t} = \frac{\partial}{\partial x}\left(D(x) \frac{\partial M_y}{\partial x}\right) + \frac{F_o}{T_o} \gamma B_1(x,t) \quad (4)$$

Where D(x) is the variable diffusion coefficient. Equation (4) is a

generalize equation of motion for the NMR flow system with a spatially variable velocity and diffusion coefficient. The behavior of the transverse magnetization or signal is depicted by the solution to equation (4). If we make the following assumption:

$$x = \exp(u_0 X) \quad (5)$$

$$\left. \begin{array}{l} D(x) = D_0 u_0^2 x^2 \\ v(x) = u_0 x \end{array} \right\} \quad (6)$$

Equation (4) becomes

$$\frac{\partial M_y}{\partial t} = D_0 \frac{\partial^2 M_y}{\partial X^2} - (1 - 2D_0 u_0) \frac{\partial M_y}{\partial X} \quad (7)$$

Where

$$D_o = \frac{v^2}{T_o} \quad (8a)$$

$$\frac{F_o}{T_o} \gamma B_1(x,t) = u_0 M_y \quad (8b)$$

Analytical solution to equation (7) is similar to those of the diffusion equation of variable diffusion coefficient (Zoppou and Knight, 1997). Hence, the solution could be written as

$$M_y(X,t;X_0) = \frac{A_{BPES}}{2\sqrt{D_0 u_0^2 \pi t}} \exp\left(\frac{-[X - X_0 - (1 - 2D_0 u_0)t]^2}{4D_0 t}\right) \quad (9)$$

The value of the constant A_{BPES} is determined using the Boubaker Polynomials Expansion Scheme (BPES) (Awojoyogbe, 2008; Zhao et al., 2008). The calculation protocol takes into account conjointly the properties of the BPES along with the already noticed (Awojoyogbe, 2008; Oyodum et al., 2009) similarity between equation (7) and the characteristic differential equation of the Boubaker polynomials.

For a component of the blood in the unit of magnetic moment being transported across the blood vessel, the value of the constant A_{BPES} is

$$A_{BPES} = \frac{1}{x_0 \exp[(u_0 t - D_0 u_0^2 t)]} = \frac{\exp[(D_0 u_0^2 t - u_0 t)]}{x_0} \quad (10)$$

Subject to the following constraint:

$$\int_0^\infty M_y(x,t)\, dx = 1 \qquad \forall t \quad (11)$$

The NMR transverse magnetization for the instantaneous release can therefore be written as:

$$M_y(x,t) = \frac{1}{2\sqrt{D_0 u_0^2 \pi t}}\left(\frac{1}{x}\right)\left(\frac{x}{x_0}\right)^{1/2D_0 \mu_0} \exp\left(-\frac{[P_{02}^2 + t^2]}{4D_0 t}\right) \quad (12)$$

This one-dimensional solution is valid quite well for the upstream and downstream of the bifurcation. For a continuous source of unit magnetic moment (Zoppou and Knight, 1997), the behavior of the NMR signal is obtained by integrating equation (12) with respect to time.

$$M_y(x,t) = \left(\frac{1}{x}\right)\left(\frac{x}{x_0}\right)^{1/2D_0\mu_0} \int_{-\infty}^{\infty} \frac{1}{2\sqrt{D_0 u_0^2 \pi t}} \exp\left(-\frac{[P_{02}^2 + t^2]}{4D_0 t}\right) dt \quad (13)$$

Where

$$P_{02}^2 = \left(\frac{1}{u_0}\ln\left(\frac{x}{x_0}\right)\right)^2$$ is the one directional perfusion function and u_0 is a constant. The expression in equations (13) gives the

behavior of the NMR signal $M_y(x,t)$ at all points for a material

or substance which is being transported (in perfusion). However, perfusing particles behave differently in different geometries. This requires that we applied some additional experimental conditions to appropriately describe equation (13) in different geometries.

THE MULTIDIMENSIONAL PERFUSION PROCESS – CYLINDRICAL GEOMETRY

Although perfusion in multi-dimension is quite rare, we may need to discuss this situation because such process can be applicable in the analysis of complex flow in regions of bifurcations. In turbulent flow for example, particles are transported in a way that is very difficult to specify the direction of the flowing particles or the direction of the resultant velocity. Hence, there is a need for point to point characterization of the fluid velocity, the diffusion coefficient and the NMR signal.

Since perfusing substances obey the advection equation, the appropriate equation to accurately describe a flow process in a cylindrical geometry based on equation (7) derived as;

$$\frac{\partial}{\partial r}\left(v_r M_y\right) + \frac{1}{r}\frac{\partial}{\partial \phi}\left(v_\phi M_y\right) + \frac{\partial}{\partial z}\left(v_z M_y\right) + \frac{\partial M_y}{\partial t} =$$

$$\frac{1}{r}\frac{\partial}{\partial r}\left(D_{r3}\frac{\partial M_y}{\partial r}\right) + \frac{1}{r^2}\frac{\partial}{\partial \phi}\left(D_\phi \frac{\partial M_y}{\partial \phi}\right) + \frac{\partial}{\partial z}\left(D_z \frac{\partial M_y}{\partial z}\right) + \frac{F_o}{T_o}\gamma B_1(\vec{r},t) \quad (14a)$$

Where

$$D_{r3} = D_r r \quad (14b)$$

Making the following assumptions (Sprawls, 2000):

$$r = \exp\left(u_{03} R\right)$$
$$\phi = \exp\left(v_{03} \Phi\right) \quad (14c)$$
$$z = \exp\left(w_{03} Z\right)$$

We can write

$$\partial r = u_{03}\exp(u_{03}R)\partial R = u_{03}r\partial R$$
$$\partial \phi = v_{03}\exp(v_{03}\Phi)\partial \Phi = v_{03}\phi\partial \Phi \left.\right\} \quad (14d)$$
$$\partial z = w_{03}\exp(w_{03}Z)\partial Z = w_{03}\partial Z$$

$$\partial r^2 = u_{03}^2 r^2 \partial R^2$$
$$\partial \phi^2 = v_{03}^2 \phi^2 \partial \Phi^2$$
$$\partial z^2 = w_{03}^2 z^2 \partial Z^2$$

If we define

$$\left.\begin{array}{l} v_r = u_{03}r \\ v_\phi = v_{03}\phi r \\ v_z = w_{03}z \end{array}\right\} \quad (14e)$$

and

$$\left.\begin{array}{l} D_{r3} = D_0 u_{03}^2 r^3 \\ D_\phi = D_0 v_{03}^2 \phi^2 r^2 \\ D_z = D_0 w_{03}^2 z^2 \end{array}\right\} \quad (14f)$$

Where and u_0, v_0, w_0, are constants. Equation (14a) becomes

$$\frac{M_y}{u_{03}^2 r}u_{03}^2 r + \frac{u_{03}r}{u_{03}^2 r}\frac{\partial M_y}{\partial R} + \frac{M_y}{v_{03}\phi r}v_{03}^2\phi r + \frac{v_{03}\phi r}{v_{03}\phi r}\frac{\partial M_y}{\partial \phi} + \frac{M_y}{w_{03}z}w_{03}^2 z + \frac{w_{03}z}{w_{03}z}\frac{\partial M_y}{\partial z} + \frac{\partial M_y}{\partial t}$$
$$= \frac{1}{u_{03}^2 r^3}3D_0 u_{03}^3 r^3\frac{\partial M_y}{\partial R} + \frac{D_0 u_{03}^3 r^3}{u_{03}^2 r^3}\frac{\partial^2 M_y}{\partial R^2} + \frac{1}{v_{03}^2\phi^2 r^2}2D_0 v_{03}^3\phi^2 r^2\frac{\partial M_y}{\partial \Phi} + \quad (15)$$
$$\frac{D_0 v_{03}^3\phi^2 r^2}{v_{03}^2\phi^2 r^2}\frac{\partial^2 M_y}{\partial \Phi^2} + \frac{1}{w_{03}^2 z^2}2D_0 w_{03}^3 z^2\frac{\partial M_y}{\partial Z} + \frac{D_0 w_{03}^2 z^2}{w_{03}^2 z^2}\frac{\partial^2 M_y}{\partial Z^2} + \frac{F_o}{T_o}\gamma B_1(\vec{r},t)$$

Giving

$$(u_{03} + v_{03} + w_{03})M_y + \frac{\partial M_y}{\partial t} = D_0\left(\frac{\partial^2 M_y}{\partial R^2} + \frac{\partial^2 M_y}{\partial \Phi^2} + \frac{\partial^2 M_y}{\partial Z^2}\right) -$$
$$(1-3D_0 u_{03})\frac{\partial M_y}{\partial R} - (1-2D_0 v_{03})\frac{\partial M_y}{\partial \Phi} - (1-2D_0 w_{03})\frac{\partial M_y}{\partial Z} + \frac{F_o}{T_o}\gamma B_1(\vec{r},t)$$

The equation of motion for NMR signals for a flow process in a cylindrical geometry can then be written as;

$$\frac{\partial M_y}{\partial t} = D_0\left(\frac{\partial^2 M_y}{\partial R^2} + \frac{\partial^2 M_y}{\partial \Phi^2} + \frac{\partial^2 M_y}{\partial Z^2}\right) - (1-3D_0 u_{03})\frac{\partial M_y}{\partial R} - \quad (16)$$
$$(1-2D_0 v_{03})\frac{\partial M_y}{\partial \Phi} - (1-2D_0 w_{03})\frac{\partial M_y}{\partial Z}$$

Provided that:

$$\frac{F_o}{T_o}\gamma B_1(\vec{r},t) = (u_{03} + v_{03} + w_{03})M_y \quad (17)$$

We seek a solution to equation (16) for an instantaneous release in the form

$$M_y(R,\Phi,Z,t) = g_{31}(R,t;R_o)g_{32}(\Phi,t;\Phi_o)g_{33}(Z,t;Z_o) \quad (18)$$

Where, g_{31}, g_{32} and g_{33} (which are not tensors) are the solutions to the one-dimensional constant coefficient advective diffusion in the transformed space.

$$\left.\begin{array}{l} g_{31}(R,t;R_o) = \dfrac{A_{31}}{2u_{03}\sqrt{\pi D_o t}} \exp\left(\dfrac{-[(R-R_o)-(1-3D_o u_{03})t]^2}{4D_o t}\right) \\[2em] g_{32}(\Phi,t;\Phi_o) = \dfrac{A_{32}}{2v_{03}\sqrt{\pi D_o t}} \exp\left(\dfrac{-[(\Phi-\Phi_o)-(1-2D_o v_{03})t]^2}{4D_o t}\right) \\[2em] g_{33}(Z,t;Z_o) = -\dfrac{A_{33}}{2w_{03}\sqrt{\pi D_o t}} \exp\left(\dfrac{-[(Z-Z_o)-(1-2D_o w_{03})t]^2}{4D_o t}\right) \end{array}\right\} \quad (19)$$

For a source of unit magnetic moment, we obtain

$$\int_0^\infty \int_0^\infty \int_0^\infty M_y(r,\phi,z,t)\,dr\,d\phi\,dz = 1 \qquad \forall t$$

The NMR transverse magnetization obtained for the instantaneous release after a long computation is;

$$M_y(r,\phi,z,t) = \dfrac{1}{8u_{03}v_{03}w_{03}(\pi D_o t)^{3/2}} \dfrac{1}{r_o\phi z}\left(\dfrac{r}{r_o}\right)^{-3/2}\left(\dfrac{r}{r_o}\right)^{1/2D_o u_{03}}$$
$$\times \left(\dfrac{\phi}{\phi_o}\right)^{1/2D_o v_{03}}\left(\dfrac{z}{z_o}\right)^{1/2D_o w_{03}} \times \exp\left(\dfrac{-[P_{03}^2+(1-D_o u_{03})^2 t^2 + 2t^2]}{4D_o t}\right) \quad (20)$$

$$P_{03}^2 = \left(\dfrac{1}{u_{03}}\ln\left(\dfrac{r}{r_o}\right)\right)^2 + \left(\dfrac{1}{v_{03}}\ln\left(\dfrac{\phi}{\phi_o}\right)\right)^2 + \left(\dfrac{1}{w_{03}}\ln\left(\dfrac{z}{z_o}\right)\right)^2$$

For the case of a continuous release of an advected substance in cylindrical geometry, we shall integrate equation (20) with respect to time:

$$M_y = \dfrac{1}{8u_{03}v_{03}w_{03}}\dfrac{1}{r_o\phi z}\left(\dfrac{r}{r_o}\right)^{-3/2}\left(\dfrac{r}{r_o}\right)^{1/2D_o u_{03}}\left(\dfrac{\phi}{\phi_o}\right)^{1/2D_o v_{03}}\left(\dfrac{z}{z_o}\right)^{1/2D_o w_{03}}$$
$$\int_{-\infty}^{\infty}\dfrac{1}{(\pi D_o t)^{3/2}}\exp\left(\dfrac{-[P_{03}^2+(1-D_o u_{03})^2 t^2 + 2t^2]}{4D_o t}\right)dt \quad (21)$$

The value of the constant A_{BPES} is determined using the Boubaker Polynomials Expansion Scheme *BPES* (Awojoyogbe, 2008; Zhao et al., 2008; Belhadj et al., 2009; Chaouachi et al., 2007; Fridjine et al., 2009; Fridjine and Amlouk, 2009; Fridjine et al., 2009; Ghanouchi et al., 2008; Gherib et al., 2008; Guezmir et al., 2009; Labiadh and Boubaker, 2007; Slama and Bessrour, 2009; Slama et al., 2009; Tabatabei et al., 2009)). The calculation protocol takes into account conjointly the properties of the *BPES* along with the already noticed (Awojoyogbe, 2008; Oyodum et al., 2009) similarity between Equation (13) and the characteristic

differential equation of the Boubaker polynomials (Slama et al., 2008; Zhao et al., 2008).

THE MULTIDIMENSIONAL PERFUSION PROCESS – SPHERICAL GEOMETRY

Within the bifurcation itself, we shall approximate the region to some spherical region (the shape actually varies). In such a spherical geometrical structure, the diffusion-advection equation describing the spatially variable perfusion process is given by,

$$\dfrac{\partial}{\partial r}(v_r M_y) + \dfrac{1}{r}\dfrac{\partial}{\partial\theta}(v_\theta M_y) + \dfrac{1}{r\sin\theta}\dfrac{\partial}{\partial\phi}(v_\phi M_y) + \dfrac{\partial M_y}{\partial t} = \dfrac{1}{r^2}\dfrac{\partial}{\partial r}\left(D_{r4}\dfrac{\partial M_y}{\partial r}\right)$$
$$+ \dfrac{1}{r^2\sin\theta}\dfrac{\partial}{\partial\theta}\left(D_{\theta 4}\dfrac{\partial M_y}{\partial\theta}\right) + \dfrac{1}{r^2\sin^2\theta}\dfrac{\partial}{\partial\phi}\left(D_\phi\dfrac{\partial M_y}{\partial\phi}\right) + \dfrac{F_o}{T_o}\gamma B_1(\vec{r},t) \quad (22)$$

Where:

$$D_{r4} = D_r r^2$$
$$D_{\theta 4} = D_\theta \sin\theta$$

Making the following assumptions [29]:

$$r = \exp(u_{04}R)$$
$$\phi = \exp(v_{04}\Phi) \qquad (23)$$
$$\theta = \exp(w_{04}\Theta)$$

And

$$\left.\begin{array}{ll} v_r = u_{04}r & \text{and} \quad D_{r4} = D_0 u_{04}^2 r^4 \\[1em] v_\theta = w_{04}\theta r & \text{and} \quad D_\phi = D_0 v_{04}^2 \phi^2 r^2 \sin^2\theta \\[1em] v_\phi = v_{04}\phi r\sin\theta & \text{and} \quad D_{\theta 4} = D_0 w_{04}^2 \theta^2 r^2 \sin\theta \end{array}\right\} \quad (24)$$

Equation (18) can be written as

$$(u_{04}+v_{04}+w_{04})M_y + \dfrac{\partial M_y}{\partial t} = D_0\left(\dfrac{\partial^2 M_y}{\partial R^2} + \dfrac{\partial^2 M_y}{\partial\Phi^2} + \dfrac{\partial^2 M_y}{\partial\Theta^2}\right)$$
$$-(1-4D_0 u_{04})\dfrac{\partial M_y}{\partial R} - (1-2D_0 v_{04})\dfrac{\partial M_y}{\partial\Phi} - (1-2D_0 w_{04})\dfrac{\partial M_y}{\partial\Theta} + \dfrac{F_o}{T_o}\gamma B_1(\vec{r},t) \quad (25)$$

Provided that

$$\dfrac{F_o}{T_o}\gamma B_1(\vec{r},t) = (u_{04}+v_{04}+w_{04})M_y \qquad (26)$$

The equation of motion for NMR signals for a flow process within process within the bifurcation is given by:

$$\frac{\partial M_y}{\partial t} = D_0\left(\frac{\partial^2 M_y}{\partial R^2} + \frac{\partial^2 M_y}{\partial \Phi^2} + \frac{\partial^2 M_y}{\partial \Theta^2}\right) - \left(1 - 4D_0 u_{04}\right)\frac{\partial M_y}{\partial R}$$

$$- \left(1 - 2D_0 v_{04}\right)\frac{\partial M_y}{\partial \Phi} - \left(1 - 2D_0 w_{04}\right)\frac{\partial M_y}{\partial \Theta} \tag{27}$$

We seek solutions to the diffusion-advection equation for an instantaneous release in the form

$$M_y(R,\Phi,\Theta,t) = g_{41}(R,t;R_o)g_{42}(\Phi,t;\Phi_o)g_{43}(\Theta,t;\Theta_o) \tag{27a}$$

Where, g_{41}, g_{42} and g_{43} (which are not tensors) are the solutions to the one – dimensional constant coefficient advective diffusion in the transformed space

$$g_{41}(R,t;R_o) = \frac{A_{41}}{2u_{04}\sqrt{\pi D_o t}}\exp\left(\frac{-[(R-R_o)-(1-4D_{04})t]^2}{4D_o t}\right)$$

$$g_{42}(\Phi,t;\Phi_o) = \frac{A_{42}}{2v_{04}\sqrt{\pi D_o t}}\exp\left(\frac{-[(\Phi-\Phi_o)-(1-2D_0 v_{04})t]^2}{4D_o t}\right) \tag{28}$$

$$g_{43}(\Theta,t;\Theta_o) = \frac{A_{43}}{2w_{04}\sqrt{\pi D_o t}}\exp\left(\frac{-[(\Theta-\Theta_o)-(1-2D_o w_{04})t]^2}{4D_o t}\right)$$

For a source of unit magnetic moment,

$$\int_0^\infty\int_0^\infty\int_0^\infty M_y(r,\phi,\theta,t)\,dr\,d\phi\,d\theta = 1 \qquad \forall t$$

The coefficients A_{41}, A_{42} and A_{43} are constants. The integral gives

$$A_{41}\exp\left(u_{04}[R_o + (1-3D_o u_{04})t]\right)A_{42}\exp\left(v_{04}[\Phi_o + (1-D_o v_{04})t]\right)$$

$$A_{43}\exp\left(w_{04}[\Theta_o + (1-D_o w_{04})t]\right) = 1$$

$$A_{41}A_{42}A_{43}r_0\phi_0\theta_0\exp\left(u_{04}t-3D_o u_{04}^2 t\right)\exp\left(v_{04}t-D_o v_{04}^2 t\right)\exp\left(w_{04}t-D_o w_{04}^2 t\right) = 1$$

An obvious choice for A_{41}, A_{42} and A_{43} would be:

$$A_{41} = \frac{1}{r_0\exp\left(u_{04}t-3D_o u_{04}^2 t\right)} = \frac{\exp\left(3D_o u_{04}^2 t - u_{04}t\right)}{r_0},$$

$$A_{42} = \frac{1}{\phi_0\exp\left(v_{04}t-D_o v_{04}^2 t\right)} = \frac{\exp\left(D_o v_{04}^2 t - v_{04}t\right)}{\phi_0},$$

$$A_{43} = \frac{1}{\theta_0\exp\left(w_{04}t-D_o w_{04}^2 t\right)} = \frac{\exp\left(D_o w_{04}^2 t - w_{04}t\right)}{\theta_0}$$

$$M_y(r,\phi,\theta,t) = \frac{1}{8u_{04}v_{04}w_{04}(\pi D_o t)^{3/2}}\frac{1}{r_0\phi z}\left(\frac{r_0}{r}\right)^2\left(\frac{r}{r_0}\right)^{1/2D_o u_{04}}$$

$$\times\left(\frac{\phi}{\phi_0}\right)^{1/2D_o v_{04}}\left(\frac{\theta}{\theta_0}\right)^{1/2D_o w_{04}}\times\exp\left(\frac{-[P_{04}^2 + (1-2D_o u_{04})^2 t^2 + 2t^2]}{4D_o t}\right) \tag{29}$$

Where,

$$P_{04}^2 = \left(\frac{1}{u_{04}}\ln\left(\frac{r}{r_0}\right)\right)^2 + \left(\frac{1}{v_{04}}\ln\left(\frac{\phi}{\phi_0}\right)\right)^2 + \left(\frac{1}{w_{04}}\ln\left(\frac{\theta}{\theta_0}\right)\right)^2$$

For the case of a continuous release of an advected substance in spherical geometry, we shall integrate equation (29) with respect to time

$$M_y = \frac{1}{8u_{04}v_{04}w_{04}}\frac{1}{r_0\phi z}\left(\frac{r_0}{r}\right)^2\left(\frac{r}{r_0}\right)^{1/2D_o u_{04}}\left(\frac{\phi}{\phi_0}\right)^{1/2D_o v_{04}}\left(\frac{\theta}{\theta_0}\right)^{1/2D_o w_{04}}$$

$$\int_{-\infty}^{t_\infty}\frac{1}{(\pi D_o t)^{3/2}}\exp\left(\frac{-[P_{04}^2 + (1-2D_o u_{04})^2 t^2 + 2t^2]}{4D_o t}\right)dt \tag{30}$$

RESULTS AND DISCUSSION

Equations (12, 13, 20, 21, 29 and 30) are the NMR transverse magnetizations and signals for the instantaneous and continuous release of advected substances in Cartesian, cylindrical and spherical geometries respectively. These NMR signals are functions of diffusion coefficient D_0 and their respective perfusion functions P_{02}^2, P_{03}^2 and P_{04}^2. The diffusion coefficient is related to the net displacement of molecules in a given time. The average distance, s, traveled relative to diffusion coefficient is given as:

$$s = \sqrt{2D_o t} \tag{31}$$

Based on equations (13, 21, 30, and 31) and applying some standard integral formulae, the NMR transverse magnetizations and signals for the continuous release of advected substances in Cartesian, cylindrical and spherical geometries can be written as:

$$\frac{M_y}{M_y(0)} = \exp(-2\beta) \tag{32a}$$

Where for example in spherical geometry,

$$M_y(0) = \frac{1}{8u_{04}v_{04}w_{04}}\frac{1}{r_0\phi z}\left(\frac{r_0}{r}\right)^2\left(\frac{r}{r_0}\right)^{1/2D_o u_{04}}\left(\frac{\phi}{\phi_0}\right)^{1/2D_o v_{04}}$$

$$\left(\frac{\theta}{\theta_0}\right)^{1/2D_o w_{04}}\frac{1}{s^3}\sqrt{\frac{\pi}{\delta^2}} \tag{32b}$$

$$\delta^2 = \frac{(1-2D_o u_{04})^2 + 2}{s^2} \tag{32c}$$

$$\beta = \frac{P_{04}^2}{s^2} = qP_{04}^2, \quad q = \frac{1}{s^2} \tag{32d}$$

In equation (32), the reduction in NMR signal $\dfrac{M_y}{M_y(0)}$ produced by the diffusion-perfusion process depends on the rate of diffusion expressed by the value of the diffusion coefficient, D_0, the perfusion function in the particular geometry and the perfusion sensitivity, q, which is determined by the average distance, s, traveled by a molecule in time t. The distance, s, depends on the diffusion coefficient for the specific tissue compartment within a voxel. A series of experiment to measure the perfusion function can be performed in which values of s, may be varied by varying n or m using equations (3, 8a and 31).

From equations (12, 20 and 29), the NMR signal intensity for the instantaneous release of advected substances in for example spherical geometry can be written as:

$$M_y = M_y(0) \exp\left(-\frac{P_{04}^2}{s^2}\right) \times \exp\left(-\frac{s^2 \delta^2 t}{4D_o}\right) \tag{33}$$

By inspection of equation (33), it can be seen that the signal intensity is a product of signal attenuation due to perfusion and signal attenuation due to diffusion. Theoretically, a series of experiment can be performed in which either s, D_o or δ is varied by varying n or m using equations (3, 8a, 31, and 32c) while keeping t constant. The real experimental conditions under which the above description of equation (33) can be used to perform the diffusion and perfusion measurements will be considered in separate studies.

Conclusion

We have obtained basic analytical expressions for the transverse magnetizations (the NMR signals) for perfusion processes in different geometrical structures and biophysical conditions based on the Bloch NMR flow equations. These analytical results are quite interesting and promising in the context of some recent works on dynamical flows (Sprawls, 2000; Awojoyogbe et al., 2010; Hassell et al., 2008; Nicolis et al., 2002). The application of these fundamental results to solve real life flow problems in which NMR-sensitive materials are transported will be presented separately. It should be mentioned that acquisition of perfusion data requires fast imaging methods based on the appropriate choice of n or m in equation (3 and 8a) because images must be acquired every few seconds to properly measure the characteristics of the bolus passage. It should be noted that, in specific tissue, the diffusion rate might be different in different directions because of the orientation of certain

tissue structures. This is a very important factor which must be taken into account when producing diffusion images. Hence, the results of diffusion-advection equation with spatially varying diffusion coefficients as discussed in this study, which is based on the fundamental Bloch NMR flow equations, can be invaluable mathematical tools to accurately understand the combined effect of diffusion and perfusion process in human physiological and pathological flow systems. The method presented in this study can have applications in functional magnetic resonance imaging (fMRI) with more accurate information. How the NMR parameters derived in the present model are linked to a practical measurement in terms of an fMRI sequence will be developed separately.

ACKNOWLEDGEMENT

The authors appreciate encouragement from Professor S.O.E Sadiku, Academic planning, Federal University of Technology, Minna, through the STEP B programme.

REFERENCES

Awojoyogbe OB, Faromika OP, Folorunsho OM, Dada M, Fuwape IA, Boubaker K (2010). Mathematical model of the Bloch NMR flow equations for the analysis of fluid flow in restricted geometries using the Boubaker polynomials expansion scheme, Curr. Appl. Phys. 10: 289-293.

Awojoyogbe OB (2004). Analytical Solution of the Time Dependent Bloch NMR Equations: A Translational Mechanical Approach. Physica A 339:437-460.

Awojoyogbe OB (2003). A Mathematical Model of Bloch NMR Equations for Quantitative Analysis of Blood Flow in Blood Vessels with Changing Cross-section II. Physica A 323: 534-550.

Awojoyogbe OB (2002). A Mathematical Model of Bloch NMR Equations for Quantitative Analysis of Blood Flow in Blood Vessels with Changing Cross-section I. Physica A 303: 163-175.

Awojoyogbe OB (2007). A quantum mechanical model of the Bloch NMR flow equations for electron dynamics in fluids at the molecular level. Phys. Scr. 5: 788-794.

Awojoyogbe OB, Boubaker K (2008). A solution to Bloch NMR flow equations for the analysis of homodynamic functions of blood flow system using m-Boubaker polynomials, Curr. Appl. Phys. 9: 278-283.

Belhadj A, Onyango O, Rozibaeva N (2009). Boubaker Polynomials Expansion Scheme-Related Heat Transfer Investigation Inside Keyhole Model, J. Thermophys. Heat Transf. 23: 639-640.

Chaouachi A, Boubaker K, Amlouk M, Bouzouita H (2007). Enhancements of pyrolysis spray disposal performance using thermal time-response to precursor uniform deposition. Eur. Phys. J. Appl. Phys. 37:105-109.

Fridjine S, Amlouk M (2009). A new parameter: An ABACUS for optimizing functional materials using the Boubaker polynomials expansion scheme, Mod. Phys. Lett. B. 23: 2179-2182.

Fridjine S, Boubaker K, Amlouk M (2009). Some electron probe X-ray microanalysis (EPMA) and (BPES)-related physical investigations on ZnSSe thin-films growth composition-related kinetics, Canad. J. Phys. 87: 653-657.

Ghanouchi J, Labiadh H, Boubaker K (2008). An Attempt to solve the heat transfer equation in a model of pyrolysis spray using 4q-Order Boubaker polynomials, Int. J. Heat Technol. 26: 49-53.

Ghrib T, Boubaker K, Bouhafs M (2008). Investigation of thermal diffusivity-micro hardness correlation extended to surface-nitrured steel using Boubaker polynomials expansion, Mod. Phys. Lett. B. 22: 2893- 2907.

Guezmir N, Ben Nasrallah T, Boubaker K, Amlouk M, Belgacem S (2009). Optical modelling of compound CuInS2 using relative dielectric function approach and Boubaker polynomials expansion scheme BPES, J. All. Comp. 481: 543-548.

Hassell DG, Mackley MR, Sahin MHJ, Harlen OG, McLeish TCB (2008). Molecular physics of a polymer engineering instability: Experiments and computation, Phys. Rev. E 77 050801.

Labiadh H, Boubaker K (2007). A Sturm-Liouville shaped characteristic differential equation as a guide to establish a quasi-polynomial expression to the Boubaker polynomials, J. Differ. Equ. Control Processes 2: 117-133.

Martinez GV, Dykstra EM, Lope-Piedrafita S, Job C, Brown MF (2002). NMR Elastometry of Fluid Membranes in the Mesoscopic Regime, Phys. Rev. E 66 050902/1 - 050902/4.

Nicolis G, Balakrishnan V, Nicolis C (2002). Moment evolution and level-crossing statistics in dichotomous and multilevel flows with time-dependent control parameters, Phys. Rev. E 65: 051109.

Oyodum OD, Awojoyogbe OB, Dada M, Magnuson J (2009). On the earliest definition of the Boubaker polynomials, Eur. Phys. J. App. Phys. 46: 21201-21203.

Segnorile HH, Barberis L, Gonzalez CE, Zamar RC (2006). Proton NMR relaxation of the dipolar quasi-invariants of nematic methyl deuterated para-azoxyanisole within the high-temperature Redfield relaxation theory, Phys. Rev. E 74, 051702.

Slama S, Bessrour J, Bouhafs M, Ben Mahmoud KB (2009). Numerical Distribution of Temperature as a Guide to Investigation of Melting Point Maximal Front Spatial Evolution During Resistance Spot Welding Using Boubaker Polynomials, Num. Heat Transf. Part A 55: 401-408.

Slama S, Boubaker K, Bessrour J, Bouhafs M (2009). Study of temperature 3D profile during weld heating phase using Boubaker polynomials expansion. Thermochimica acta. 482: 8-11.

Slama S, Bouhafs M, Ben Mahmoud KB, Boubaker A (2008). Polynomials Solution to Heat Equation for Monitoring A3 Point Evolution During Resistance Spot Welding, Int. J. Heat Technol. 26: 141-146.

Sprawls P (2000). Magnetic Resonance Imaging: Principle, Methods, and Techniques. Medical Physics Publishing: Madison, Wisconsin pp. 137-144.

Tabatabaei S, Zhao T, Awojoyogbe O, Moses F (2009). Cut-off cooling velocity profiling inside a keyhole model using the Boubaker polynomials expansion scheme, Heat Mass Transf. 45: 1247-1251.

Valfouskaya A, Adler PM (2005). Nuclear-magnetic-resonance diffusion simulations in two phases in porous media, Phys. Rev. E 72, 056317.

Zhao TG, Wang YX, Ben Mahmoud KB (2008). Limit and uniqueness of the Boubaker-Zhao polynomials imaginary root sequence, Int. J. Math. Comp. 1: 13-16.

Zoppou C, Knight JH (1997). Analytical solution of the spatially variable co-efficient advective-diffusion equation in one, two and three dimensions. Mathematics Research Report. Mathematical Sciences Institute; the Australian National University. MRR 56-97.

Zoppou C, Knight JH (1997). Analytical Solutions for Advection and Advection Diffusion Equations with spatially variable Coefficients. J. Hydraulic Eng. 123 (2): 144 -14.

Comparative study of inhibition of drug potencies of c-Abl human kinase inhibitors: A computational and molecular docking study

D. Kshatresh Dubey*, K. Amit Chaubey, Azra Parveen and P. Rajendra Ojha

Biophysics Unit, Department of Physics, DDU Gorakhpur University, Gorakhpur 273009, India.

Structural studies suggest that the c-Abl protein kinase domain exists in two conformations; an active and an inactive form. There are many inhibitors which bind this tyrosine kinase in both forms. Many of these kinase inhibitors are in clinical trials too. The inhibition potency of these inhibitors is a common topic of discussion. In the present study we have taken a library of eight different inhibitors and docked those using GLIDE. After GLIDE docking we have also calculated induced fit results. The validity of the docking scores was compared to the post-docking score calculated by the Molecular mechanics - gernalised Boltzman/ surface area (MM-GB/SA) approach. During this process, Imatinib and Nilotinib showed very similar scores and binding energy. A comparative study of all eight inhibitors suggest that Imatinib and Nilotinib have the best binding scores and hence, they can be considered as the best drugs relative to PHA, VX6, PD3, PD5, P17 and Dasatinib. Our findings provide further rationale for considering kinase conformation in the design of kinase inhibitors.

Key words: Docking, scoring, binding affinity.

INTRODUCTION

The development of protein crystallography over a last few decades has revolutionized advances in drug discovery. This technique enables us to visualize the precise positions of the individual atoms of a molecule which causes disease, thus revealing the mechanism of disease, which then assists us in drug discovery (Blundell et al., 2002). Comparative crystallographic studies of protein kinases have uncovered the secrets of irregularities in cell growth. A protein kinase is an enzyme that modifies other proteins by chemically adding phosphate groups to them. This phosphorylation usually results in a functional change of the target protein by changing enzyme activity (Goodshell, 2005). Thus, kinases cause changes in a cell's metabolism. Hence, deregulated protein kinase activities, occurring via mutation, can cause diseases like cancer due to uncontrolled changes in a cell. A protein kinase inhibitor is an enzyme that blocks the activity of these kinases by fitting into a binding site. In theory these inhibitors can be

used to treat the irregularities and uncontrolled changes in cellular function that occur. During the last few decades there have been many groups who have investigated the conformation and molecular structure of many kinases. These studies reveal that among more than 500 different protein kinases, c-Abl kinase is a type of kinase which transmits messages to the cell for adding to their neighboring cells, to grow or to move to a new location. Since this kinase plays an important role in the structure and the function of organisms, its signaling should be carefully regulated. If it is not, then the total balance of the cell is destroyed resulting in diseases like chronic myelogenous leukemia (CML) (Deininger et al., 2000).

Tyrosine kinases are structurally separated into two domains, an N-terminal domain and a C-terminal domain. The N-terminal domain is made of largely of β sheets while the C-terminal domain is largely comprised of α helices. The N-terminal domain also contains an α helix, which is often known as the αC helix. The binding site for ATP is situated between these two domains. Between these two domains there is an activation loop (245 - 254) which contains a tyrosine residue, which after

---

*Corresponding author. E-mail: kshatresh@gmail.com.

phosphorylation, electrostatically interacts with its neighboring arginine residue producing the open conformation of the kinase. This open loop conformation is suitable for peptide substrates binding. The α helical loop also plays an important role in the kinase's mechanism. This helix presents a glutamate residue that forms a salt bridge with a lysine residue from the N lobes and this pairing coordinates the phosphate groups of ATP. In the inactive form of the kinase, it projects outward preventing the formation of the salt bridge (Sicheri et al., 1997). The N lobes of the kinase also contain a flexible glycine residue, which plays an important role in kinase activity (Nagar, 2007).

These regulatory elements imply that disruption of any one of them can result in changes in the activity of this protein kinase. This presents an important concept for designing inhibitors which bind to the ATP cleft (Al-Obeidi et al., 2000; Atwal et al., 2004). Fortunately many inhibitors have been developed as targets of this cleft (Zhang et al 2009). Imatinib is a type of ATP competitive inhibitor, which binds to the cleft between the N- and C-terminal domains of the kinase. It has been approved for use in patients affected with CML. The crystallographic structure of c-Abl protein kinase bound to Imatinib (Cowan-Jacob et al., 2007) revealed that after binding of Imatinib, the activation loop is folded inward (Nagar et al., 2003; Nagar, 2007). This inverted conformation prevents the aspartate residue from ligating with magnesium. Due to its strong binding behavior, Imatinib has been used in clinical trials. Besides Imatinib, Nilotinib is also a highly potent, very selective and active inhibitor against these types of kinases (Rix et al., 2007; Redaelli et al., 2008). Due to its increased inhibition, it was also used in clinical trials. A related inhibitor, Dasatinib, has also been studied; it has been found to be more effective for the Src/Abl kinase inhibition with potent anti proliferative activity against hematological malignancies harboring activated BCR-ABL (Tokaraski et al., 2006). A study shows that Dasatinib blocks the migration and the invasion of human cells without affecting proliferation and survival (Buettener et al., 2008). Another study shows that PHA 739358 also plays an important role as an Abl kinase inhibitor and binds the pan aurora as well as Abl kinase (Carpinelli et al., 2007). The role of the aurora kinase inhibitor, VX680 (Weisberg et al., 2006), as well its complex structure, have also been studied. It has been identified as a potent and selective small molecule inhibitor, which suppresses tumor growth (Young et al., 2006). By applying similar strategies, other chemo type inhibitors have been developed and their crystallographic studies have been done. These inhibitors have been predicted for therapeutic use by binding to the tyrosine kinases. The chemical structures of these inhibitors are shown in Figure 1.

In the present work we have investigated the c-Abl human kinase receptor from the 2hyy complex (Cowan-Jacob et al., 2007) and the above inhibitors, that is, Imatinib, Nilotinib, Dasatinib, PHA, PD3, PD5, P17, VX680

from different complexes. These drugs have already predicted for kinase inhibition. Computational docking of the inhibitors into kinases gives an accurate insight of the behavior of this molecule. We have calculated docking scores for each drug and each target complex separately, which theoretically explains binding of these inhibitors with the c-Abl human tyrosine kinases.

METHODOLOGY

Simulation was performed using SCHRODINGER Inc software package (Schrodinger Inc, New York). The initial structure was taken from the protein data bank (pdb id 2HYY, www.pdb.org), which was used as the starting coordinates for the simulation studies.

Structure preparation

The coordinates for all proteins were obtained from RCSB protein data bank (PDB code 2hyy). The imported structure was a tetramer in which all the units contain same binding site and ligand, so we have considered only one unit of the complex with a single ligand for our studies. The complex was prepared by protein preparation wizard software, where hydrogens were added automatically and refinement of the structure was also done. Since water molecules in the crystallographic complex were not critical to the functioning of the protein-ligand interaction, all of the crystallographic waters were deleted and the bond orders were re-assigned. Then the systems were minimized to a RMSD (Root Mean Square Deviation) 0.30 angstrom. The complex structure contains some missing residues, which were added through the prime application of the Schrödinger Suite.

Ligand preparation

Ligands were obtained from complex structures and we prepared them using the Ligprep module of the Maestro software. The crystallographic ligands did not have correct bond orders so we modified the bond orders manually according to their pdb data. Each ligand was subjected to a full minimization in the gas phase with the OPLS (Optimized Potential for liquid simulations) force field (Jorgenson et al., 1996) to eliminate the bond length and bond angles biased from the crystal structure. Under any physiological condition, a molecule can exist in a variety of protonation (ionization) states, so here we have produced multiple structures for each ligand with different combinations of ionized states based on the ionizable group present. This operation was done by help of ionizer in Ligprep. Since most proteins function in the pH range of 5 - 9, we have restricted the ionization to this range.

Since the crystal structure contains only one ligand structure but there is a chance that one of the tautomeric forms interacts more strongly with the binding site relative to the other forms, we have taken into account this concept and generated all of the other possible tautomeric states of one inhibitor.

Ligand docking and scoring

Prepared ligand and receptor were used as the initial coordinates for docking purposes. We have used c-Abl tyrosine kinase as the target receptor. The principle ligand can be docked by two methods: (1) Assuming that the ligand is flexible and the receptor is rigid and (2) Assuming that the ligand is rigid and the receptor is flexible. So here we have used both strategies of ligand docking. In

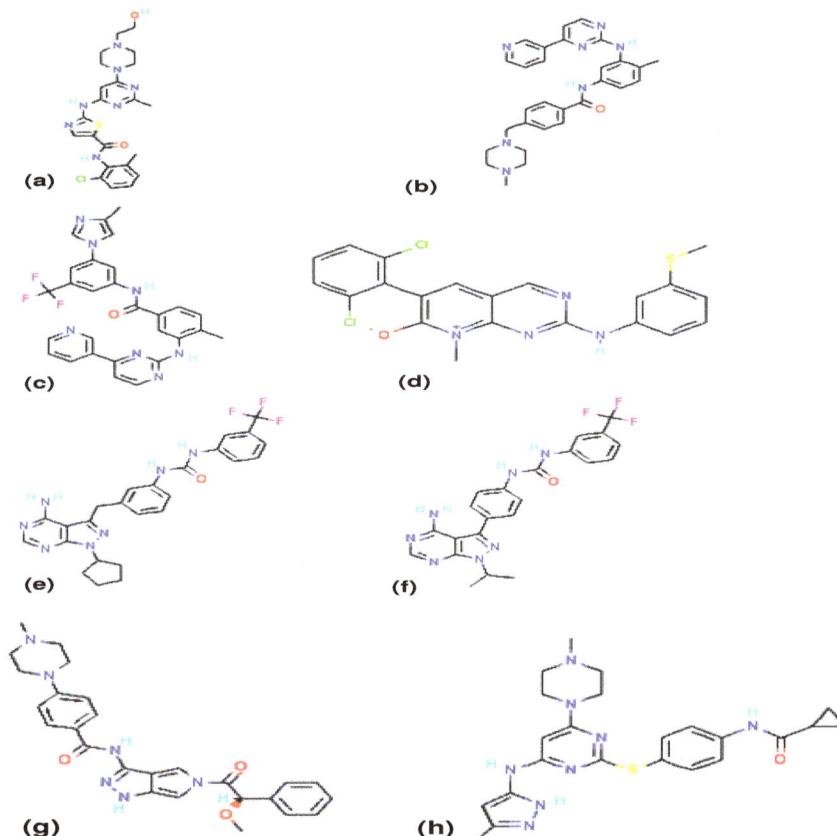

Figure 1. Chemical structure of different kinase inhibitors in their minimized positions. Non-carbon hydrogens are colored in green. The structures were generated using Chimera (a) Dasatinib, (b) Imatinib, (c) Nilotinib, (d) P17, (e) PD3, (f) PD5, (g) PHA and (h) VX-6.

both processes we have used GLIDE for docking (Friesner et al., 2004; Halgren et al., 2004).

The first stage for ligand docking was the receptor grid generation; for that purpose we have used the kinase protein structure complexed with Imatinib. During the grid generation, no Vander Waal radius sampling was done and the partial charge cutoff (Cho et al., 2007) has been taken as 0.25 and no constraints were applied (Sherman et al., 2006). The location of Imatinib was taken as binding site for docking of all of the other ligands. The ligand docking calculations were done in the standard precision mode of GLIDE. During the docking process, the receptor was treated as fixed while ligand was flexible. In the minimization of ligands, we have used a distance-dependent dielectric constant with a value of 2.0 and a conjugate gradient algorithm with small 100 steps. All of the inhibitors were passed through a scaling factor of 0.80 and partial charge cutoff of 0.15. After docking, we have performed post-docking minimization to improve the geometry of the poses. The post-docking minimization specifies a full force-field minimization of those poses which are considered for the final scoring. After docking, the results were used for binding energy calculations and docking scores (Bissantz et al., 2000). The MM-GBSA approach was used to predict the free energy of binding for a receptor and a set of ligands (Nu et al., 2006; Gohlke et al., 2000, 2002).

For the second process, we have assumed that the ligand was fixed while the receptor was flexible (that is, we have used an induced fit docking protocol in the first stage of this protocol). Softened potential docking was performed to generate 20 initial poses. The softened potential docking consisted of scaling the Vander Wall radii by a factor of 0.5; this process was done to give an extra cavity to ligand to be fit in the binding site (Moitessier et al., 2006). All of the docking calculations were performed using the standard Precision mode of Glide. For each of 20 poses from the initial softened potential docking step, a full cycle of protein refinement was performed. For refinement we have also used the OPLS_2005 parameter set and the surface Generalized Born implicit solvent model (Zhang et al., 2000). After convergence to a low energy solution, an additional minimization was performed to make all of the residues in the backbone and the side chains to be relaxed. The sum of molecular mechanics and solvation energy was calculated and those complexes having energy within the range of 30 kcal/ mol, were accepted for docking score.

RESULTS AND DISCUSSION

Virtual docking

We have applied the GLIDE docking method to seven inhibitors of tyrosine protein kinases to build a binding affinity model for the c-Abl human tyrosine kinase receptor that was then used to compute the free energy of binding for this kinase.

The ligand preparation procedure generated different

Table 1. Average Vander waals (vdw), electrostatic (coul) and Site energy (site) after GLIDE docking[a].

Inhibitors	E_{vdw}	E_{coul}	E_{site}	Docking score	Energy
Imatinib 1	-64.432	-11.216	-0.059	-11.953	-75.649
Imatinib 2	-64.165	-08.431	-0.063	-10.662	-72.596
Imatinib 3	-62.186	-16.276	-0.117	-12.302	-78.462
Imatinib 4	-61.354	-13.318	-0.099	-10.788	-74.673
Imatinib 5	-62.890	-18.320	-0.167	-13.136	-81.210
Imatinib 6	-60.489	-15.283	-0.163	-11.484	-75.773
Imatinib 7	**-65.248**	**-21.593**	**-0.153**	**-13.400**	**-86.841**
Imatinib 8	-59.592	-20.087	-0.244	-11.587	-79.680
Nilotinib 1	-67.135	-11.497	-0.121	-12.825	-78.633
Nilotinib 2	**-66.141**	**-13.755**	**-0.136**	**-12.914**	**-79.897**
Nilotinib 3	-65.086	-08.909	-0.108	-11.171	-73.995
Nilotinib 4	-64.384	-11.402	-0.133	-11.477	-75.786
Dasatinib	-57.271	-15.059	-0.174	-09.910	-72.276
PHA 1	-39.277	-07.407	-0.172	-04.622	-46.685
PHA 2	-46.271	-03.350	-0.032	-04.790	-49.622
PHA 3	-42.689	-12.486	-0.306	-05. 311	-59.195
PHA 4	-50.365	-08.830	-0.041	-07.629	-55.175
VX6 1	-36.666	-02.802	0.000	-05.391	-39.468
VX6 2	-34.512	-12.396	-0.302	-04.384	-46.908
PD17	-51.248	-04.064	0.000	-10.533	-55.314
PD3	-49.847	-06.670	-0.234	-06.780	-56.517
PD5	-49.545	-02.818	-0.040	-07.498	-52.363

[a] The energy data are written according to the structure generated by the docking program. The best docked poses are written in bold letter. All energies are given in Kcal/mol.

training sets of the inhibitors whose scoring function is given in Table1. These different structures were found by using different orientations of the inhibitors and different positions of the hydrogens. The results of docking and scoring are given in Table 1. According to the table we see that among all the energy parameters the largest contribution for binding energy comes from Vander Waals interactions. The cavity energy term is very small, which indicates that there is a very low energy penalty when the ligand is buried in the cavity. The low rms deviation (0.30 Angstrom) indicates that the receptor has approximately the same structure as it does in the crystal structure. We then calculated docking scores for all other inhibitors. It is clear that among all of the 26 generated structures of the inhibitors, Imatinib and Nilotinib have very good docking scores. During the ligand preparation we have got eight different training sets of Imatinib. All of the training sets of Imatinib have different docking energies as well as docking scores. This observation shows that the inhibition of any ligand depends not only on the cavity site but also on the ionization and the tautomeric state of the inhibitors. Among all the eight training sets, Imatinib 7 had the best docking score, and the binding energy of this training set was also maximum. In best docked position, Imatinib forms five hydrogen bonds with THR-315, GLU-286, ASP-381, and ILE- 360 and MET-318

residues of the c-Abl kinase receptor (Figure 2a). The hydrogen binding with the ASP-381 plays an important role in kinase inhibition because the aspartate residue is the main constituent of the DFG loop of the tyrosine kinase receptor (Nagar, 2007). In the active kinase, the conformation of the aspartate residue of the DFG motif is oriented towards the bound ATP and is capable of ligating a critical magnesium ion bound to the phosphate group of ATP. After docking, Imatinib forms hydrogen bonds with the aspartate of the DFG loop, which negates its ability to bind to ATP. Thus, the activity of kinase may be reduced. Hence the mechanism of Imatinib inhibition has been clarified by computational docking. Nilotinib also has a very good docking score, which is slightly less than Imatinib. The docked structure shows that Nilotinib has approximately the same bound structure with the receptor (Figure 2b). Hence, we may conclude that the mechanism of inhibition of Nilotinib and Imatinib are similar. Nilotinib exhibits four different structures based on its different ionization and tautomeric states. When we compare the results of Table 1, it is obvious that the Vander Waals energy contribution for Nilotinib and Imatinib are approximately the same but there is a considerable difference in their columbic interactions.

This difference occurs due to the fluorine atoms present in the Nilotinib structure.

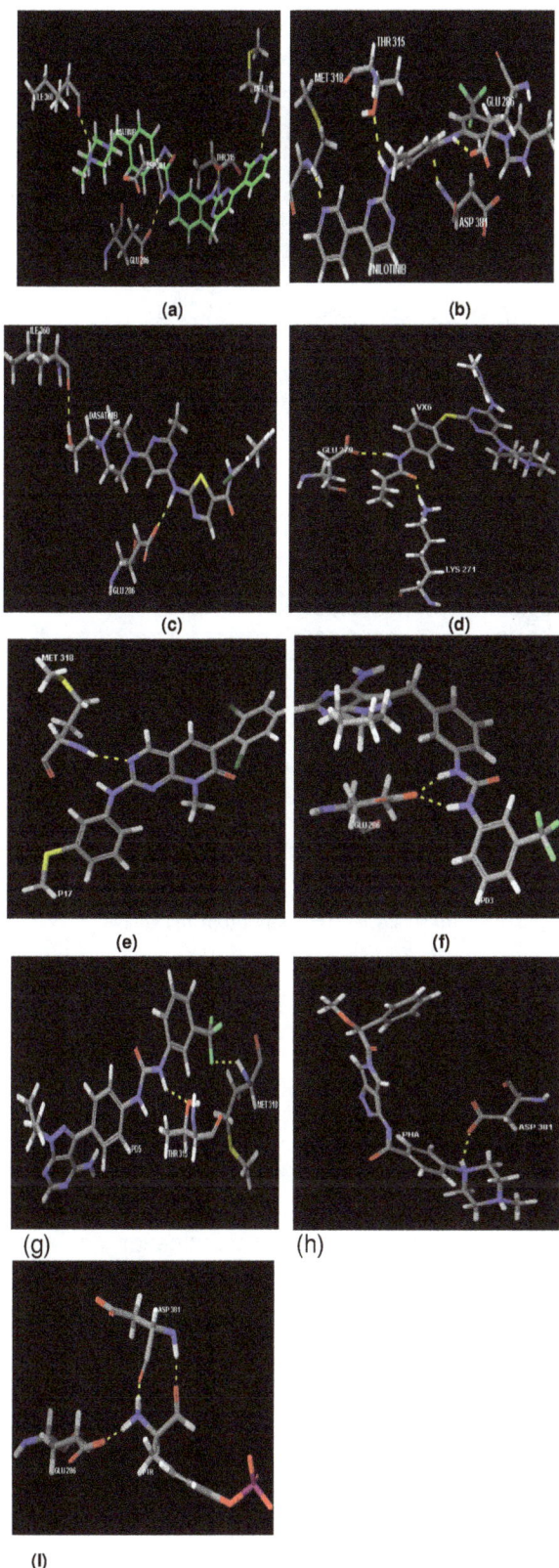

(a)

(b)

(c)

(d)

(e)

(f)

(g)

(h)

(I)

Figure 2. Structures of the different kinase inhibitors bound to the protein 2hyy. Only residues that undergo significant movement or are hydrogen bonded to the ligand are shown (a) binding of Imatinib, (b) binding of Nilotinib, (c) binding of Dasatinib, (d) binding of VX6, (e) binding of P17, (f) binding of PD3, (g) binding of PD5, and (h) binding of PHA.

Table 1 shows that, besides well known inhibitors like Imatinib and Nilotinib, there are also inhibition possibilities with Dasatinib, PHA, VX6, and PD5 and PD3. Interactions of these inhibitors with the c-Abl kinase receptor are shown in Figure 2c - I. The inhibition potential of these kinase inhibitors is in same sequence as written above. These inhibitors have significant docking scores and binding energies by MM-GB/SA approach. The binding energies of these inhibitors are shown in Table 2.

Induced fit results

In virtual docking (Taylor et al., 2002), the ligands are docked into binding site of the receptor where the receptor is held rigid and the ligand is free to move. But the c-Abl kinase receptor shows a critical hinging and on-off state, hence it may undergo side chain or backbone movement. These changes may allow alterations in the receptor so that it more closely conforms to the shape and binding modes of the ligands. Thus, we have also taken into account receptor flexibility (Broughton et al., 2000) by induced fit docking. In induced fit docking, we obtained 18 different poses for Imatinib, 19 different poses for Nilotinib, 19 for Dasatinib, 15 poses for VX680, 10 poses for PD3, 14 poses for PD5 and 14 different poses for P17. The results of the induced fit docking are given in Table 3, which displays the best docked poses. From the results of the induced fit docking, it is clear that there are some considerable changes in the docking scores and energies of the docked complexes. By comparing the results of flexible receptor docking and rigid receptor docking, we see that there is no noticeable change for Imatinib. In each case the docking score is approximately same. Structure analysis also shows no change in hydrogen bonding. The similarity in the results implies that Imatinib binds to receptor in an ideal manner and after the docking, it does not allow further flexibility in the binding site. But for other inhibitors, like Nilotinib, Dasatinib, PHA, PD3, PD5, P17, and VX6, there are noticeable changes in docking scores as well as in hydrogen bonding. These differences in the results may also be explained due to the strategy adopted in induced fit methods. The strategy in the induced fit method is to first dock the ligands into a rigid receptor using a softened energy function such that steric clashes, do not prevent at least one ligand pose from assuming a conformation close to the correct one. Then there is a sampling of the receptor degree of freedom and a minimization of the receptor-inhibitor complex for many different receptor poses and it is attempted to identify low free energy conformation of the each complex. A second round of the ligand docking is then performed on the refined protein structure, this time using a hard potential function to further sample ligand conformational space within the refined protein environment (Sherman et al., 2006).

Table 2. Binding energy[a] data for different inhibitors[b].

Ligand	dG(1)c	dG(2)	G_comp	G_lig	G_rec	Lstrain
VX 6 1	-05.07	-15.02	86218.83	-161.89	86385.80	09.95
VX6 2	-19.22	-22.75	86205.09	-161.49	86385.80	03.53
PTR2	71.49	61.03	86141.17	-316.12	86385.80	10.46
PTR1	76.23	73.83	86146.49	-315.54	86385.80	02.41
PHA3	-30.39	-38.60	86377.54	022.14	86385.80	08.20
PHA4	-33.22	-39.49	86374.10	021.53	86385.80	06.27
PHA2	-31.74	-38.22	86424.70	070.64	86385.80	06.48
PHA1	29.58	-37.23	86427.09	070.87	86385.80	07.65
PD5	-36.37	-41.49	86249.33	-100.10	86385.80	05.12
PD3	-34.96	-46.03	86270.14	-080.70	86385.80	11.07
Nilotinib2	-63.05	-73.84	86157.15	-165.60	86385.80	10.79
Nilotinib1	-67.79	-73.71	86178.41	-139.59	86385.80	05.91
Nilotinib4	-48.80	-60.16	86155.83	-181.16	86385.80	11.35
Nilotinib3	-54.13	-59.98	86176.73	-155.53	86385.80	05.84
Imatinib7	-67.37	-76.03	86134.53	-183.90	86385.80	08.66
Imatinib5	-70.65	-82.34	86183.64	-131.50	86385.80	11.69
Imatinib3	-73.02	-84.28	86186.79	-125.99	86385.80	11.27
Imatinib1	-46.62	-57.37	86262.92	-076.26	86385.80	10.75
Imatinib8	-47.96	-60.06	86138.57	-199.26	86385.80	12.10
Imatinib6	-55.91	-67.80	86182.36	-147.52	86385.80	11.89
Imatinib4	-57.00	-69.03	86186.68	-142.12	86385.80	12.03
Imatinib2	-31.52	-42.47	86261.67	-092.61	86385.80	10.96
Dasatinib	-51.83	-34.72	86158.09	-175.87	86385.80	10.

[a] All energy values are given in the Kcal/mol. [b] Binding energy data are not arranged in the best posed position. [c] dG1 and dG2 are binding energy in different condition of complex, which are calculated by using following formula. dG(1) = E_complex(minimized) - (E_ligand(minimized) + E_receptor). dG(2) = E_complex(minimized) - (E_ligand(from minimized complex) + E_receptor).

Table 3. Results of the induced fit docking[a].

Inhibitors	Docking energy	Docking score
Imatinib	-86.073	-13.313
Nilotinib	-86.630	-14.178
Dasatinib	-68.274	-11.220
PD5	-64.244	-09.951
PD3	-69.341	-10.866
VX6	-77.340	-13.066
PHA	-46.843	-05.311
P17	-65.392	-12.238
PTR	-44.743	-07.912

[a] docking energy is written only for the best docked poses. All energy values are given in Kcal/mol.

Finally, a composite scoring function is applied to rank the complexes, accounting for the receptor ligand interaction energy as well as strain and solvation energy. The validity of the induced fit method is already proved for flexible proteins as well as where little or no conformational changes in the receptor are required to

properly dock the ligand.

Binding energy calculations

In the induced fit calculations, improvement in the ranking of the known ligand and better discrimination among the rest of compounds in the database was achieved by taking into account the ligand – receptor solvation energy (Jacobson et al., 2002, 2004). Hence in this methodology, the MM-GB/SA calculations have been performed also to enhance the docking scoring (Zhang et al., 2001). The MM-GB/SA results for docking all of the ligands are given in Table 2. Table 2 shows that the ligand strain energy is different for different inhibitors. The ligand strain energy is the difference of (1) the binding energy dG1 calculated from the difference of the energy of minimized complex and sum of the energies of the minimized ligand and receptor and (2) the binding energy dG2 calculated as difference of energy of the minimized complex to the sum of the energy of the ligand from minimized structure and energy of the receptor (Given as formula in the legend of Table 2).

Activity of inhibitors relative to Imatinib

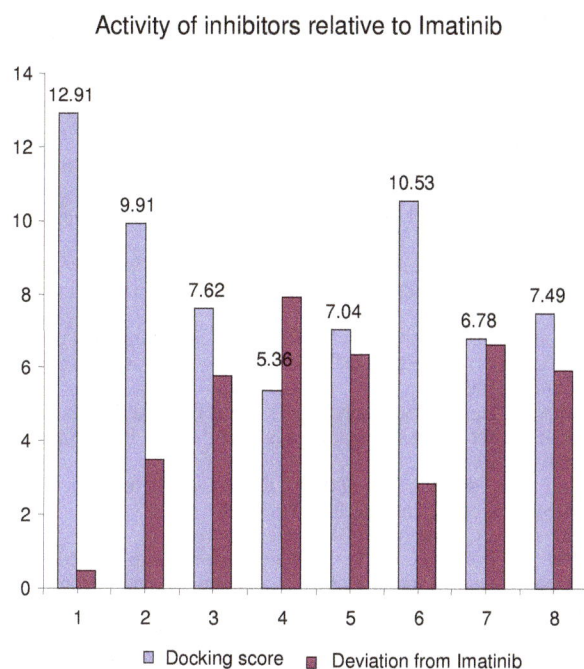

Figure 3. Comparative scoring of different inhibitors with respect to Imatinib. Here Imatinib is taken as the standard unit of kinase inhibition and the other ligands are compared to it. 1- Nilotinib, 2-dasatinib, 3-PHA, 4,-VX6, 5-PD17, 6- PD3, 7-PD5.

Comparing the results of dG1 and dG2, we see very considerable differences between these two free energies. In the first case when we minimized the ligand in the absence of receptor, the ligand adjusts its conformation independently. It does not show the exact position of ligand in the complex. But the minimization of the ligand into the complex represents ligand stabilization with respect to the receptor; hence there is a noticeable difference between the free energies in both cases. It is also obvious from the Table 2 that the free energy in the second case also gives better results than the first case. This shows that ligands in second case, bind the receptor more accurately than the first case. Hence the second case represents the better position of the ligand.

The first three training sets of Imatinib exhibit high free energy, whereas, the first two ligands of Nilotinib have more reliable free energies. Comparing these results with Table 1 we see that inhibitors which have good docking scores also have reliable free energies. Table 1 show that Imatinib, Nilotinib computationally behave as the best inhibitors because they have good docking scores and binding energies (Sotriffer et al., 2002; Taylor et al., 2002). Figure 3 shows a comparative graph of the other inhibitors with respect to Imatinib. It is clear from the graph that Nilotinib and Dasatinib are the best inhibitors compared to Imatinib. Also, comparing the results of Tables 1 and 2, it clear that by GLIDE docking methods we have approximately same free energy results for

Nilotinib and Imatinib.

Conclusion

The glide score can be used as a semi-quantitative descriptor for the ability of ligands to bind to a specific conformation of the protein receptor. Generally speaking for low glide score good ligand affinity to the receptor may be expected. According to the glide score and MM-GB/SA binding energy, the results of the inhibition for the c-Abl human tyrosine kinase receptor may be arranged in the following manner: Imatinib> Nilotinib > P17 > Dasatinib > PHA > PD5 >PD3> > VX6. Conformational analyses of different docked complexes also show that residues ASP-381, THR-315, GLU-286, MET-318, ILE-360, LYS-281and GLU-279 play important role in this kinase's activity.

Docking studies performed by GLIDE has confirmed that above inhibitors fit into the binding pocket of the c-Abl kinase receptor. From the results we may observe that for successful docking, intermolecular hydrogen bonding and liphophilic interactions between the ligand and the receptor are very important. We can also explain why the glide score and MM-GB/SA results for Imatinib are higher than others. The main reason for this are penalties for close intra-ligand contacts. Using the results, we have described a structural-based model of inhibition of c-Abl tyrosine kinase. Docking results show that Imatinib, Nilotinib, Dasatinib, PHA, PD5, PD3, PD17, VX6 inhibitors may penetrate deeply into binding site. The results also suggest that Imatinib and Nilotinib bind with the tyrosine residue of the DFG loop and compete with ATP binding.

A comparison of the induced fit and virtual docking gives the role of protein flexibility. It is obvious from the results that a combined method of soft docking and side chain optimization gives better results. It is also clear that an average distribution of docking free energy ranging from 2 kcal/mol or more, is sufficient to mis-rank a potential drug candidate as a weak binder. However, by combining the MM-GB/SA and relaxed complex methods we are able to show the best ranked binding modes.

ACKNOWLEDGMENTS

The authors thank the Department of Physics, of DDU Gorakhpur University for providing the computational facilities. The authors also thank DST, for issuing the grant for purchasing the computational equipment.

REFERENCES

Al-Obeidi FA, Lam KS (2000). Development of inhibitors for protein tyrosine kinases. Oncogene.19: 5690-5701.
Atwell S, Adams JM, Badger J, Buchanan MD, Feil IK, Froning KJ, Gao X, Hendle J, Keegan K, Leon BC, Muller-Deickmann HJ, Nienaber

VL, Noland BW, Post K, Rajashankar KR, Ramos A, Russell M, Burley SK, Buchanan SG (2004). A novel mode of Gleevec binding is revealed by the structure of spleen tyrosine kinase. J .Biol.Chem., 279: 55827-55832.

Bissantz C, Folkers G, Rognan D (2000). Protein based virtual screening of chemical database. 1. Evaluation of different docking /scoring combinations, J. Med. Chem., 43: 4759-4767.

Broughton HB (2000). A Methods for including protein flexibility in protein-ligand docking: Improving tools for database mining and virtual screening. J. Mol. Graphics Modell., 18: 147-257.

Buettner R, Mesa T, Vulture A, Lee F, Jove R (2008). Inhibition of Src family kinases with dasatinib blocks migration and invasion of human melanoma cells. Mol.Cancer Res., 6: 1766-1774.

Carpinelli P, Ceruti R, Giorgini ML, Cappella P, Gianellini L, Croci V, Degrassi A, Texido G, Rocchetti M, Vianello P, Rusconi L, Storici P, Zugnoni P, Arrigoni C, Soncini C, Alli C, Patton V, Marsiglio A, Ballinari D, Pesenti E, Fancelli D, Moll J (2007). PHA-739358, a potent inhibitor of Aurora kinases with a selective target inhibition profile relevant to cancer. Mol.Cancer Res., 6: 3158-3168.

Cowan-Jacob SW, Fendrich G, Floersheimer A, Furet P, Liebetanz J, Rummel G, Rheinberger P, Centeleghe M, Fabbro D, Manley PW (2007). Structural biology contributions to the discovery of drugs to treat chronic mylogenous leukemia. Acta Crystallographica. D63: 80-93.

Deininger MWN, Goldman JM, Melo JV (2000). The molecular Biology of chronic myeloid leukamia. Blood. 96: 3343-3356.

Friesner RA, Banks JL, Murphy RB, Halgren TA, Klicic JJ, Mainz DT, Repasky MP, knoll EH, Shelly M, Perry JK, Shaw DE, Francis P, Shenkin PS (2004). Glide: A New approach for rapid accurate docking and scoring. 1 Method and assessment of docking accuracy. J. Med. Chem., 47: 1739-1749.

Gohlke H, Klebe G (2002). Approaches to the description of the binding affinity of small molecule ligand to macromolecular receptors. Angew. Chem. Int. Ed., 41:2644-2676.

Gohlke H, Hendlich M, Kelbe G (2000). Knowledge based scoring function to predict protein ligand interactions. J. Mol. Biol., 295: 337-356.

Goodsell DS (2005). The molecular perspective: c-Abl Tyrosine kinases. The Oncologist, 10: 758-759.

Halgren TA, Murphy RB, Friesner RA, Beard HS, Frye LL, Pollard WT, Banks JL (2004). Glide: A New approach for rapid accurate docking and scoring. 2 enrichment factors in database screening. J. Med. Chem., 47:1750-1759.

Jacobson M P, Kaminski GA, Friesner RA, Rapp CS (2002). Force Field Validation Using Protein Side Chain Prediction. J. Phys. Chem. B., 106:11673–11680.

Jacobson MP, Pincus DL, Rapp CS, Day TJF, Honig B, Shaw DE, Friesner RA (2004). A Hierarchical Approach to All-Atom Protein Loop Prediction", Proteins: Structure. Function and Bioinformatics, 55: 351-367.

Jorgenson WL, Maxwell DS, Tirado-Rives J (1996). Development and testing of the OPLS all atom force field on conformational energetic and properties of organic liquids. J. Am. Chem. Soc., 118: 11225-11236.

Moitessier N, Therrien E, Henessian S (2006). A method for induced fit docking scoring and ranking of flexible ligands. Application to peptidic and pseudopeptidic βsecretase (BACE 1) inhibitors. J. Med. Chem., 49: 5885-5894.

Nagar B (2007). c – Abl Tyrosine Kinase and inhibition by the cancer Drug Imatinib. j. Nut. 137: 1518S-1523S.

Nagar B, Hantschel O, Young MA, Scheffzek K, Veach D, Bornmann W, Clarkson B, Superti-Furga G, Kuriyan J (2003). Structural basis for the autoinhibition of c-Abl tyrosine kinase Cell (Cambridge,Mass). 112: 859-71.

Rix U, Hantschel O, Durnberger G, Rix LLR, Planyavsky M, Fernbach NV, Kaupe I, Bennett KL, Valent P, Colinge J, Kocher T, Furga GS (2007).

Chemical proteomic profiles of the BCR-ABL inhibitors Imatinib, Nilotinib and Dasatinib reveals novel kinases and non kinases targets. Blood. 110: 4055-4063.

Sherman W, Day T, Jacobson MP, Friesner RA, Farid R (2006). Novel procedure for modeling ligand/receptor induced fit effects. J. Med. Chem. 49: 534-553.

Sicheri F, Moarefi I, Kuriyan J(1997). Crystal structure of the Src- family tyrosine kinaseHck. Nature.385: 602-609.

Schrödinger LLC. New York; Glide user manual.

Sotriffer CA, Gohlke H, Klebe G (2002). Docking into knowledge based potential fields; a comparative evaluation of drug score. J. Med. Chem.45:1967-1970.

Taylor RD, Jewsbury PJ, Essex JW (2002). A review of protein small molecule docking methods. J. Comp. Aided Mol. Des. 16: 151-166.

Tokarski JS, Newitt JA, Chang CYJ, Cheng JD, Wittekind M, Kiefer SE, Kish K, Lee FYF, Borzillerri R, Lombardo LJ, Xie D, Zhang Y, Klei HE (2006). The structures of Dasatinib bound to activated ABL kinases domain Elucidates its inhibitory Activity against Imatinib-Resistant ABL Mutants. Cancer Res 66: 5790-5797.

Weisberg E, Manley PW, Breitenstein W, Brueggen J, Cowan-Jacob SW, Ray A, Huntly B, Fabbro D, Fendrich G, Hall-Meyers E, Kung AL, Mestan J, Daley GQ, Callahan L, Catley L, Cavazza C, Azam M, Neuberg D, Wright RD, Gilliland DG, Griffin JD, Young MA, Shah NP, Chao LH, Seeliger M, Milanov ZV, Biggs WH, Treiber DK, Patel HK, Zarrinkar PP, Lockhart DJ, Sawyers CL, Kuriyan J (2006). Structure of the kinase domain of an imatinib-resistant Abl mutant in complex with the Aurora kinase inhibitor VX-680. Cancer Res. 66: 1007-14.

Zhang J, Yang PL, Gray NS (2009). Targeting cancer with small molecule kinases inhibitors. Nature. 9:28-39.

Zhang LU, Gallicchio E, Friesner RA, Levy RM (2001) Solvent Models for ProteinLigand Binding: Comparison of Implicit Solvent Poisson and Surface Generalized Born Models with Explicit Solvent Simulat. J. Comp Chem, 22,591-607.

18

Chlorosoma: How can it contribute to photosynthesis of green bacteria?

A. Y. Borisov

A. N. Belozersky Institute of Physico-Chemical Biology in M. V. Lomonosov Moscow State University. Vorob'ev hills, 119992 Moscow. E-mail: borissov@belozersky.msu.ru.

Numerous kinetic data for excitation energy decay in bacteriochlorophylls as well as extremely short lifetimes of excitations in dominating chlorosoma pigment exclude the possibility of an efficient migration of this energy from chlorosoma to the main intra-membrane photosystem of some green bacteria. Author discusses the main purpose and molecular mechanism which account for this inconsistency in huge, 3D chlorosoma and substantiates the conditions of cell growing under which this energy migration from chlorosoma may become efficient.

Key words: Green bacteria, chlorosoma, energy migration and trapping.

INTRODUCTION

Green bacteria have light-harvesting organelles – chlorosomes located on the cytoplasmic side of the inner cell membranes. In sulfur bacteria chlorosomes may have up to 10.000 - 20.000 bacteriochlorophyll (BChl) c (or d, e) molecules which are enveloped into thin lipid capsule (Blankenship et al., 1995; Blankenship, Matsuura 2003). BChl-c concentration in chlorosomes is extremely high, its pigments are not bound to proteins, but aggregated themselves, therefore the red shift of their absorption bands, up to 720 - 750 nm is reasonably attributed to pigment-pigment interactions (Van Grondelle et al., 1994). So called base plates are located between chlorosoma and photosynthetic membrane (Blankenship et al., 1995; Blankenship and Matsuura, 2003). It contains BChl-a, carotenoids and proteins. A modest red shift of its absorption peak to about 795 nm suggests that these molecules are apparently monomers like B800 ones in purple bacteria.

Many green bacteria contain up to 1000 - 2000 BChl-c

Abbreviations: **BChl**, bacterio/chlorophyll; **EE**, electronic excitation.

BChl-c (C740) → BChl-a (B795) → BChl-a (B866) → BChl-a (P860)

chlorosoma   base plate   membrane |

Antenna system | RC

(in chlorosoma), 30 – 50 BChl-a (in base plate) and about 60 - 70 BChl-a molecules (in membrane) per reaction center (RC) (Blankenship and Matsuura, 2003). This pigment system seems to represent a classical type of cascade-like array, which should funnel excitations from the light-harvesting antenna to the RC traps.

The lifetime of photoexcited BChl-c was first measured in green bacterium Chloribium limicola as, $\tau^* = 35 \pm 15$ ps (Borisov et al., 1977). It was supposed in this work that such short lifetime reflects an efficient transfer of electronic excitations (EEs) to the major BChl-a in the membrane. Later this lifetime was measured with greater precision and often turned to be within 10 - 15 ps (see Holzwarth et al., 1990; Miller et al., 1991; Lin et al., 1991; Savikhin et al., 1994) and reviews (Blankenship et al., 1995; Van Grondelle et al., 1994). Many researchers made efforts to prove the activity of the above presented model in EE funneling, either by revealing the fluorescence of the main BChl-a B866, induced by photoexcitation of chlorosoma C740 pigment, or by monitoring the increase of C740 lifetime caused by disconnecting of chlorosoma from the main BChl-a apparatus, but without real success. Several non evident suppositions were suggested in literature pursuing the aim to reconcile such data with the above presented migration model.

Below a hypothetical notion is suggested which apparently explains the reasons of these troubles and

Strong non-photochemical excitation quenching in 3D chlorosoma

What kind of mechanism may be responsible for EE quenching in chlorosoma pigment? The answer is borrowed from physics. It was demonstrated in classical experiments by Nobel prize winner J. Perrin and S. Vavilov with rhodamin and methilene blue that their fluorescence emissions were quenched by about two orders when these molecules have formed associates, mostly dimers (Monograph Terenin, 1967). Later many similar works appeared. Chlorophylls are not exceptions in this virtue. Fluorescence lifetime (τ_{fl}) of BChl-*a* dissolved in a number of organic solutes occurred to be within 2 - 4 ns. In photosynthetic tissues it is much shorter, for example in RC special pairs of purple bacteria the rate constant for the sum of wasteful losses approximates to $(0,3\ ns)^{-1}$ (Shuvalov, 2000) although these BChl pairs are not real dimers, but closely positioned molecules. The mechanism of such EE quenching was well established. When chromophores of dyes make direct contacts, their outer orbits got distorted. This effect has two consequences: (a) a tremendous increase of singlet-triplet intercombination rate constant, (b) the splitting of the lower singlet excited level into "*n*" ones (*n* - the number of interacting molecules) hampers greatly EE deactivation into oscillations. Both these mechanisms increase greatly the sum of EE deactivation constants especially in higher molecular associates which phenomena manifest themselves in a strong decrease of EE lifetime.

In photosynthesis, the fine structures of photosystems of all known bacteria, alga and plants are organized by specific transmembrane polipeptides. They arrange Chl and BChl antenna complexes in such a way, that although having the majority of closest intermolecular distances within 9 - 12 Å, they have no direct contacts. In order to illustrate the importance of this unique virtue of transmembrane proteins, one should remember that such imterchromophore spacing corresponds to mean concentrations of 0.1 - 0.03 M/l (!) while in solutions, the formation of dimers starts at concentrations ~ $3\ 10^{-4}$ M/l. The only exception is chlorosoma of green bacteria in which BChl-*c* molecules apparently form direct contacts. It was proved in Betti et al. (1982) that in non polar solutions "...Chl-*c* forms aggregates whose spectra resemble those in chlorosomes". It is appropriate to refer to unpublished experiments which the author has performed in late professor Tumerman' laboratory in which in slightly acidic aqueous solutions, Chl-*a* molecules readily adsorbed on the surface of ZnO crystal powder and on fine-dispersed drops of stearine.

In the beginning its fluorescence lifetime was about 3.5 ns, but it drop drastically, when the adsorbed dye molecules formed more than 2 - 3 layers on the surfaces of these bearers. Apparent explanation: Chl-*a* molecules were well separated from each other in a single or two layers on hydrophobic surfaces, because the dimensions of their p-electron clouds (by definition they envelop 90% electron density) are 14 × 15 Å and no quenching centers were formed. It appears likely that 1 - 2 next Chl layers also could keep some kind of pseudo-crystalline structure with low quenching ability.

In the light of above said, it seems reasonable to come to the conclusion. The efficient EE migration is hardly possible from the multilayer BChl-*c* complexes quenched at the rate $(10 - 15\ ps)^{-1}$ like we often observe in huge chlorosoma.

DISCUSSION

But why could such huge chlorosama appear? The following explanation may be suggested. When green bacteria get into the media with some unfortunate parameters, the weakly persistent structure in the C740 outer layers cannot survive and hence get broken. So, a number of quenching centers appear in it which leads to a decrease of the portion of EEs delivered from it to the base plate and then to the major B866 antenna. This diminution of the EE income activates the natural regulatory mechanism to synthesize additional BChl-*c* molecules but in such situation, they only increase the mass of heaving up quenchers on the outer side of the chlorosoma. This leads to a further quenching of EEs in it and thus decreases C740 lifetime to the level of 10 - 15 ps. Possibly redox titration of chlorosoma (both *in vivo* and in isolated ones, as it was demonstrated in some works (Karapetian et al., 1980; Wang et al., 1990; Blankenship et al., 1993) renders such an example. Such redox titrations down to Em about - 1.46 V (Blankenship et al., 1993) produce such unfavorable conditions that the fine structure of chlorosoma got damaged which leads to a creation of numerous quenching centers especially in the outer layers of BChl-*c*. If this notion is reasonable, the fluorescence lifetime of BChl-*c* should be high in intact chlorosoma and should fall down to the unfortunately low value in the course of titration. This version requires lifetime control: the value of C740 fluorescence lifetime at least of 2 - 3 hundreds of picoseconds should be used as the criterion of chlorosoma intactness.

Possibly, enormously huge chlorosoma are specific for widespread regime of cell growing: in pursuit for more biomass, scientists keep up bacteria to very high optical densities. Thus, in the final days, most shadowy cells get enormously weak light. As a result, such cells continue to synthesize surplus of BChl-*c* unless its outer layers in chlorosoma loses its crystalline structure and produce many quenching associates. Guessing, the energy migration from chlorosoma to B795 may be rather high if green bacteria are collected from the reactor before their optical density exceeds ~ 0.5 OD.

Excitons in chlorosoma

One may reasonably expect that light absorbed by C740 creates excitons (Blankenship and Matsuura, 2003; Van Grondell et al., 1994) which expands over tens of chlorosoma molecules, provided they do have crystalline structure. Bearing in mind that the transition dipoles of C740 are parallel within 15 - 20° (Betti et al., 1982; Fetisova et al., 1987; Van Amerongen et al., 1988; Griebenov et al., 1991) and assuming high degree of their crystallinity, the values of efficient dipoles of such excitons may be considerably greater than that of monomer BChl-c molecule. The rate constant values between chlorosoma C740 and C795 of the baseplate may thus increase close to the vector sum of individual C740 molecules involved. Unfortunately, even 30-fold increase of these rate constants can not help to elevate considerably the efficiency of C740 → C795 migration, given the distance between C740 and C795 exceeds 45 - 50 Å. It should be also noted, that the values of EE lifetime of the order of 10 - 15 ps apparently excludes the presence of good crystalline structure in C740.

Base plate

As it concerns the base plates of green bacteria, the efficiency of EE delivery from their BChl-a B795 to the major BChl-a, B866 must be high, because there exist two important advantages as compared with EE transfer from chlorosoma.

(a) Contrary to huge chlorosoma the usual ratio of B795 to B866 is only 0.5 - 0.7. Thus, the entropy barrier of about 70-90 meV, which exists in chlorosoma with 1000 – 2000 BChl-c per about 60 B866 is eliminated. This circumstance ensures 20 - 40-fold gain in the migration constants from its B795 to the major B866.
(b) As in B800 monomers of purple bacteria, the rate constant of wasteful losses in monomers B795 must be of similar order ~ $(1.5 - 2 \text{ ns})^{-1}$ [compare with $(\sim 15 \text{ps})^{-1}$ in some chlorosoma], which makes 100 - 150-fold gain in favor of EE delivery from the base plate pigment to B866. Our modeling within the use of T. Ferster' theory proved that the efficiency of B795* → B866 EE transfer reaches 70 and 50% for the distances between C795 and B866 pigments as long as 45 Å and 48 Å respectively.

REFERENCES

Betti JA, Blankenship RE, Nagarajan LV, Dickinson LC, Fuller RC (1982).Antenna organization and evidence for the function of a new antenna pigment species in the green photosynthetic bacterium *Chloroflexus aurantiacus*, Biochim. Biophys. Acta, 680: 194-201.

Blankenship RE, Cheng PL, Causgrove TP, Brune DC, Wang SH, Choh JU, Wang J (1993). Redox regulation of energy transfer efficiency in antennas of green Photosynthetic bacteria. Photochem. Photobiol., 57: 103-107.

Blankenship RE, Matsuura K (2003). Antenna complexes from green bacteria. In light-harvesting antennas in photosynthesis. In: Purple bacteria, Green LR, Parson WW (Eds). Kluwer Acad. Publ.The Netherlands. pp. 195-217.

Blankenship RE, Olson JM, Miller M (1995). Antenna complexes from green bacteria In: Anoxygenic photosynthetic bacteria, Blankenship RE, Madigan MT, Bauer CE (ed-s), Kluwer Acad. Publ. The Netherlands. pp. 399-435.

Borisov AY, Fetisova ZG, Godik VI (1977). Energy transfer in photoactive complexes from green bacterium *Ch. Limicola*. Biochim. Biophys. Acta, 461: 500-509.

Fetisova ZG, Freiberg AM, Timpmann KE (1987).Investigation by picosecond polarized fluorescence spectroscopy of energy transfer in living cells of the green bacterium *Chlorobium limicola*. FEBS Lett., 199: 234-236.

Griebenov K, Holzwarth AR, Van Mourik F, Van Grondelle R (1991). Pigment organization and energy transfer in green bacteria. 2. Circular and linear dichroism of protein containing and protein-free chlorosomes from *Chloroflexus aurantiacus* strain Ok-70-f1. Biochim. Biophys Acta, 1058:194-202.

Holzwarth AR, Miller MG, Griebenov K (1990). Picosecond energy transfer Kinetics between pigment pools in different preparations of chromosomes from green Bacterium *Chloroflexus aurantiacus*, J. Photochem. Photobiol. Acta, 65: 61-71.

Karapetian NV, Swarthoff T, Rijgersberg CP, Amesz J (1980).In: Current Research in Photosynthesis (M. Baltshevsky edr) V. II, Kluwer Acad. Publ. Dordrecht. pp 17-24.

Lin S, Van Amerongen H, Struve WS (1991). Ultra fast pump-probe spectroscopy of bacteriochlorophyll-c antennae in bchl-a containing chlorosomes from *Chloroflexus Aurantiacus*. Biochim. Biophys. Acta, 1060: 13-24

Miller MG, Cox RP, Gilbro T (1991).Energy transfer kinetics in chromosome from Chloroflexus aurantiacus, studies using picosecond's absorption spectroscopy. Biochim. Biophys. Acta, 1057: 187-194.

Savikhin S, Zhu Y, Lin S, Blankenship RE, Struve WS (1994).Femtosecond Spectroscopy of chlorosoma antennas from green photosynthetic bacteria *Chloroflexus aurantiacus*, J. Phys. Chem., 98:10322-10334.

Shuvalov VA (2000). The conversion of solar radiation in the primary act of charge separation in photosynthetic reaction centers. Nauka edition. Moscow. pp. 3-47.

Terenin AN (1967). Fotonics of dye molecules, Nauka edition, Leningrad. pp. 465-476.

Van Amerongen H, Vasmel YH, van Grondelle R (1988). Linear dichroisn of chlorosomes from *Chloroflexus aurantiacus* in compressed gel and electric fields. J.Biophys. 54: 65-76.

Van Grondelle R, Dekker JP, Gilbro T, Sundstrem V (1994). Energy transfer and trapping in photosynthesis, Biochim. Biophys Acta, 1187: 1-65.

Wang J, Brune DC, Blankenship RE (1990). Effects of oxidants and reductants on the efficiency of excitation transfer in green photosynthetic bacteria. Biochim. Biophys. Acta,1015: 457-463.

19

Spectroscopic approach of the interaction study of ceftriaxone and human serum albumin

Abu Teir M. M.[1] , Ghithan J.[1], Abu-Taha M. I.[1], Darwish S. M.[1] and Abu-hadid M. M[2].

[1]Department of Physics, Faculty of Science, Al-Quds University, Jerusalem, Palestine.
[2]Department of Immunology, Faculty of Medicine, Al-Quds University, Jerusalem, Palestine.

Under physiological conditions, interaction between ceftriaxone and human serum albumin was investigated by using fluorescence spectroscopy and ultra violet (UV) absorption spectrum. From spectral analysis, ceftriaxone showed a strong ability to quench the intrinsic fluorescence of human serum albumin (HSA) through a static quenching procedure. The binding constant (k) is estimated as $K=1.02 \times 10^3$ M^{-1} at 298 K. Fourier transform infrared spectroscopy (FT-IR) spectroscopy with Fourier self-deconvolution technique was used to determine the protein secondary structure and drug binding mechanisms. The observed spectral changes indicated the formation of H-bonding between ceftriaxone and HSA molecules at higher percentage for α-helix than for the β-sheets.

Key words: Ceftriaxone, amide I-III, binding mode, binding constant, protein secondary structure, Fourier transform IR, UV-spectroscopy, Flurosence spectroscopy.

INTRODUCTION

In recent years, many investigations on the binding of drugs and natural products to human serum albumin (HSA) were carried out (Il'ichev et al., 2002; Qing et al., 2011; Ahmad et al., 2006; Liu et al., 2009). Ceftriaxone, as a possible ligand, was not studied in details upon its binding reaction with HSA. It has been reported that Ceftriaxone binds to HSA (Guowen et al., 2011, Bibiana et al., 1996).

Ceftriaxone, a cephalosporin, is bound reversibly to defatted human serum albumin from adults, with a first stoichiometric binding constant of 6×10^4 M^{-1}, as found by equilibrium dialysis at pH 7.4, 37°C (Roberton et al., 1989). So far, none of the investigations determine in details the Ceftriaxone-HSA binding constant and the effects of Ceftriaxone complexation on the protein.

Ceftriaxone belongs to a group of antibiotics called the cephalosporins. It is a parenteral cephalosporin that displays a broad spectrum of activity against Gram-negative and Gram-positive pathogens; its chemical structure is shown in Figure 1 (Bilirubin et al., 1989). Ceftriaxone is indicated for a wide variety of infections. These include infections of the lower respiratory and urinary tracts, bacterial septicemia, skin and skin structure infections, bone and joint infections, pelvic inflammatory disease, uncomplicated gonorrhea, intra-abdominal infections, acute bacterial otitis media, meningitis, as well as surgical prophylaxis (Quaglia et al., 1997).

The drug is widely used because of its broad spectrum of antibacterial activity, infrequent side-effects, and long serum half-life and it has recently been recommended as the drug of choice for use in newborn infants exposed to Neisseria gonorrhea during delivery (Bibiana et al., 1996).

It has been reported that ceftriaxone is transported through the blood-brain barrier (Scott et al., 2008; Reynold et al., 1987) and it is widely accepted that some conformational diseases, such as Alzheimer's and prion diseases can be the result of misfolding due to lack of stability in parallel β-sheets (Locht et al., 1990). Recent reports suggest that somedrugs may accelerate these

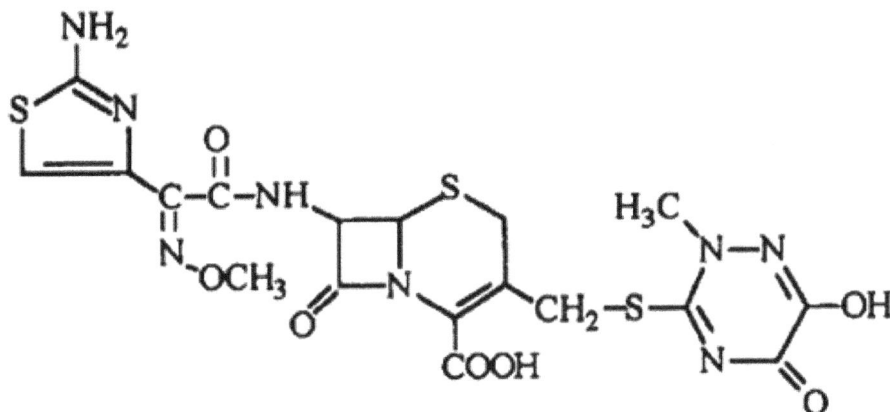

Figure 1. Chemical structure of ceftriaxone.

neurodegenerative diseases (Rabia et al., 2009). This emphasizes the needs to understand protein folding and sheets stability in some proteins complexes with drugs.

The widespread use of ceftriaxone and its transport through the blood-brain barrier makes it necessary to study structural changes of ceftriaxone -protein complexes to understand the biological effects and functions of ceftriaxone in the body. Thus, the study of ceftriaxone - protein interaction is of great interest in the field of biophysics, life sciences and clinical medicine.

HSA is the most abundant protein in human plasma, which is synthesized in the liver (Fengling et al., 2006) and is able to bind and thereby transport various compounds such as fatty acids, hormones, bilirubin, tryptophan, steroids, metal ions, therapeutic agents and a large number of drugs. HSA serves as the major soluble protein constituent of the circularity system, it contributes to colloid osmotic blood pressure, it can bind and carry drugs which are mainly poorly soluble in water (Peters et al., 1985). HSA accounts for approximately 60% of the total plasma protein corresponding to a concentration of 40 mg/ml in the blood (~0.6 mM) (Peters et al., 1985). The three dimensional structure of HSA was determined through x-ray crystallographic measurements (He et al., 1992). This globular protein consists of a single polypeptide chain of 585 amino acids (Hal et al., 2000), which have a molecular weight of 66 kDa (Bian et al., 2004). HSA composed of three homologous domains I, II and III, each containing two sub-domains, A and B, each having six and four α-helices, respectively (Curry et al., 1999). The tertiary structure of HSA is stabilized by 17 disulphide bridges giving it a heart shaped molecule (He et al., 1992; Bian et al., 2004).

It has been shown that distribution, free concentration, and metabolism of various drugs can be significantly altered as a result of their binding to HSA (Artali et.al, 2005; Kang et al., 2004). The binding properties of albumin depend on the three dimensional structure of the binding sites, which are distributed all over the molecule.

Strong binding can decrease the concentrations of free drugs in plasma, whereas weak binding can lead to a short lifetime or poor distribution or both. Its remarkable capacity to bind a variety of drugs results in its prevailing role in drug pharmacokinetics and pharmacodynamics (Kandagal et.al, 2007).

Multiple drug binding sites have been reported for HSA by several groups of researchers (Bhattacharyya et al, 2006; Simard et al., 2006; Ulrich .et al., 2006). The principal regions of ligand binding sites of HSA are located in hydrophobic cavities in sub domains IIA and IIIA, which corresponds to site I and site II, respectively. Site I is dominated by strong hydrophobic interaction with most neutral, bulky, heterocyclic compounds, while site II mainly by dipole-dipole, van der Waals, and/or hydrogen-bonding interactions with many aromatic carbo-xylic acids. HSA contained a single tryptophan residue (Trp-214) in domain IIA and its intrinsic fluorescence is sensitive to the ligands bound nearby (Krishnakumar et al., 2002; Muravchick et al., 1995). Therefore, it is often used as a probe to investigate the binding properties of drugs with HSA.

The interaction of ceftriaxone sodium (CS), a cephalosporin antibiotic, with the major transport protein, bovine serum albumin (BSA), was investigated using different spectroscopic techniques such as fluorescence, circular dichroism (CD), and UV-vis spectroscopy (Cao et al., 2012). The mechanism of the interaction between bovine serum albumin (BSA) and ceftriaxone with and without zinc (II) (Zn^{2+}) was studied employing fluorescence, ultraviolet (UV) absorption, circular dichroism (CD), and synchronous fluorescence spectral methods. The intrinsic fluorescence of BSA was quenched by ceftriaxone in a static quenching mode, which was authenticated by Stern-Volmer calculations.

The binding constant, the number of binding sites, and the thermodynamic parameters were obtained, which indicated a spontaneous and hydrophobic interaction between BSA and ceftriaxone regardless of Zn^{2+} (Liu et al.,

2012).

In this study, we have investigated the interaction of ceftriaxone with HSA by means of FT-IR, UV/VIS, and fluorescence spectrophotometer. Infrared spectroscopy provides measurements of molecular vibrations due to the specific absorption of infrared radiation by chemical bonds. It is known that the form and frequency of the Amide I band, which is assigned to the C=O stretch-ing vibration within the peptide bonds is very characteristic for the structure of the studied protein.

From the band secondary structure, components peaks (α-helix, β-strand) can be derived and the analysis of this single band allows elucidation of conformational changes with high sensitivity (Abu et al., 2011). This work will be limited to the mid-range infrared, which covers the frequency range from 4000 to 400 cm^{-1}. This wavelength region includes bands that arise from three conformational sensitive vibrations within the peptide backbone (Amides I, II and III) of these vibrations. Amide I is the most widely used and can provide information on secondary structure composition and structural stability (Cui et al., 2008; Kang et al., 2004; Rondeau et al., 2007).

One of the advantages of infrared spectroscopy is that it can be used with proteins that are either in solution or in thin film. In addition there is a growing body of literature on the use of infrared to follow reaction kinetics and ligand binding in proteins, as will as a number of infrared studies on protein dynamics.

The identification of the binding sites in albumin was also performed using probes for the so-called sites I, II, bilirubin and fatty acids binding sites. Albumin showed two types of binding sites for cefoperazone and ceftria-xone, while for cefsulodin it showed a single type of bind-ing site (Pico et al., 1996).

The mechanism of the interaction between human serum albumin (HSA) and ceftriaxone has been studied by using UV, fluorescence and FTIR spectroscopy. Further-more temperature dependent conformation of HSA struc-ture at different concentrations of Ceftriaxone has been studied by FT-IR spectroscopy.

MATERIALS AND METHODS

Human serum albumin (HSA, 96-99% purity) and Ceftriaxone disodium salt hemi(heptahydrate) in powder form were pur-chased from Sigma Aldrich chemical company and used without further purification.

Preparation of stock solutions

HSA was dissolved in phosphate buffer saline, at physiological pH 7.4 at (80 mg/ml) concentration. Ceftriaxone with molecular weight of 66.5 kDa was dissolved in phosphate buffer saline (1.21 mg/ml), the solution was placed in ultrasonic water path (SIBATA AU-3T) for 6 h to ensure that all the amount of Ceftriaxone was completely dissolved. The final concentrations of HSA- Ceftriaxone complexes were prepared by mixing equal volume of HSA and Ceftriaxone stock solution. HSA concentration in all samples were fixed at 40 mg/ml (~0.6 mM). However, the concentration of Ceftriaxone in the final protein drug solutions was decreased gradually to attain the desired drug concentrations of 0.687, 0.910, 1.14, 1.37, 1.60, 1.83 and 2.06 mM. The solution of Ceftriaxone and HSA were incubated for 1 h (at 25°C) before spectro-scopic measurements were taken.

Fluorescence

The fluorescence measurements were performed by a Nano-Drop ND-3300 Fluorospectrometer at 25°C. The excitation had been done at the wavelength of 360 nm and the maximum emission wavelength was at 439 nm. The excitation source comes from one of three solid-state light emitting diodes (LED's). The excitation source options include: UV LED with maximum excitation of 365 nm, Blue LED with excitation of 470 nm, and white LED from 500 to 650 nm excitation. A 2048-element CCD array detector covering 400-750 nm, was connected by an optical fiber to the optical measurement surface. The emission spectra were recorded for free HSA 40 mg/ml (~0.6 mM) and for its complexes with ceftriaxone solutions with the concentrations of (0.687, 0.910, 1.14, 1.37, 1.60, 1.83, 2.06 mM). Repeated measurements were done for all samples and no significant differences were observed.

FT-IR spectroscopy experimental procedures

The FT-IR measurements were obtained on a Bruker IFS 66/S spectrophotometer equipped with a liquid nitrogen-cooled MCT detector and a KBr beam splitter. The spectrometer was conti-nuously purged with dry air during the measurements. The absorption spectra were obtained in the wave number range of 400-4000 cm^{-1}. A spectrum was taken as an average of 60 scans to increase the signal to noise ratio, and the spectral resolution was at 4 cm^{-1}. The aperture used in this study was 8 mm, since we found that this aperture gives best signal to noise ratio. Baseline correction, normalization and peak areas calculations were per-formed for all the spectra by OPUS software. The peak positions were determined using the second derivative of the spectra. The infrared spectra of HSA, and Ceftriaxone-HSA complex were obtained in the region of 1000-1800 cm−1. The FT-IR spectrum of free HSA was acquired by subtracting the absorption spectrum of the buffer solution from the spectrum of the protein solution. For the net interaction effect, the difference spectra [(protein and Ceftriaxone solution) - (protein solution)] were generated using the featureless region of the protein solution 1800-2200 cm^{-1} as an internal standard (Surewicz et al., 1993). The accuracy of this subtraction method is tested using several control samples with the same protein or drug concentrations, which resulted into a flat base line formation. The obtained spectral differences were used here, to investigate the nature of the drug-HSA interaction. We had also used ELAB 12/05 thermo system to directly and simultaneously determine the thermo-dependent structural changes of drug-protein complexes.

RESULTS AND DISCUSSION

Analysis of fluorescence quenching of HSA by Ceftriaxone

Fluorescence spectroscopy is one of the most widely used spectroscopic techniques in the fields of bioche-mistry and molecular biophysics today (Royer et al.,

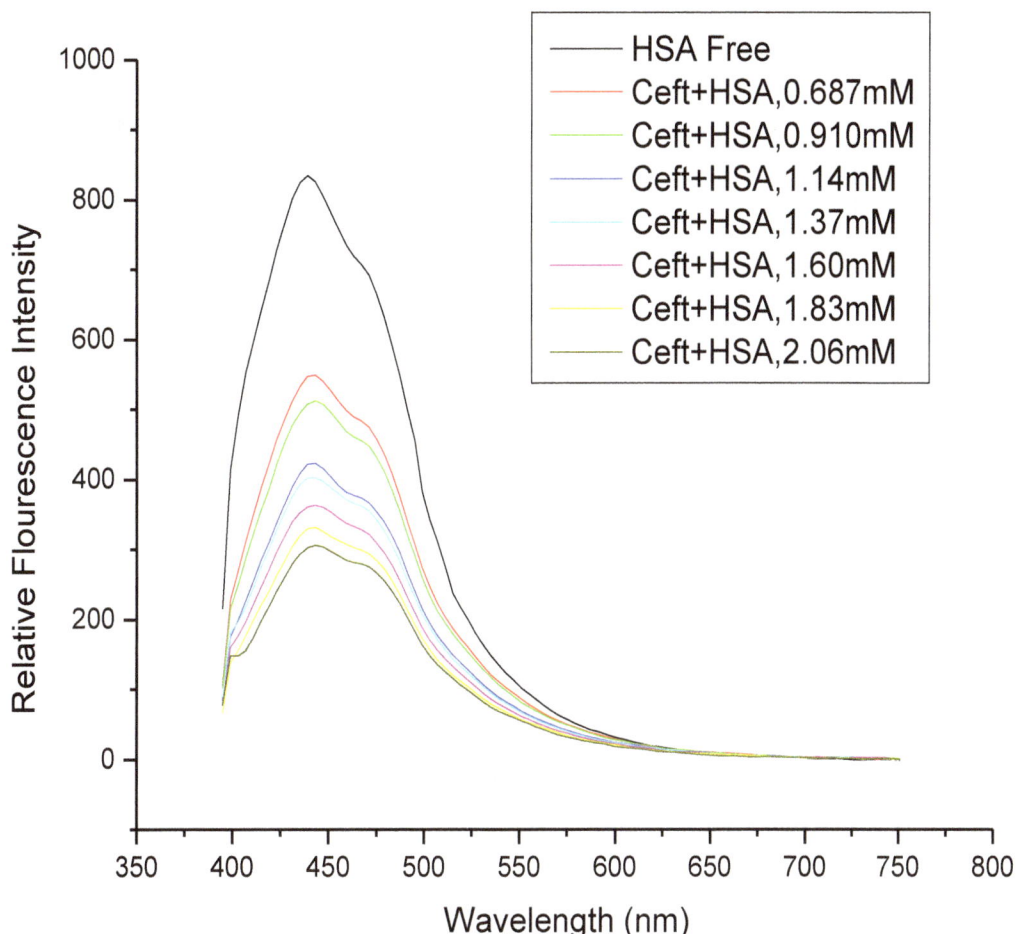

Figure 2. Fluorescence emission spectra of HSA in the absence and presence of ceftriaxone at different concentrations.

1995). Fluorescence measurements can give some information on the binding mechanism of small molecule substances to protein, including binding mode, binding constants, binding sites and intermolecular distances (Liu et al., 2004). The fluorescence of HSA comes from tryptophan, tyrosine and phenylalanine residues. Actually, the intrinsic fluorescence of HSA is almost contributed by tryptophan alone (Sulkowska et al., 2002).

Figure 2 shows the fluorescence spectra of HSA in the presence of various concentrations of ceftriaxone. It is obvious that HSA fluorescence intensity gradually decreased with the increase of Ceftriaxone concentration, while the peak position shows little or no change upon increasing the concentration of Ceftriaxone to HSA, indicating that Ceftriaxone binds to HSA. Under the same condition, no fluorescence of Ceftriaxone was observed hich indicates that Ceftriaxone could quench the auto fluorescence of HSA which confirm that Ceftriaxone interacts with HSA, leading to a change in the microenvironment around the tryptophan residue exposing it to the polar solvent (Wang et al., 2007; Cui et al., 2007).

Two mechanisms, namely dynamic quenching and static quenching are responsible for the fluorescence quenching. The possible quenching mechanism can be deduced from the Stern-Volmer plot. The dynamic quenching process can be described by the Stern-Volmer equation (Tian et al., 2003)

$$\frac{F_0}{F} = 1 + K_q \tau_0 [Q] = 1 + K_{sv}[Q] \cdot \tag{1}$$

Where, F_0 and F are the fluorescence intensities of HSA in the absence and presence of the quencher, respectively. K_q is the quenching rate constant of the biomolecule, K_{SV} is the Stern-Volmer dynamic quenching constant, and $K_{SV} = K_q\tau_0$. τ_0 is the average lifetime of the biomolecule without quencher. The value of τ_0 of the biopolymer was $10^{-8}s^{-1}$ (Chen et al., 1990), and [Q] is the concentration of quencher ceftriaxone.

The Stern-Volmer quenching constant Ksv indicates the sensitivity of the fluorophore to a quencher. Linear curves were plotted according to the Stern-Volmer

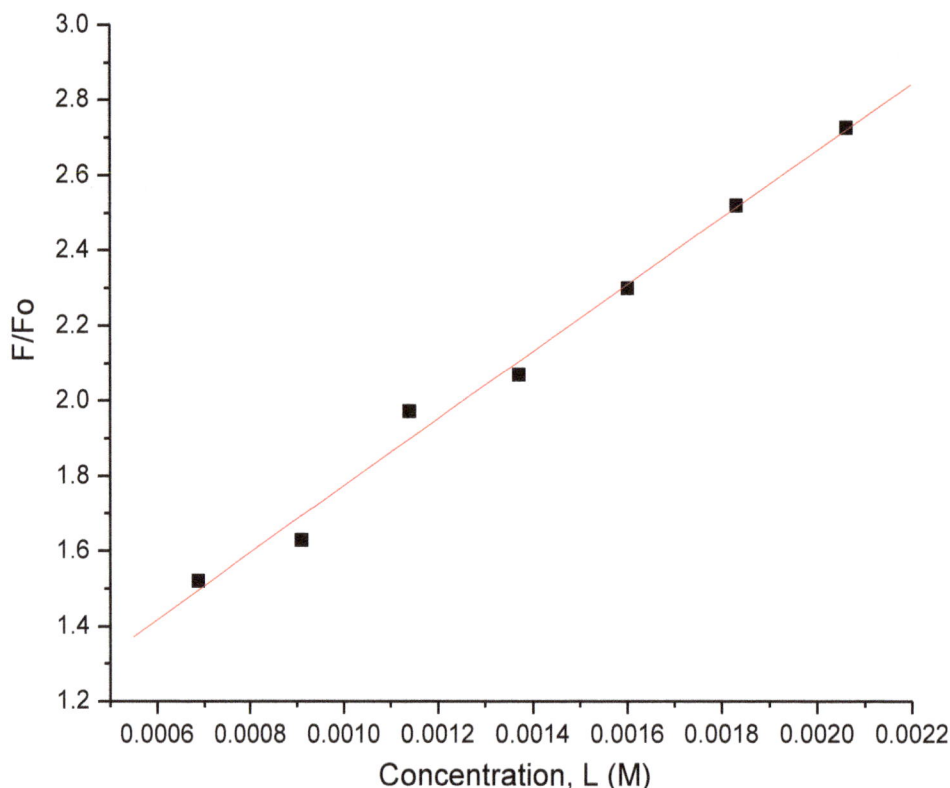

Figure 3. The Stern-Volmer plot for Ceftriaxone-HSA complexes.

equation as shown in Figure 3 for ceftriaxone- HSA complexes. The Stern-Volmer quenching constant Ksv was obtained by the slope of the curve obtained in Figure 3, and its value equals 8.92×10^2 L mol^{-1}.

From the equation above, the value of Ksv=Kqτ_0, can be used to calculate the value of Kq using the fluorescence life time of 10^{-8} s for HSA, to obtain Kq value of (8.92×10^{10} L mol^{-1}s^{-1}) for ceftriaxone- HSA complexes. Generally, the maximum dynamic quenching constant, Kq of various kinds of quenchers with biopolymer was 2.0×10^{10} L mol^{-1}s^{-1} (Lakowicz et al., 1973).

Obviously, the values of kq were greater than that of the maximum dynamic quenching constant. This suggested that the fluorescence quenching was not the result of dynamic quenching, but the consequence of static quenching (Chen et al., 1990, Wang et al., 2008).

When static quenching is dominant, the modified Stern-Volmer equation could be used (Lakowicz, et al.1999).

$$\frac{1}{F_0 - F} = \frac{1}{F_0 K L} + \frac{1}{F_0}$$

$$(2)$$

Where, K is the binding constant of ceftriaxone with HSA, and can be calculated by plotting 1/(F0-F) vs. 1/L (Figure 4). The value of K was obtained from the values of slope

and intercept, respectively. K equals the ratio of the intercept to the slope.

The value of K was 1.02×10^3 M^{-1} at room temperature, which is consistent with the value obtained by Bibiana et al., (1996) and Patrick et al. (1990). *In vitro* protein binding studies were conducted to examine the interaction between ceftriaxone (CEF), probenecid (PROB) and diazepam (DIAZ). The presence of PROB and DIAZ at concentrations equal to molar albumin concentration caused a decrease in CEF affinity from 3.7 x 10^4 M^{-1} (control) to 1.1 x 10^4 (PROB) and 2.6 x 10^4 (DIAZ) M^{-1}, but not in binding capacity in pooled human plasma (Stoeckel et al., 1990).

The value obtained is indicative of a weak Ceftriaxone-HSA interaction with respect to the other drug-HSA complexes with binding constants in the range of 10^5 and 10^6 M^{-1} (Kragh-Hansen et al., 1981). The highly effective quenching constant in this case has lead to a lower value of binding constant between the drug and HSA due to an effective hydrogen bonding between ceftriaxone and HSA.

FT-IR spectroscopy

FT-IR spectroscopy is a powerful technique for the study of hydrogen bonding (Li et al., 2006), and has been identified as one of the few techniques that is established

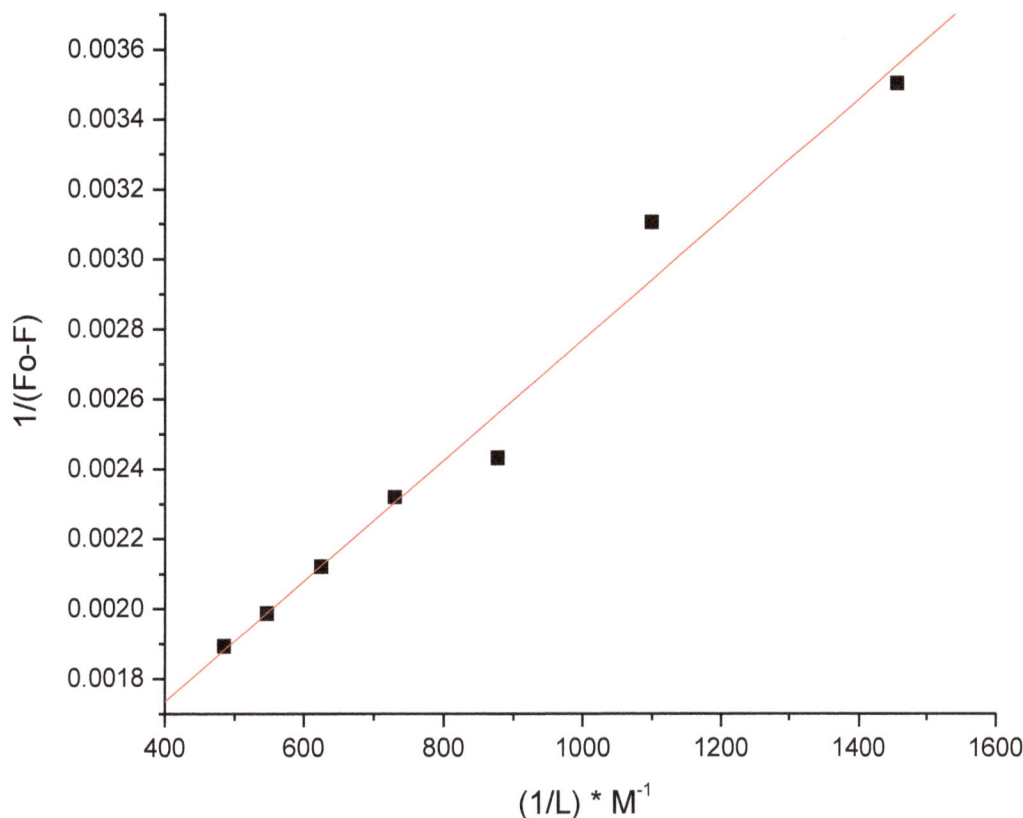

Figure 4. The plot of 1/(Fo-F) vs (1/L) for ceftriaxone- HSA complexes.

in the determination of protein secondary structure at different physiological systems (Sirotkin et al., 2001; Arrondo et al., 1993). The information on the secondary structure of proteins could be deduced from the infrared spectra. Proteins exhibit a number of amide bands, which represent different vibrations of the peptide moiety. The amide group of proteins and polypeptides presents characteristic vibrational modes (amide modes) that are sensitive to the protein conformation and largely been constrained to group frequency interpretations (Ganim et al., 2006).

The modes most widely used in protein structural studies are amide I, amide II and amide III. Amide I band ranging from 1700 to 1600 cm^{-1} and arises principally from the C=O stretching (Vandenbussche G et al., 1992), has been widely accepted to be used (Workman et al., 1998). The amide II band is primarily N-H bending with a contribution from C-N stretching vibrations; amide II ranging from 1600 to 1480 cm^{-1} while amide III band ranging from 1330 to 1220 cm^{-1} which is due to the C-N stretching mode coupled to the in-plane N - H bending mode (Arrondo et al., 1993; Jackson et al., 1991).

The second derivative of free HSA is shown in Figure 5A, where the spectra is dominated by absorbance bands of amide I and amide II at peak positions 1656 and 154 4cm^{-1}, respectively. Figure 5B, shows the spectrum of ceftriaxone- HSA complexes with different concentrations of ceftriaxone.

The peak positions of amide I bands in HSA infrared spectrum shifted as listed in Table 1: 1637-1642 cm^{-1}, 1655-1658 cm^{-1}, 1683-1679 cm^{-1}, 1693-1692 cm^{-1} after interaction with ceftriaxone. In addition, peaks at 1613 and 1626 cm^{-1} had disappeared and also a new peak at 1663 cm-1 had appeared after the interaction of ceftriaxone with HSA. In amide II, the peak positions have shifted as follows: 1496 to 1494 cm^{-1}, 1512 to 1513 cm^{-1}, 1544 to 1536 cm^{-1} and 1580 to 1584 cm^{-1}. In addition, a new peaks at 1552 and 1568 cm^{-1} appeared after the interaction of ceftriaxone with HSA. In the Amide III region little or no change of the peak positions has been observed. The changes of these peak positions and peak shapes implied that the secondary structures of HSA had been changed by the interaction of ceftriaxone with HSA. The minor changes in peak positions can be attributed to the effect of the newly formal H-bonding between ceftriaxone molecules with the protein. It is suggested that, the shift to a higher frequency for the major peak in amide I region (1656-1658 cm^{-1}) came as a result of stabilization by hydrogen bonding by having the C-N bond assuming partial double bond character due to a low of electrons from the C=O to the C-N bond (Jackson et al., 1991).

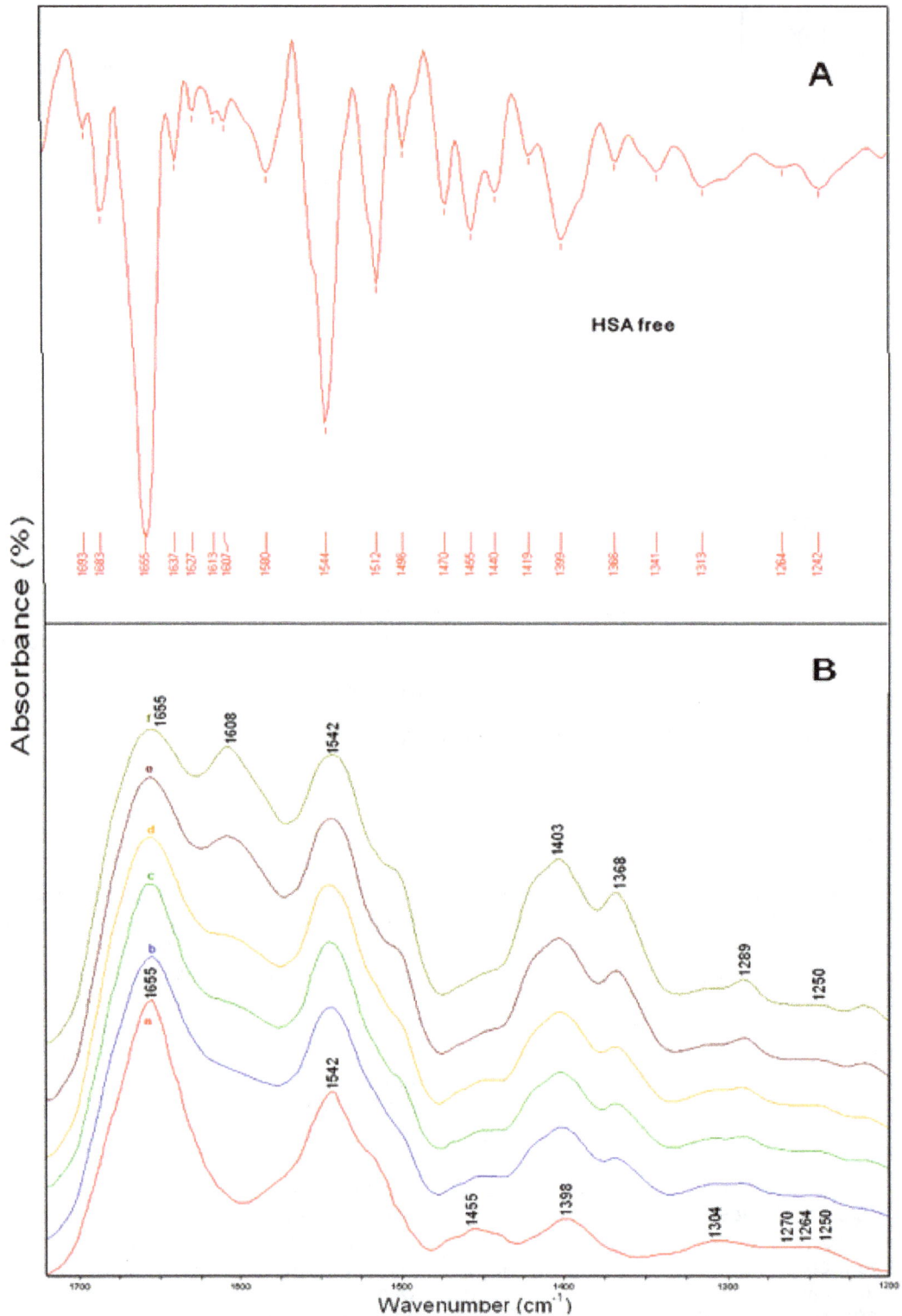

Figure 5. The spectra of (A) HSA free (second derivative) and (a, b, c, d, e, and f) HSA-Ceftriaxone IR absorption spectrum with Ceftriaxone concentrations (0.0, 0.687, 0.910, 1.14, 1.37, and 1.60 mM) respectively.

Determination of the secondary structure of HSA and its ceftriaxone complexes was carried out on the basis of the procedure described by Byler et al. (1986). In this work, a quantitative analysis of the protein secondary structure for the free HSA and ceftriaxone- HSA complexes in dehydrated films is determined from the shape of amides

Table 1. Band assignment in the absorption spectra of HSA with different ceftriaxone concentrations for amide I, II and III regions.

Bands	HSA Free	HSA-Ceft. 0.687mM	HSA-Ceft. 0.910 mM	HSA-Ceft. 1.14 mM	HSA-Ceft. 1.37 mM	HSA-Ceft. 1.60 mM
Amide I (1610-1700 cm^{-1})	1613					
	1626	1627	1625	1629		
	1637	1641	1641	1642	1642	1642
	1655	1654	1655	1656	1657	1658
		1662	1662	1662	1662	1663
	1683	1680	1679	1679	1678	1679
	1693	1693	1693	1692	1692	1692
Amide II (1480-1600 cm^{-1})	1496	1497	1498	1498	1499	1494
	1512	1513	1513	1514	1514	1513
	1544	1545	1546	1536	1535	1536
		1552	1551	1551	1551	1552
		1569	1568	1568	1568	1568
	1580	1586	1585	1585	1586	1584
Amide III (1220-1330 cm^{-1})	1242	1242	1242	1242	1241	1241
	1263	1263	1264	1261	1261	1260
	1303	1288	1288	1288	1288	1288
	1313	1312	1312	1311	1311	1312

I, II and III bands. Infrared Fourier self-deconvolution with second derivative resolution and curve fitting procedures, were applied to increase spectral resolution and therefore to estimate the number, position and area of each component bands. The procedure was in general carried out considering only components detected by second derivatives and the half widths at half height (HWHH) for the component peaks are kept around 5 cm^{-1}; the above procedure was reported in our recent publications (Abu et al., 2011; Darwish et al., 2010; Abu et al., 2012).

The component bands of amides I, II, and III regions were assigned to a secondary structure accor-ding to the frequency of its maximum raised after Fourier self deconvolution have been applied for amide I band ranging from 1610 to 1700 cm^{-1} generally assigned as follows: 1610-1627 cm^{-1} are generally represented to parallel β-sheet, 1627-1643 cm^{-1} represent random coil, 1643-1672 cm^{-1} represent α-helix, 1672-1787 cm^{-1} represent turn structure, and 1687-1700 cm^{-1} represent β-antiparallel. For amide II ranging from 1480 to 1600 cm^{-1}, the absorption band was assigned in the following order: 1485-1502 cm^{-1} represent parallel β-sheet, 1502-1529 cm^{-1} represent random coil, 1529-1563 cm^{-1} represent α-helix, 1563-1587 cm^{-1} represent turn structure, and 1587-1600 cm^{-1} represent β-antiparallel.

For amide III ranging from 1220 to 1330 cm^{-1} was assigned as follows: 1220-1255 cm^{-1} represent parallel β-sheet, 1255-1301 cm^{-1} represent random coil, 1301-1317 cm^{-1} represent turn structure, and 1317-1330 cm^{-1} repre-sent α-helix. Most investigations have concentrated on

Amide I band assuming higher sensitivity to the change of protein secondary structure (Vass et al., 1997). How-ever, it has been reported that amide II and amide III bands have high information content and could be used for prediction of proteins secondary structure (Oberg et al., 2004, Jiang et al., 2004, Liu et al., 2003).

Based on the above assignments, the percentages of each secondary structure of HSA were calculated from the integrated areas of the component bands in amide I, amide II and amid III, respectively. Table 2 shows the content of each secondary structure of HSA before and after the interaction with ceftriaxone at different concen-trations. The percentage values for the components of amide I of free HSA are consistent with the results of other recent spectroscopic studies.

Figure 6 illustrates the variation of relative intensities with concentration change for the α-helix, parallel and antiparallel β-sheets in HSA free and ceftriaxone-HSA complexes. The results exhibited a reduction of α-helical structures so that of amide II and amide III showed similar trends in their percentage values to that of amide I.

The decrease of α-helix percentage with the increase of ceftriaxone concentrations is evident in the calculations and this trend is consistent in the three Amide regions. However, for the parallel and antiparallel β-sheet, relative percentage increased with increasing ceftriaxone concentrations and the relative intensity increased for the antiparallel β-sheets remarkably having a faster rate increase than the parallel β-sheets.

The reduction of α-helix intensity percentage in favor of

Table 2. Secondary structure determination for amide I, II and III regions for HSA and its ceftriaxone complexes.

Band	HSA Free	HSA-Ceft. 0.687 mM	HSA-Ceft. 0.910 mM	HSA-Ceft. 1.14 mM	HSA-Ceft. 1.37 mM	HSA-Ceft. 1.60 mM
Amide I						
B-sheet parallel (1610-1627 cm^{-1})	10	11	12	13	14	16
Random (1627-1643 cm^{-1})	11	5	4	5	5	7
α-helix (1643-1672 cm^{-1})	68	62	60	45	39	27
Turns (1672-1687 cm^{-1})	8	16	16	24	25	32
B-sheet Anti-parallel (1687-1700 cm^{-1})	3	7	8	13	18	20
Amide II						
B-sheet parallel (1485-1502 cm^{-1})	8	22	24	23	29	33
Random (1502-1529 cm^{-1})	27	9	6	9	4	2
α-helix (1529-1563 cm^{-1})	52	55	55	52	46	43
Turns (1563-1587 cm^{-1})	13	14	13	13	13	14
B-sheet Anti-parallel (1587-1600 cm^{-1})	1	2	3	4	7	9
Amide III						
B-sheet (1220-1255 cm^{-1})	30	20	14	16	14	12
Random (1255-1301 cm^{-1})	19	20	21	20	21	22
Turns (1301-1317 cm^{-1})	29	36	41	41	47	49
α-helix (1317-1330 cm^{-1})	23	25	24	23	19	17

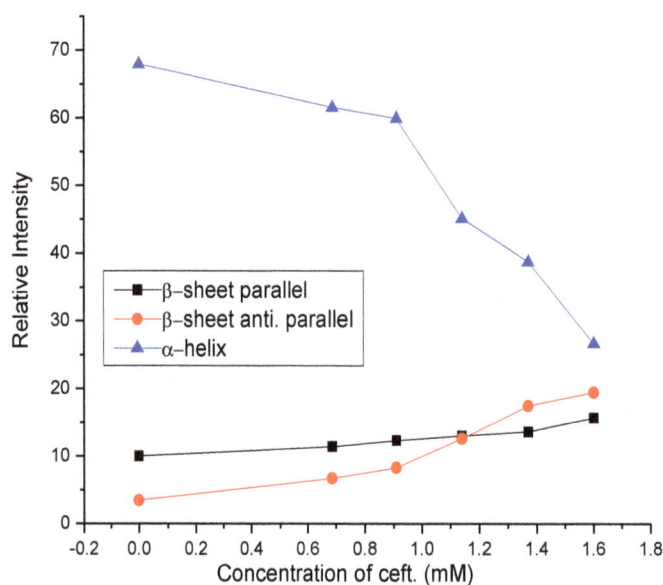

Figure 6. Relative intensities of HSA secondary-structure Amide I components as a function of increasing ceftriaxone concentration. The isolated symbols on the right correspond to the respective secondary structure components.

the increase of β-sheets are believed to be due to the unfolding of the protein in the presence of ceftrixone as a result of the formation of H bonding between HSA and the antibiotic. The newly formed H-bonding result in the C-N bond assuming partial double bond character due to a flow of electrons from the C=O to the C-N bond which decreases the intensity of the original vibrations (Krimm et al., 1986); the hydrogen bonds in α-helix are formed inside the helix and parallel to the helix axis, while for β-sheet, the hydrogen bonds take position in the planes of β-sheets as the preferred orientations especially in the anti-parallel sheets, so the restrictions on the formation of

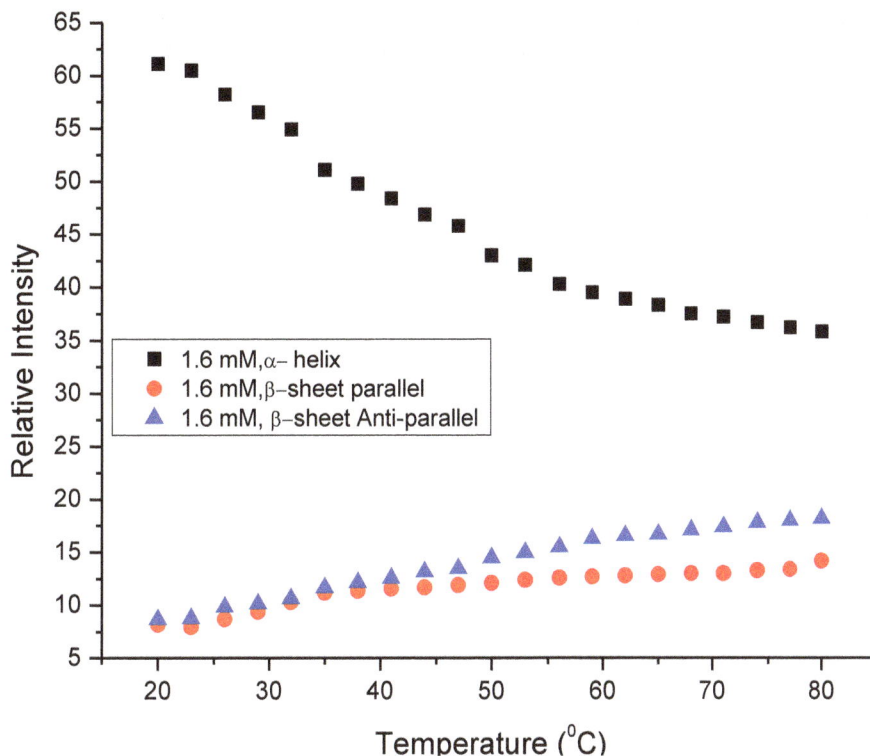

Figure 7. Relative intensities of HSA secondary-structure components at different concentrations ceftriaxone as a function of increasing temperatures. The isolated symbols on the left correspond to the respective secondary structure components.

hydrogen bonds in β-sheet relative to the case in α-helix explains the larger effect on reducing the intensity percentage of α-helix to that of β-sheet. Similar conformational transitions from an α-helix to β-sheet structures were observed for the protein unfolding upon protonation and heat denaturation (Surewicz et al., 1987; Holzbaur et al., 1996).

The variation of intensities with temperature change for α-helix, parallel and antiparallel β-sheets in ceftriaxone-HSA complexes are shown in Figure 7. The relative intensity increases as -temperature and ceftriaxone concentration increases- for the antiparallel β-sheets while it remarkably decreases for α-helix and the rate of increase for antiparallel β-sheets remarkably faster than the parallel β-sheets. The difference in behavior for the two types of β-sheets can be explained by the different amino acids and their preferred secondary structural arrangements in these β-sheets. It has been reported, that the two forms of β-sheets have different thermodynamic propensities scale (Kim et al., 1993).

The slight gradual increase in intensity of the parallel β-sheets is believed to be mainly due to the unfolding of the by fluorescence spectroscopy and by FTIR spectroscopy. From the fluorescence study, we determined values for the binding constant and the quenching constant for ceftriaxone-HSA complexes. The results indicate that the

protein as a result of the formation of H-bonding between HSA and ceftriaxone. The newly formed H-bonding result in the C-N bond assuming partial double bond character due to electrons flow from the C=O to the C-N bond which decreases the intensity of the molecular vibrations (Fabian et al., 1993).

On the other hand, an increase of intensity implies more stability and less conversions of the C=O bond as a result of the interaction with ceftriaxone. The parallel arrangement is less stable because the geometry of the individual amino acid molecules forces the hydrogen bonds to occur at an angle, making them longer and thus weaker.

Contrarily, in the anti-parallel arrangement, the hydrogen bonds are aligned directly opposite to each other, making stronger and more stable bonds.

Conclusion

The binding of ceftriaxone to HSA has been investigated

intrinsic fluorescence of HSA was quenched by ceftriaxone through static quenching mechanism. Analysis of the FTIR spectra reveals that HSA-ceftriaxone interaction results in major protein secondary structural changes in the compositions of α-helix to that of the β-sheets.

The obtained data show an increase in the intensity of the absorption band of the antiparallel βsheets as temperature increases. Higher concentrations of ceftriaxone provide an increase of the absorption band intensity for both the antiparallel and parallel β-sheets.

The intensity analysis is in support of higher stability for the anti-parallel β-sheets compared to that of parallel β-sheets. These variations in behavior are mainly due to the differences in the intrinsic properties of parallel and antiparallel-β sheets. More experimental studies are needed to determine the degree of stability of β-sheets and how does it affect protein folding.

ACKNOWLEDGEMENT

This work was supported by the German Research Foundation DFG Grant No. DR228/24-2.

REFERENCES

Abu Teir MM, Ghithan SJH, Darwish S, Abu-Hadid MM (2011). "Study of Progesterone interaction with Human Serum Albumin: Spectroscopic Approach," J. Appl. Biol. Sci. 5 (13):35-47.

Abu Teir MM, Ghithan SJH, Darwish S, Abu-hadid MM (2012). Multi-spectroscopic investigation of the interactions between cholesterol and human serum albumin. J. Appl. Biol. Sci. 6(3):45-55.

Ahmad B, Parveen S, Khan RH (2006). Effect of albumin conformation on the binding of ciprofloxacin to human serum albumin. Biomacromolecules 7:1350-1356.

Arrondo JL, Muga A (1993). Quantitative studies of the structure of proteins in solution by Fourier-transform infrared spectroscopy. Prog. Biophys. Mol. Biol. 59:23-56.

Artali R, Bombieri G, Calabi L, Del Pra A (2005). A molecular dynamics of human serum albumin binding sites. II Farmaco 60:485-495.

Bian Q, Xu LC, Wang SL, Xia YK, Tan LF, Chen JF, Song L, Chang HC, Wang XR (2004). Study on the relation between occupational fenvalerate exposure and spermatozoa DNA damage of pesticide factory workers. Occup. Environ. Med. 61:999-1005.

Bibiana N, Beatriz Farruggia, Guillermo Pico (1996). A Comparative study of the binding characteristics of ceftriaxone, cefoperazone and cefsudolin to human serum albumin. Biochem. Mol. Biol. Int. 40(4):823-831.

Bilirubin RB, Alex R (1989). Ceftriaxone binding to human serum albumin. Mol. Pharmacol. 36:478-483.

Brodersen R, Robertson A (1989). Ceftriaxone binding to human serum albumin: competition with birirubin. Mol. Pharnmacol. 36: 478-483.

Byler M and Susi H (1986). Examination of the secondary structure of proteins by deconvolved FTIR spectra. Biopolymers. 25:469-487.

Chen GZ, Huang XZ, Xu JG, Zheng ZZ, Wang ZB (1990). Method of Fluorescence Analysis, Science Press, Beijing,. 112-119.

Cui F, Qin L, Zhang G, Liu X, Yao X, Lei B (2008). A concise approach to 1,11-didechloro-6-methyl-40-O-demethyl rebeccamycin and its binding to human serum albumin: Fluorescence spectroscopy and molecular modeling method. Bioorg. Med. Chem.16:7615-7621.

Cui F, Wang J, Cui Y, Yao X, Qu G, Lu Y (2007). Investigation of interaction between human serum albumin and N6-(2-hydroxyethyl) adenosine by fluorescence spectroscopy and molecular modelling. Luminescence. 22:546-553.

Curry S, Brick P, Franks NP (1999). Fatty acid binding to human serum albumin: new insights from crystallographic studies. Biochem. Biophys. Acta 1441:131-140.

Darwish SM, Abu sharkh SE, Abu Teir MM, Makharza SA, Abu hadid MM (2010). Spectroscopic investigations of pentobarbital interaction with human serum albumin. J. Molecular Structure. 963:122-129.

Fabian H, Schultz C, Naumann D, Landt O, Hahn U, Saenger WJ (1993). Secondary Structure and temperature-induced unfolding and refolding of ribonuclease T1 in aqueous solution: A Fourier transform infrared spectroscopic study. Mol. Biol. 232:967-981.

Ganim Z, Tokmakoff A (2006). Spectral Signatures of Heterogeneous Protein Ensembles Revealed by MD Simulations of 2DIR Spectra. Biophys. J. 91 : 2636.

Guowen Zhang N, Nan Z, Lin W (2011). Probing the binding of vitexin to human serum albumin by multi spectroscopic techniques. J. Lumin 131:880-887.

Hal CP, Thomas PA, Ronald AH, Bliss SP (2000). Simmondsin and wax ester levels in 100 high-yielding jojoba clones. Industrial Crops Prod. 12:151-157.

He XM, Carter DC (1992). Atomic structure and chemistry of human serum albumin. Nature 358:209-215.

Holzbaur IE, English AM, Ismail AA (1996). FTIR study of the thermal denaturation of horseradish and cytochrome c peroxidases in D2O. Biochemistry 35:5488-5494.

Il'ichev AL, Gut LJ, Williams DG, Hossain MS, Jerie PH (2002). Area-wide approach for improved control of oriental fruit moth Grapholita molesta (Busck) (Lepidoptera: Tortricidae) by mating disruption. Gen. Appl. Entomol. 31:7-15.

Jackson M, Mantsch HH (1991). Protein secondary structure from FT IR spectroscopy with dihedral angles from three-dimensional Ramachandran plots. J. Chem. 69:1639-1643.

Jiang M, Xie MX, Zheng D, Liu Y, Li XY, Chen X (2004). Spectroscopic studies on the interaction of cinnamic acid and its hydroxyl derivatives with human serum albumin. J. Mol Structure. 692:71-80.

Kandagal PB, Shaikh SMT, ManjunathaDH, Seetharamappa J, Nagaralli BS (2007). Spectroscopic studies on the binding of bioactive phenothiazine compounds to human serum albumin. J. Photochem. Photobiol. A-Chem. 189(1):121-127.

Kang J, Liu Y, Xie MX, Li S, Jiang M, Wang YD (2004). Interactions of human serum albumin with chlorogenic acid and ferulic acid. Biochimica et Biophysica Acta.1674:205-214.

Kang S, Wu Y, X Li (2004). Effects of statin therapy on theprogression of carotid atherosclerosis: a systematic review and meta-analysis. Atherosclerosis 177(2):433-442.

Kim CA, Berg JM (1993). Thermodynamic beta-sheet propensities measured using a zinc-finger host peptide. Nature 362:267-270.

Kragh-Hansen U (1981). Molecular aspects of ligand binding to serum albumin. Pharmacol. Rev. 33:17- 53.

Krimm S, Bandekar J (1986). Vibrational spectroscopy and conformationof peptides, polypeptides and proteins. Adv. Protein Chem. 38:181-364.

Kumar VK, Ramasamy R (2002). Spectral and normal coordinate analysis of 6-methoxypurine. Indian J. Pure Appl. Phys. 40:252.

Lakowicz JR (1999). Principles of fluorescence spectroscopy. Kluwer Academic Publishers/Plenum Press, Dordrecht/New York.

Lakowicz JR, Weber G (1973). Quenching of protein fluorescence by oxygen. Detection of structural fluctuations in proteins on the nanosecond time scale Biochemistry 12 : 4161.

Li Y, He WY, Dong YM, Sheng F, Hu ZD (2006). Human serum albumin interaction with formononetin studied using fluorescence anisotropy, FT-IR spectroscopy and molecular modeling methods. Bioorg. Med. Chem. 14:1431-1436.

Liu JQ, Tian JN, He WY, Xie JP, Hu ZD, Chen XG (2004). J. Pharm. Biomed. Anal. 35:671.

Liu XP, Du YX, Sun W, Kou JP, Yu BY (2009). Study on the interaction of levocetirizine dihydrochloride with human serum albumin by molecular spectroscopy. Spectrochim. Acta Part A 74:1189.

Liu Y, Xie MX, Kang J, Zheng D (2003). Studies on the interaction of total saponins of panax notoginseng and human serum albumin by Fourier transform infrared spectroscopy. Spectrochim. Acta. Part A. 59:2747-2758.

Locht F, Dorcbe G, Anbert G, Boissier C, Bertrand AM, Branon J (1990). The penetration of ceftriaxone into human brain tissue. J. Antimicrob. Chemother. 26:81-86.

Mcnamara P, TruebV, Stoeckel K (1990). Ceftriaxone binding to human serum albumin. Inderect displacement by probenecid and diazepam. ochem. Pharmacol. 40:1247-1253.

Muravchick S, Smith DS (1995). Parkinsonian symptoms during emergence from general anesthesia. Anesthesiology 82:305-307.

Nerli B, Farruggia B, Picó G (1996). A comparative study of the binding characteristic of of ceftriaxone, cefoperazone and cefsulodin to human
serum albumin. Biochem. Mol. Biol. Int. 40:823-831.

Oberg KA, Ruysschaert JM, Goormaghtigh E (2004). The optimization of protein secondary structure determination with infrared and circular dichroism spectra. Eur. J. Biochem. 271:2937-2948.

Pan J, Ye Z, Cai X, Wang L, Cao Z (2012). Biophysical study on the interaction of ceftriaxone sodium with bovine serum albumin using spectroscopic methods. J. Biochem. Mol. Toxicol. 26:487-492.

Patrick J, Mcnamara VrenyTruebs, Klaus Stoeckel (1990). Ceftriaxone binding to human serum albumin indirect displacement. Biochem. Pharmacol. 40(6):1247-1253.

Peters T (1985). Serum albumin. Adv. Protein Chem. 37:161-245.

Peters T Jr. (1985). Ado. Protein Chem. 37:161-245.

Qing Y, Xi-min Z, Xing-guo C (2011). Combined molecular docking and multi-spectroscopic investigation on the interaction between Eosin B and human serum albumin. J. Lumin. 131:880-887.

Quaglia MG, Bossu E, Dell'Aquila C, Guidotti M (1997). Determination of the binding of a β_2-blocker drug, frusemide and ceftriaxone to serum proteins by capillary zone electrophoresis. J. Pharm. Biomed. Analys. 15:1033-1039.

Rabia S, Shiori T, Leonid B, Sylvie D, Yves F, Dufr E, Vasanthy N, Erik G, Jean-Marie R, Vincent R (2009). Antiparallel β-sheet: a signature structure of the oligomeric amyloid β-peptide Emilie CERF. Biochem. J. 421:415-423.

Reynold Spector (1987). Ceftriaxone transport through the blood-brain barrier. J. Infect. Dis. 156(1):209-211.

Rondeau P, Armenta S, Caillens H, Chesne S, Bourdon E (2007). Assessment of temperature effects on b-aggregation of native and glycated albumin by FTIR spectroscopy and PAGE: Relations between structural changes and antioxidant properties. Archives Biochem. Biophys. 460:141-150.

Royer CA (1995). Approaches to teaching fluorescence spectroscopy. Biophys. J. 68:1191-1195.

Scott VM, William AP Jr, Kristin KJ, Lori K, Joseph AP (2008). Safety of ceftriaxone sodium at extremes of age. Expert Opin. Drug Saf. 7(5):515-523.

Simard AR, Soulet D, Gowing G, Julien JP, Rivest S (2006). Bone marrow-derived microglia play a critical role in restricting senile plaque formation in alzheimer's disease. Neuron 49:489-502.

Sirotkin VA, Zinatullin AN, Solomonov BN, Faizullin DA, Fedotov VD (2001). Calorimetric and Fourier transform infrared spectroscopic study of solid proteins immersed in low water organic solvents. Biochimica et Biophysica Acta. 1547:359-369.

Sudip Bhattacharyya, Tod ES, Jean H, Swank, Craig BMarkwardt (2006). RXTEObservationsof1A1744361: correlated spectral and timing behavior. Astrophys. J. 652:603-609.

Sulkowska A (2002). Interaction of drugs with bovine and human serum albumin. J. Mol. Struct. 614:227-232.

Surewicz WK, Mantsch HH, Chapman D (1993). Determination of protein secondary structure by Fourier transform infrared spectroscopy: A critical assessment. Biochemistry 32:389-394.

Surewicz WK, Moscarello MA, Mantsch HH (1987). Secondary structure of the hydrophobic myelin protein in a lipid environment as determined by Fourier- transform infrared spectrometry. J. Biol. Chem. 262:8598-8609.

Tian JN, Liu JQ, Zhang JY, Hu ZD, Chen XG (2003). Chem. Pharm. Bull. 51:579.

Ulrich H, Cleber AT, Arthur AN 2006). DNA and RNA apta-mers: from tools for basic research towards therapeutic applications. Comb. Chem. High Throughput Screen 9:619-32.

Vandenbussche G, Clercx A, Curstedt T, Johansson J, Jornvall H, Ruysschaert JM (1992). Structure and orientation of the surfactant associated protein C in a lipid bilayer. Eur. J. Biochem. 203:201-209.

Vass E, Holly S, Majer Z, Samu J, Laczko I, Hollosi M (1997). FTIR and CD spectroscopic detection of H-bonded folded polypeptide structures. J. Mol. Struct. 408/409:47-56.

Wang C, Wu Q, Li C, Wang Z, Ma J, Zang X, Qin N (2007). Interaction of tetrandrine with human serum albumin: a Fluorescence quenching study. Analyt. Sci. 23:429-433.

Wang T, Xiang B, Wang Y, Chen C, Dong Y, Fang H, Wang M (2008). Spectroscopic investigation on the binding of bioactive pyridazinone derivative to human serum albumin and molecular modeling. Colloids Surf. B. 65:113-119.

Workman JR (1998). Applied Spectroscopy: Optical Spectrometers, Academic Press, San Diego. pp. 21-53.

Yue Q, Shen T, Wang C, Gao C, Liu J (2012). Study of the interaction of bovine serum albumin with ceftriaxone and the inhibition effect of Zinc (II). Int. J. Spectrosc. Article ID 284173, doi: 10.1155/2012/284173.

Zhang FL, Jespersen KG, Bjorstrom C, Svensson M, Andersson MR, Sundstrom V, Magnusson K, Moons E, Yartsev A, Inganas O (2006). Influence of guest solvents on the morphology and performance of solar cells based on polyfluorene copolymer/fullerene blends. Adv. Funct. Mater. 16:667-674

A thermodynamic investigation of bovine carbonic anhydrase II interaction with cobalt ion at 300 and 310K

G. Rezaei Behbehani[1]*, A. Divsalar[2, 3], A. A. Saboury[2] and Z. Rezaei[4]

[1]Chemistry Department, Imam Khomeini International University, Qazvin Iran.
[2]Institute of Biochemistry and Biophysics, University of Tehran, Tehran, Iran.
[3]Department of Biological Sciences, Tarbiat Moallem University, Tehran, Iran.
[4]Chemistry Department, Islamic Azad University, Gachsaran, Iran.

A thermodynamic study on the interaction of bovine carbonic anhydrase II, CAII, with cobalt ions was studied by using isothermal titration calorimetry (ITC) at 27 and 37°C in tris buffer solution at pH = 7.5. The heats of Co^{2+}+CAII interaction are reported and analysed in terms of the new solvation theory. It was indicated that there are three identical and non-cooperative sites for Co^{2+}. The binding of a cobalt ion is exothermic with dissociation equilibrium constants of 81.306 and 99.126 µM at 27 and 37°C respectively. The binding of cobalt ions can cause some changes in structure of enzyme, which results in a decrease in the activity and stability of the enzyme.

Key words: Bovine carbonic anhydrase, cobalt ion, isothermal titration calorimetry.

INTRODUCTION

The carbonic anhydrase, CA, is ubiquitous zinc enzymes, presents in Archaea, prokaryotes and eukaryotes, being encoded by three distinct, evolutionarily unrelated gene families: the α–CA, β-CA and the γ-CA (Sly and Hu, 1995; Lyer et al., 2006; Sarraf et al., 2004; Supuran et al., 2001). CA is one of the fastest enzymes known, with a maximal turnover rate for CO_2 hydration of ~10^6 s^{-1} at 25°C, which catalyze the reversible hydration of CO_2 to form HCO_3^- and protons according to the following reaction: $CO_2 + H_2O \leftrightarrow H_2CO_3 \leftrightarrow HCO_3^- + H^+$. The first reaction catalyst by carbonic anhydrase and second reaction occurs in stantaneously, which is probably the reason why the activation of CA has not been much studied. In contrast, inhibition of CA has been widely investigated and several crystal structures of CA complexes with inhibitor molecules have been reported (Sly and Hu, 1995; Lyer et al., 2006; Sarraf et al., 2004; Supuran et al., 2001; Bertini et al., 1983). Seven distinct isozymes are presently known in higher vertebrates, though their physiological function is not completely known.

CAII is novel as a metal protein due to its unusually high affinity for zinc, so that the CAII+Zn^{2+} dissociation constant is 1 - 10 pM (Sly and Hu, 1995). The role of highly conserved aromatic residues surrounding the zinc binding site of human carbonic anhydrase II (CAII) in determining the metal ion binding specificity of this enzyme has been previously examined by mutagenesis (Lindskog and Nyman, 1964; Sarraf et al., 2005). Residues F93, F95, and W97 are located along a β-strand containing two residues that coordinate zinc, H94 and H96, and these aromatic amino acids contribute to the high zinc affinity and slow zinc dissociation rate constant of CAII. Substitutions of these aromatic amino acids with smaller side chains enhance the copper affinity (up to 100-fold) while decreasing the affinity of both cobalt and zinc, thereby altering the metal binding specificity up to 10^4-fold. Furthermore, the free energy of the stability of native CAII, determined by solvent-induced denaturation, correlates positively with increased hydrophobicity of the amino acids at positions 93, 95, and 97 as well as with cobalt and zinc affinity (Stadie and O'Brien, 1933; Nishino et al., 1999). Conversely, increased copper affinity correlates with decreased protein stability. Although CAII is loaded with zinc in its physiologically relevant format, it can bind a number of other metal ions in the zinc binding site, such as Co^{2+}, Ni^{2+}, Cu^{2+}, Cd^{2+}, Hg^{2+} and Pb^{2+} with various affinities (Supuran et al., 2001; Sarraf et al., 2005). Some Zn (II) and Cu (II) metal complexes of sulfonamides

*Corresponding author. E-mail: grb402003@yahoo.com.

incorporating polyaminopolycarboxylated tails have also been reported, which indeed showed very good in vitro CA inhibitory activity against isoforms CA I, II, and IV. Rami et al reported the preparation and inhibition assay of some Cu (II) complexes of aromatic/heterocyclic sulfonamides incorporating EDTA and DTPA tails. In addition, such copper (II) derivatives with potent CA IX/XII inhibitory activity might also be important for developing positron emission tomography (PET) imaging agents for tumor hypoxia (Winum et al., 2007; Alzuet et al., 1998). In this paper the effect of the cobalt ion on the structure and stability of the CAII, in addition to some investigations on the binding parameters of Co^{2+} to the enzyme has been considered.

MATERIALS AND METHODS

Erythrocyte bovine carbonic anhydrase was obtained from Sigma. Copper sulfate was obtained from Merck. The buffer solution used in the experiments was 50 mM Tris, pH = 7.5, which was obtained from Merck. All the experiments were carried out in 300 and 310 K. The experiments were performed with the 4-channel commercial microcalorimetric system, Thermal Activity Monitor 2277, Thermometric, Sweden. Each channel is twin heat conduction calorimeter (multijuction thermocouple plates) positioned between the vessel holders and the surrounding heat sink. Both sample and reference vessels were made from stainless steel. The limited sensitivity for the calorimeter is 0.1 μ cal. cobalt nitrate solution (5 mM) was injected by use of a Hamilton syringe into the calorimetric titration vessel, which contained 1.8 mL CA, 30 μM, in Tris buffer (30 mM), pH = 7.5. Thin (0.15mm inner diameter) stainless steel hypodermic needles, permanently fixed to the syringe, reached directly into the calorimetric vessel. Injection of copper nitrate solution into the perfusion vessel was repeated 30 times and each injection included 20μL cobalt nitrate solution. The calorimetric signal was measured by a digital voltmeter that was part of a computerized recording system. The heat of injection was calculated by the "Thermometric Digitam 3" software program. The heat of dilution of the cobalt solution was measured as described above except CAII was excluded. Also, the heat of dilution of the protein solution was measured as described above except that the buffer solution was injected to the protein solution in the sample cell. The enthalpies of copper and protein solutions dilution were subtracted from the enthalpies of cobalt nitrate solutions in CAII solutions. The determined enthalpies for Co^{2+}+CAII interactions, were listed in Table 1 (in μJ). The micro-calorimeter was frequently calibrated electrically during the course of the study.

RESULTS AND DISCUSSION

It has been shown previously (Rezaei et al., 2006; Rezaei and Saboury, 2007; Rezaei et al., 2008; Rezaei and Saboury, 2008; Rezaei et al., 2008) that the enthalpies of interactions of biopolymers with ligands (Co^{2+}+CAII in this case) in the aqueous solvent (Co^{2+}+water in the present case) mixtures, can be reproduced via the following equation:

$$q = q_{max} x'_B - \delta_A^\theta (x'_A L_A + x'_B L_B) - (\delta_B^\theta - \delta_A^\theta)(x'_A L_A + x'_B L_B) x'_B \quad (1)$$

The parameters δ_A^θ and δ_B^θ are the indexes of the CAII stability as a result of interaction with Co^{2+} in the low and high Co^{2+} concentrations respectively. If the binding of ligand at one site increases the affinity for ligand at another site, the macromolecule exhibits positive co-operativity. Conversely, if the binding of ligand at one site lowers the affinity for ligand at another site, the protein exhibits negative cooperativity. If the ligand binds at each site independently, the binding is non-cooperative. p < 1 or p > 1 indicate positive or negative cooperativity of macromolecule for binding with ligand respectively; p = 1 indicates that the binding is non-cooperative. x'_B can be expressed as follows:

$$x'_B = \frac{px_B}{x_A + px_B} \quad (2)$$

x_B is the fraction of the Co^{2+} needed for saturation of the binding sites, and $x_A = 1 - x_B$ is the fraction of unbounded Co^{2+}. We can express x_B fractions, as the total Co^{2+} concentrations divided by the maximum concentration of the Co^{2+} upon saturation of all CAII as follows:

$$x_B = \frac{[Co^{2+}]_T}{[Co^{2+}]_{max}} \qquad x_A = 1 - x_B \quad (3)$$

$[Co^{2+}]_T$ is the total concentration of cobalt and $[Co^{2+}]_{max}$ is the maximum concentration of the cobalt upon saturation of all CAII. In general, there will be "g" sites for binding of Co^{2+} per CAII molecule. L_A and L_B are the relative contributions of unbounded and bounded Co^{2+} to the enthalpies of dilution with the exclusion of CAII and can be calculated from the enthalpies of dilution of Co^{2+} in buffer, q_{dilut}, as follow:

$$L_A = q_{dilut} + x_B \left(\frac{\partial q_{dilut}}{\partial x_B} \right)$$

$$L_B = q_{dilut} - x_A \left(\frac{\partial q_{dilut}}{\partial x_B} \right) \quad (4)$$

The heats of Co^{2+}+CAII interactions, q, were fitted to Equation 1 over the whole range of Co^{2+} compositions. In this procedure, the only adjustable parameter (p) was changed until the best agreement between the experimental and calculated data was approached (Figure 1).

The optimized δ_A^θ and δ_B^θ values are recovered from the coefficients of the second and third terms of equation

Table 1. The heats of Co^{2+}+CAII interactions, q, at 300 K (O) and 310 K (Υ). qdilut are the heats of dilution of $Co(NO_3)_2$ with water. Precision is ±0.100 µJ or better.

$[Co^{2+}]$ / mM	[CAII] / µM	q (O) / µJ	q_{dilut} (O)/ µJ	q (Υ) / µJ	q_{dilut} (Υ)/ µJ
0.055	29.670	-635.3	-424.8	-563.5	-398.3
0.109	29.348	-1065.4	-790.4	-954.2	-742
0.161	29.032	-1350.6	-1089.6	-1222.6	-1021.5
0.213	28.723	-1543.7	-1344	-1410.7	-1260
0.263	28.421	-1679.4	-1562.4	-1546.7	-1464.7
0.312	28.125	-1778.5	-1740	-1648.3	-1631.2
0.361	27.835	-1853.4	-1895.2	-1726.5	-1776.7
0.408	27.551	-1911.7	-2028.8	-1788.2	-1902
0.454	27.273	-1958.2	-2140.8	-1838	-2008
0.500	27.000	-1996.1	-2232	-1879	-2092.5
0.544	26.733	-2027.6	-2316.5	-1913.2	-2171.3
0.588	26.470	-2054.1	-2390.1	-1942.2	-2240.3
0.631	26.214	-2076.7	-2456.5	-1967.1	-2302.5
0.673	25.962	-2096.2	-2510.9	-1988.6	-2353.5
0.714	25.714	-2113.2	-2561.8	-2007.4	-2401.7
0.755	25.472	-2128.1	-2605	-2024	-2442.2
0.794	25.233	-2141.3	-2644.8	-2038.8	-2479.7
0.833	25.000	-2153.1	-2681.6	-2052	-2513.2
0.872	24.771	-2163.7	-2715.2	-2063.9	-2544.7
0.909	24.545	-2173.2	-2745	-2074.6	-2572.4
0.946	24.324	-2181.8	-2771.4	-2084.3	-2597.2
0.982	24.107	-2189.7	-2794.8	-2093.2	-2618.9
1.018	23.894	-2196.9	-2816.4	-2101.3	-2640.2
1.053	23.684	-2203.5	-2836.2	-2108.8	-2658.9
1.087	23.478	-2209.6	-2854.6	-2115.7	-2676.6
1.121	23.276	-2215.2	-2870.4	-2122.1	-2691.6

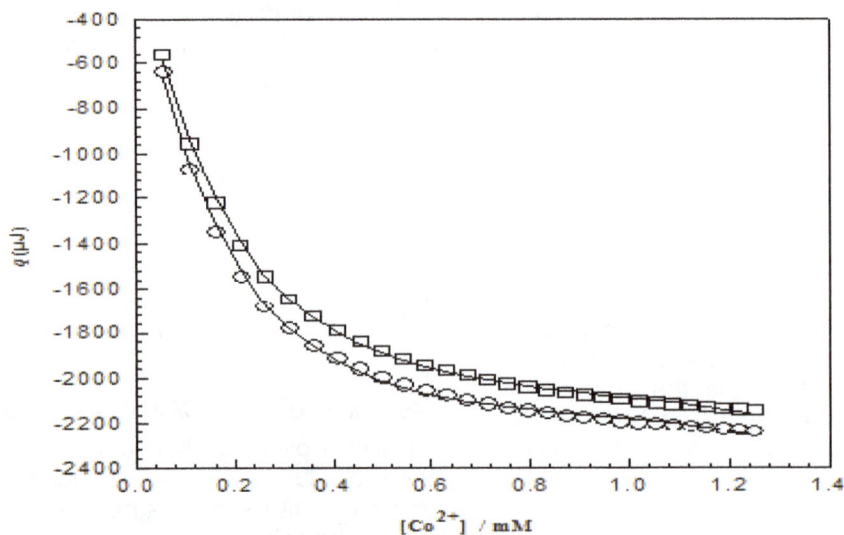

Figure 1. Comparison between the experimental heats for Co^{2+}+CAII interactions at 300 K (O) and 310 K (Υ) and calculated data (lines) via Equation 1.

coefficients of the second and third terms of Equation 1. The agreement between the calculated and experimental results (Figure 1) is excellent, and gives considerable support to the use of Equation 1.

Φ is the fraction of CAII molecule undergoing complexation with Co^{2+} which can be expressed as follows:

$$\Phi = \frac{q}{q_{max}} \qquad (5)$$

q_{max} represents the heat value upon saturation of all CAII. The appearance equilibrium constant values, K_a, as a function of free concentration of Co^{2+}, $[Co^{2+}]_F$, can be calculated as follows:

$$K_a = \frac{\Phi}{(1-\Phi)[Co^{2+}]_F} = \frac{\Phi}{(1-\Phi)[Co^{2+}]_T(1-x_B)} \qquad (6)$$

The standard Gibbs free energies as a function of Co^{2+} concentrations can be obtained as follows:

$$\Delta G^o = -RT LnK_a \qquad (7)$$

The standard Gibbs energies, ΔG^o, at different temperatures calculated from Equation 7 have shown graphically in Figure 2. $T\Delta S$ values were calculated using ΔG^o values at different temperatures and have shown in Figure 3.

Consider a solution containing a ligand (Co^{2+}) and a macromolecule (CAII) that contains "g" sites capable of binding the ligand. If the multiple binding sites on a macromolecule are identical and independent, the ligand binding sites can be reproduced by a model system of monovalent molecules $[(CAII)_g \rightarrow g(CAII)]$ with the same set of dissociation equilibrium constant, K_d, values. Thus, the reaction under consideration can be written:

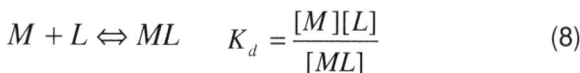

$$M + L \Leftrightarrow ML \qquad K_d = \frac{[M][L]}{[ML]} \qquad (8)$$

If α is defined as the fraction of free binding sites on the biomacromolecule, M_0 is the total biomacromolecule concentration, and L_0 is the total ligand concentration, then the free concentrations of monovalent molecule [M] and ligand [L] as well as the concentration of bound ligand [ML] can be deduced as follows:

$$[ML] = g(1-\alpha)M_0 \qquad (9)$$

$$[L] = L_0 - [ML] = L_0 - g(1-\alpha)M_0 \qquad (10)$$

$$[M] = gM_0 - [ML] = gM_0 - g(1-\alpha)M_0 = \alpha gM_0 \qquad (11)$$

Substitution of free concentrations of all these components in Equation (8) gives:

$$K_d = (\frac{\alpha}{1-\alpha})L_0 - \alpha gM_0 \qquad (12)$$

or

$$\alpha M_0 = (\frac{\alpha}{1-\alpha})\frac{1}{g}L_0 - \frac{K_d}{g} \qquad (13)$$

The value of $1-\alpha$ as the fraction of occupied binding sites on the biomacromolecule:

$$1-\alpha = \frac{q}{q_{max}} \qquad (14)$$

Where q represents the heat value at a certain L_0 and q_{max} represents the heat value upon saturation of all biomacromolecules. If q and q_{max} are calculated per mole of biomacromolecule then the molar enthalpy of binding for each binding site (ΔH_{bin}) will be:

$$\Delta H_{bin} = \frac{q_{max}}{g}$$

The combination of Equations 13 and 14 yields:

$$\frac{\Delta q}{q_{max}}M_0 = (\frac{\Delta q}{q})L_0\frac{1}{g} - \frac{K_d}{g} \qquad (15)$$

Where $\Delta q = q_{max} - q$. Therefore, the plot of $\frac{\Delta q}{q_{max}}M_0$

versus $\frac{\Delta q}{q}L_0$ should be a linear plot with a slope of $\frac{1}{g}$

and a vertical-intercept of $\frac{K_d}{g}$.

The linearity of the plot has been examined by different estimated values for q_{max} to reach the best value for the correlation coefficient. The best linear plot with the correlation coefficient value ($r^2 \approx 1$) was obtained using -2398 µJ and -2333 µJ (equal to -22.204 kJ/mol and -21.602 kJ/mol at 27 and 37°C respectively). The values of g and

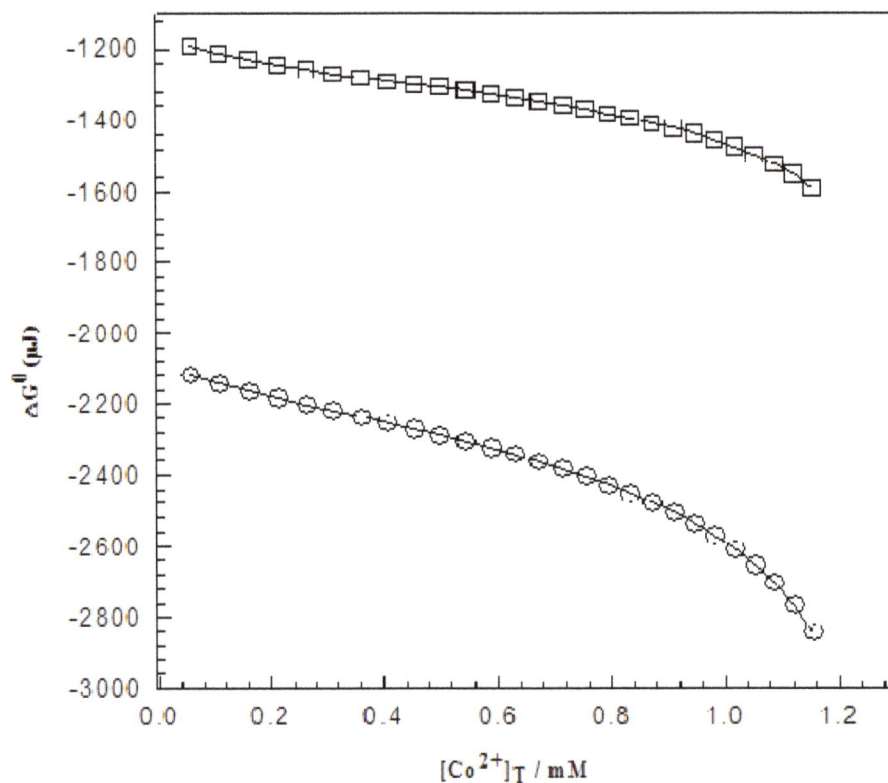

Figure 2. Comparison between the experimental Gibbs free energies at 300 K (O) and 310 K (Υ) for Co^{2+}+CAII interactions in and calculated data (lines) via Equations 7 and 8.

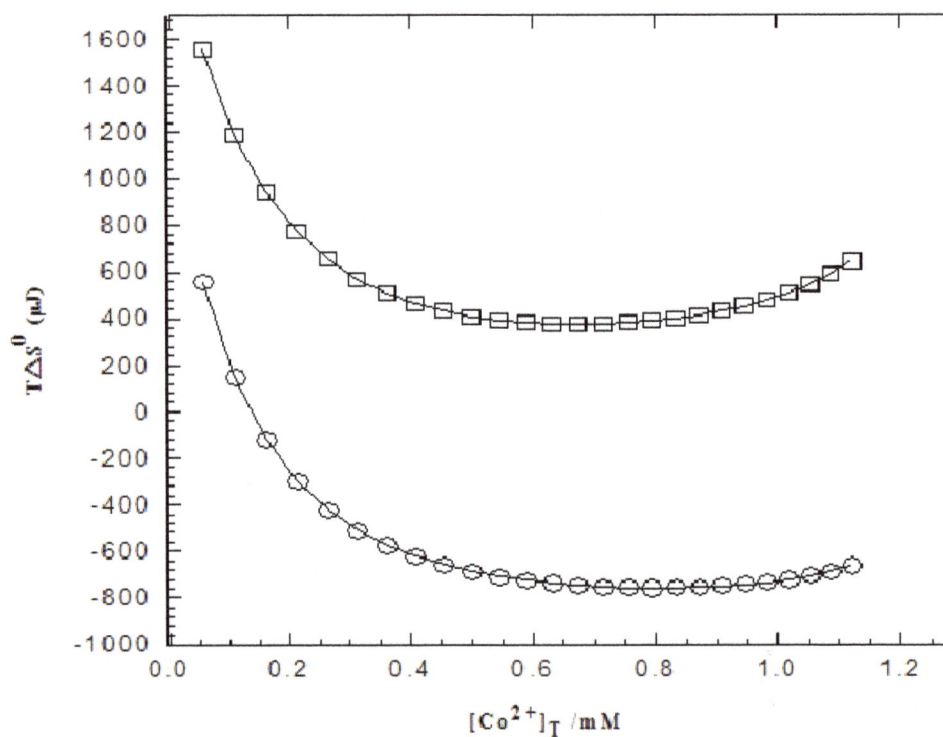

Figure 3. Comparison between the experimental entropies at 300 K (O) and 310 K (Υ) for Co^{2+}+CAII interactions in and calculated data (lines) via equations 7 and 8.

values of g and K_d, obtained from the slope and vertical-intercept plot, are listed in Table 2. The calorimetric method described recently allows obtaining the number of binding sites (g), the molar enthalpy of binding site (ΔH_{bin}) and the dissociation equilibrium constant (K_d) for a set of biomacromolecule binding sites. The lack of a suitable value for q_{max} to obtain a linear plot of $\square$ ($\Delta q/q_{max}$) $M0$ vs. ($\square\Delta q/q$) $L0$ may be attributed to the existence of non-identical binding sites or the interaction between them. Using this method shows that there is a set of three identical and non-interacting binding sites for cobalt ions. Binding parameters for Co^{2+}+CAII interactions using the new model are listed in Table 2.

It is possible to introduce a correlation between change in δ_A^θ and increase in the stability of proteins. The δ_A^θ value reflects the hydrophobic property of CAII, leading to the enhancement of water structure. The small δ_A^θ values recovered from Equation 1 suggest that there are no significant changes in CAII structure as a result of it interaction with Co^{2+} ion. In the high concentration of Co^{2+}, the positive value of δ_B^θ (3.680) reflects stabilization of the CAII structure in the high concentration of Co^{2+} ion. P = 1 value shows the non-cooperativity for the interaction of Co^{2+} ions with CAII including specific interactions with native CAII structure. These results are consistent with the association equilibrium constants due to specific interactions in the certain sites on CAII ($K_1 = K_2 = K_3 = 87.150$ mM^{-1}), underlying the existence of some partially unfolded intermediate forms of CAII, which form Co^{2+}+CAII complexes. The positive values of δ_B^θ suggest that specific interactions, defined here as preferential interactions between Co^{2+} ion and the native folded state of CAII, are dominant.

REFERENCES

Sly WS, Hu PY (1995). Humancarbonic anhydrases an carbonic anhydrase deficies. Annu. Rev. Biochem 11:375-401.
Lyer R, Barrese AA,Parker CN, Tripp BC (2006). J.Biomol. Screen. 11:782.
Sarraf NS, Saboury AA, Ranjbar B, Moosavi-Movahedi AA (2004). Structural and functional changes of bovine carbonic anhydrase as a consequence of temperature. Acta Biochemica Polonica. 51:665-671
Supuran CT, Scozzafava A, Mastrolorenzo A (2001). Bacterial pro-

teases: current therapeutic use and future prospects for the development of new antibiotics. Exp. Op. Ther. Patents, 2:221-259
Bertini I, Lanini G, Luchinat C (1983). Equilibrium species in cobalt (II) carbonic anhydrase. J. Am. Chem. Soc., 105:5116-5118
Lindskog S, Nyman PO (1964). Metal-binding properties of human erythrocyte carbonic anhydrases. Biochim.. Biophys. Acta, 85: 462-474
Sarraf NS, Mamaghani-Rad S, Karbassi F, Saboury AA (2005). Thermodynamic Studies on the Interaction of Copper Ions with Carbonic Anhydrase. Bull. Korean Chem. Soc., 26:1051-1056
Stadie WC, O'Brien H(1933). The catalysis of the hydration of carbon dioxide and dehydration of carbonic acid by an enzyme isolated from red blood cells. J. Biol. Chem., 103:521-529
Nishino S, Ishikawa Y, Nishida Y(1999). Interaction between a copper (II) compound and protein investigated in terms of the capillary electrophoresis method. Inorg. Chem. Com., 2: 438-441
Winum JY, Rami M, Montero JL, Scozzafava A, Supuran CT(2007). Carbonic anhydrase IX: A new druggable target for the design of antitumor agents. Med. Res. Rev., 28:445-463
Alzuet G, Casanova J, Borrfis j, Garcfa-Granda S, Gutidrrez-Rodrfguez A, Supuran CT (1998). Copper complexes modelling the interaction between benzolamide and Cu-substituted carbonic anhydrase. Crystal structure of Cu (bz) (NH₃)₄ complex. Inorg. Chim. Acta, 273:334-338
Rezaei Behbehani G, Ghamamy S, Waghorne WE (2006). Enthalpies of transfer of acetonitrile from water to aqueous methanol, ethanol and dimethylsulphoxide mixtures at 298.15 K. Thermochim. Acta, 448:37-42
Rezaei Behbehani G, Saboury AA (2007). Using a new solvation model for thermodynamic study on the interaction of nickel with human growth hormone. Thermochim. Acta 452: 76-79
Rezaei Behbehani G, Saboury AA, Taleshi E (2008). A Comparative Study of the Direct Calorimetric Determination of the Denaturation Enthalpy for Lysozyme in Sodium Dodecyl Sulfate and Dodecyltrimethylammonium Bromide Solutions. J. Solution Chem. 37:619-629
Rezaei Behbehani G, Saboury AA, Takeshi E (2008). Determination of partial unfolding enthalpy for lysozyme upon interaction with dodecy ltrimethylammonium bromide using an extended solvation model. J. Mol. Recognit. 21:132-135
Rezaei Behbehani G, Divsalar A, Saboury AA, Gheibi N (2008). A new approach for Thermodynamic Study on binding some metal ions with human growth hormone. J. Solution Chem.

pH uniquely modulates protein arginine methylation

Wen Xie[1], George Merz[2] and Robert B. Denman[3]*

[1]Division of Hematology and Medical Oncology, Department of Medicine, Weill Medical College of Cornell University, New York, NY 10065, USA.
[2]Department of Developmental Neurobiology, New York State Institute for Basic Research in Developmental Disabilities, 1050 Forest Hill Road, Staten Island, NY 10314, USA.
[3]Department of Molecular Biology, New York State Institute for Basic Research in Developmental Disabilities, 1050 Forest Hill Road, Staten Island, NY 10314, USA.

Protein arginine methyltransferases (PRMTs) function in the alkaline milieu of the nucleus and at neutral pH of the cytosol. Accordingly, several PRMTs are broadly active over a range of pHs. We investigated the effect altering pH had on protein arginine methylation using a variety of defined substrates, recombinant PRMTs and cell extracts. We demonstrate that pH-induced alterations in the extent of methylation and the methyl-product formed depend both on the particular substrate assayed and the PRMT that modifies it. We also find that transient intracellular alkalinization of mouse embryonic P19 neurons by NH_4Cl results in sustained changes in substrate methylation. Altogether our results are consistent with a hypothesis in which altered substrate methylation resulting from pH-induced changes of PRMT activities coupled with low levels of demethylation may provide the long-term tag(s) necessary for the formation and maintenance of "molecular memory".

Key words: Arginine methylation, protein arginine methyltransferase, pH, P19 cells, insect cell lysate, SmD1 peptide.

INTRODUCTION

pH homeostasis is crucial to life (Garcia-Moreno, 2009). Because of this organisms have evolved complex molecular systems that both sense pH e.g. carbonic anhydrase (CA) (Seksek and Bolard, 1996; Tresguerres et al., 2010) and the capsacin receptor (TRPV1) (Dhaka et al., 2009) aswell asmaintain intracellular and intraorganellar pH; the latter include: the Na^+/H^+ antiporter (Murer et al., 1976), the acid pumping VATPase (Tabares and Betz, 2010), a group of Cl^-/HCO_3^- exchangers (Chesler, 2003) and Na^+ coupled bicarbonate transporters (NCBTs) (Majumdar and Bevensee, 2010).

Changes in intracellular or intraorganellar pH affect the charge state of many amino acid side chains and thus can profoundly alter the activities of enzymes. Indeed, the fact that every enzyme has an optimal pH at which it functions is a basic principle of biochemistry (Fersht, 1977). The pH of the nucleus is among the most basic of all of the organelles (Seksek and Bolard, 1996). Thus, it is not surprising that enzymes operating in this milieu would tend to have pH optima that are more basic. Protein methyltransferases (PMTs) methylate a variety of nuclear proteins including histones (Izzo and Schneider, 2010), splicing factors (Chen et al., 2010; Deng et al., 2010), spliceosomal components (Miranda et al., 2004), transcription factors (Kowenz-Leutz et al., 2010), and mRNA export complex components (Hung et al., 2010). In addition, several protein methyltransferases modulate their own activities via auto methylation (Chin et al., 2007; Frankel et al., 2002; Kuhn et al., 2010). In keeping with this, Zhang et al have shown that protein lysine

*Corresponding author. E-mail: rbdenman@yahoo.com.

methyltransferases (PKMTs) and protein arginine methyltransferases (PRMTs) are quite active at alkaline pH (Cheng et al., 2005; Zhang and Cheng, 2003). This feature assists in catalysis as the substrate amino lysine and amino arginine groups of these enzymes must be deprotonated in order to initiate a nucleophilic attack on the Ado-Met methyl group. Nevertheless, PRMTs are also found in the cytoplasm, which is pH-neutral (Garcia-Moreno, 2009), where they methylate RNA binding proteins (Dolzhanskaya et al., 2006; Stetler et al., 2006; Xie and Denman, 2011), spliceosomal components (Friesen et al., 2001; Meister et al., 2001), and receptors (Chen et al., 2004; Infantino et al., 2010). In this case it is likely that hydrophobic groups surrounding the methyltransferase active site play a role in lowering the pKa of the target amino substituent group so that it can stay in the deprotonated state required for its methylation.

To date, the effect pH has on protein arginine methyltransferase activity has not been systematically evaluated. Here we have begun to address this problem by assessing *in vitro* protein arginine methylation at neutral and basic pH, approximating the cytosol/peroxisome and the mitochondria/nucleus environments, respectively (Abad et al., 2004; Garcia-Moreno, 2009; Jankowski et al., 2001; Seksek and Bolard, 1996). We demonstrate that the pH at which *in vitro* methylation reactions occur significantly impacts the type and extent of methylation in a substrate-specific manner. We also find that intracellular alkalinization results in the modification of protein arginine methyltransferases activity ex vivo, implying that protein arginine methylation is functionally coupled to local pH changes.

MATERIALS AND METHODS

Buffers and chemicals

HMTase Buffer is 50 mM Tris-HCl pH 9.0, 1 mM PMSF and 0.5 mM DTT as previously described (Denman, 2006; Dolzhanskaya et al., 2006). PBS and SNARF-1AM were purchased from Invitrogen. Protease inhibitors were purchased from Roche. ω-N^G-Monomethylarginine (MMA), ω-N^G, N^G-asymmetric dimethylarginine (aDMA), ω-N^G, N^G-symmetric dimethylarginine (sDMA) and S-adenosyl-homocysteine (SAH) were obtained from Sigma. [^{3}H]-S-Adenosyl-L-methionine 15 Ci/mmol was purchased from MP Biomedical, Inc.

Proteins, peptides and antibodies

Recombinant PRMT1 and recombinant CARM1/PRMT4 were purchased from Millipore. Recombinant PRMT5 and PRMT7 were obtained from Origene. Histone H3 and H4 were purchased from Roche Applied Biosciences. Myelin basic protein (mouse) was obtained from Sigma. CIRP overexpression lysate and NOLA1 protein were purchased from GenWay. *FMR1* exon 15-17 truncation mutants, FMRP$_{Ex15a}$ and FMRP$_{Ex15b}$ were prepared as

previously described (Dolzhanskaya et al., 2006; 2008). Biotinylated-SmD1peptide, biotin-KREAVAGRGRGRGRGRGRGRGRGGPRR, was synthesized by CPC Scientific.Antibodies to PRMT1, PRMT4, PRMT5 and PRMT7 were purchased from Millipore. Anti-Hsp70c mAb was obtained from Stressgen. Anti-dimethylarginine antibody, ASYM24, was purchased from Millipore.

In vitro translation systems

TNT rabbit reticulocyte lysate (RRL) *in vitro* translation lysate was purchased from Promega; the insect cell *in vitro* translation kit (ICL) was purchased from Qiagen.

Protein preparation and Western blotting

Proteins were prepared from cultured cells as previously described (Sung et al., 2003). Western blotting was performed as described (Sung et al., 2003). Rabbit polyclonal antibodies to ASYM24, PRMT1, PRMT4, PRMT5 and PRMT7 were used at a 1:300, 1:1000, 1:1000, 1:500 and 1:500 dilution of primary antibody, respectively. HSP70c was detected using a 1:5000 dilution. A 1:5,000 dilution of HRP-conjugated goat anti-rabbit secondary antibody or HRP-conjugated goat anti-mouse secondary antibody (Pierce) was used for detection. Blots were blocked for 1 h at room temperature in PBS supplemented with 3% non-fat dry milk and probed overnight in fresh buffer with the corresponding primary antibody at 4°C. Blots were developed using the PicoTag system (Pierce).

In vitro methylation assays

In vitro methylation by the endogenous methyltransferases (MTs) of the proteins in cultured cell lysates (30 μg), or *in vitro* translation lysates (30-50 μg) or various substrate proteins (1 μg) was performed in either HMTase-Buffer (50 mM Tris-HCl pH 9, or pH 7.5, 0.5 mM DTT) or PBS supplemented with 1 μCi of ^{3}H-SAM in a total volume of 20 μl. *Because ^{3}H-SAM is reconstituted in weak acid the actual pH of these reactions is 8.3 and 7.0, respectively.In vitro* methylation by recombinant PRMTs were performed by denaturing the cell lysates (30 μg), or *in vitro* translation lysates (30-50 μg) for 3 min at 75°C and then incubating them in either HMTase-Buffer (50 mMTris-HCl pH 9, 0.5 mM DTT) or PBS supplemented with 1 μCi of ^{3}H-SAM and the indicated PRMT(1 μl, 0.1 μg) in a total volume of 20 μl. The methyltransferase dependence of the incorporated ^{3}H was demonstrated by the addition of 1000 μM S-adenosyl-homocysteine (SAH) to the reactions. Incubations were allowed to proceed for 2 h at 30°C. Following the incubation, an equal volume of 2x Laemelli buffer was added to the reaction mixtures and the proteins were resolved in duplicate on SDS-polyacrylamide gels. The gels were fixed, in 30% methanol, 10% acetic acid overnight. After removing the fixing solution the gels were either soaked in En^{3}Hance (Perkin Elmer) for 1 h and then water for 30 min. and then dried and subjected to fluorography as described, or stained with Coomassie Brilliant Blue (Dolzhanskaya et al., 2006). *In vitro* methylation of the biotinylated-SmD1 peptide (100 pmols) by the endogenous methyltransferases (MTs) of the proteins in cultured cell lysates (30 μg), or *in vitro* translation lysates (30-50 μg) or recombinant PRMTs (0.2 μg) was performed in either HMTase-Buffer (50 mM Tris-HCl pH 9, 0.5 mM DTT) or PBS

Table 1. PRMT Primers.

Name	Sequence	Size (bp)
PRMT1-F	5'GAGGCCGCGAACTGCATCAT3'	374, 428, 545
PRMT1-R	5'TGGCTTTGACGATCTTCACC3'	
PRMT3-F	5'GCAGTTGCTGGGTACTTTGATA3'	247
PRMT3-R	5'TCACTGGAGACTGTAAGTCTGG3'	
PRMT4-F	5'GCTGTGGCTGGAATGCCTACT3'	109, 179, 306, 584
PRMT4-R	5'TCCCTGGGCACCTGAGGACCT3'	
PRMT5-F	5'CACGAAGGCCAGAACATCTG3'	220
PRMT5-R	5'ATGTTCTACACCTTCTGTGC3'	
PRMT7-F	5'CAACAGCCTATGCAATCCAAGGGCAC3'	320
PRMT7-R	5'CTCAATAAGAGATCAGCTCAAGGTG3'	
PRMT8-F	5'GACATTTACACTGTGAAGACGG3'	434, 375
PRMT8-R	5'GATACAGATGTTTCACACAGCTG3'	

supplemented with 1 µCi of ^{3}H-SAM in a total volume of 20 µl. Incubations were allowed to proceed for 2 h at 30°C. Subsequently, 150 pmol of Tetralinkavidin resin (Promega) was added and the biotinylated peptide was captured at 4°C for 20 min. on a rotating mixing platform. The resin was centrifuged at 14K for 2 min. and the supernatant was removed. The resulting pellet was washed twice with 100 µl of buffer and the bound radioactivity counted in 10 ml of Filtron X (National Diagnostics). Competition reactions were performed using either 1000 µM SAH or 36 mM biotin (Sigma). Background values were measured in mock reactions containing all of the components except the methyltransferases(s) and were subtracted from the sample values.

Methylarginine amino acid analysis

In vitro methylation reactions (20 µl) containing 1 µCi of ^{3}H-SAM were hydrolyzed with 250 µl of 6 N HCl at 110°C for 21 h in a sealed glass ampule. The hydrolyzed amino acids were dried in an oven after opening the ampule. Ten microliters of water was added to the dried residue, and then 1 µl of the solution was applied to each lane of a Silica 60 TLC plate (Whatman) along with 1.0 µg(0.5 µl) ofeach of the standards ω-N^G-monomethylarginineacetatesalt (ω-MMA; SigmaM7033), asymmetric ω-N^G,N^G-dimethylarginine hydrochloride (aDMA; Sigma D4268) and ω-N^G, N^G-symmetric dimethylargininedi(p-hydroxyazobenzene-p'-sulfonate) salt (sDMA; Sigma D0390) for amino acid analysis by TLC. Sodium citrate buffer (0.35 M Na$^+$, pH5.27) was used to separate the amino acids (Miranda et al., 2004). Under these conditions aDMA and sDMA are not completely resolved and therefore, we consider the composite peak their sum, which we designate dimethylarginine (DMA). Color was developed with a ninhydrin spray (Sigma) following incubation of the plate at 55°C for 15 min.The plates were scanned with an Epson Perfection V300 scanner and the resulting image data converted to a ninhydrin intensity distributions using UN-SCAN-ITGel 6.1 (Silk Scientific Software). MMA and DMA spots were subsequently scraped into liquid scintillation vials containing 5 ml Filtron X (National Diagnostics) and were subsequently counted.

RNA Isolation and RT-PCR

RNA was extracted from cells and tissues using RNAeasy Mrna mini columns (Qiagen). Purified RNA was eluted in 25 µl of DEPC-treated H_2O and quantified spectrophotometrically. One microgram of RNA was used to prepare first strand cDNA (Invitrogen). Aliquots of the cDNAs were amplified in a Perkin Elmer GeneAmp 9700 thermocycler using the primers in Table 1 and the following cycling parameters: 94°C for 30 s, 55°C for 30 s and 72°C for 1 min. A range of cDNA concentrations and cycles was used to insure accurate quantification as previously described (Xie et al., 2009). Reaction products were resolved on 1–2% TAE agarose gels along with appropriate size markers and plasmid controls. Gels were imaged using a Scion CFW-1308M mega pixel camera and captured using FOTO/Analyst PC Image software version 9.04 (FOTODYNE).The resulting image files were digitized and analyzed using UN-SCAN-IT gel version 6.1 (Silk Scientific, Inc).

Cell culture

P19 cells were cultured in ω-MEM supplemented with 10% fetal bovine serum. Differentiation of rapidly growing cultures into neuronal and glial cells was accomplished *via* the method of Jones-Villeneuve (Jones-Villeneuve et al., 1982). All of the results presented here correspond to cells cultured for 6-8 days in the presence or absence of 0.1 µM retinoic acid.

Retinoic differentiated P19 cells were cultured on 40 mm coverslips.Twenty four hours later the cells were loaded with the pH-sensing fluorescent dye SNARF-1AM, 4 µM 30 min, 37°C, according to the manufacturer's instructions. Subsequently, the dye was washed out and the cover slips mounted in a Sykes-Moore chamber (Bellco Glass Inc., Vineland, NJ). The cells were maintained at 37°C with an air curtain incubator (NevTek, Williamsville, Va). Baseline images of the cells were acquired using a Nikon 90i microscope coupled to a Nikon C1 three-laser scanning confocal system (NIKON Instruments Inc., Melville, NY) by exciting at 488 nm and recording at 580 nm and 640 nm. The cells were treated

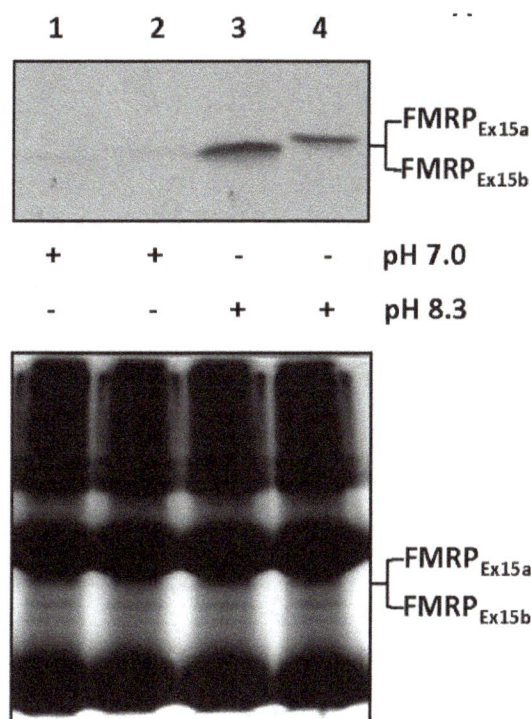

Figure 1A. Increasing pH facilitates FMRP *in vitro* methylation. Equal amounts of FMRPE$_{x15a}$ and FMRPE$_{x15b}$ were methylated with 30 µg of rabbit reticulocyte lysate at pH 7.0 and pH 8.3 as indicated. The reactions were resolved by SDS-PAGE and subject to fluorography.

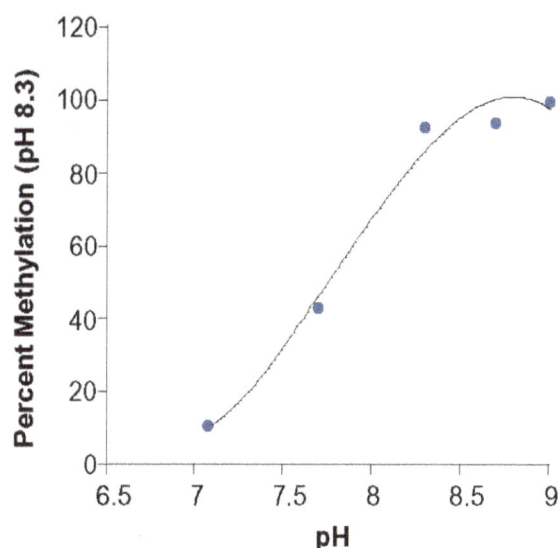

Figure 1B. Increasing pH facilitates FMRP *in vitro* methylation. Relative methylation of FMRPE$_{x15a}$ at various pHs. Fluorgram intensities were compared to the intensity at pH 8.3, which was arbitrarily set to 100%. Values plotted are the average of two experiments whose means differ by less than 10%.

or not treated with NH$_4$Cl (100 mM) and timelapse images recorded at 580 nm and 640 nm in 1 min intervals over the course of 12-20 min. The images were thresholded using Image-Pro (version 5.1) to define the cells and the ratio of fluorescence intensities, 640 nm/580 nm, within the cells was computed. The results were exported to a Microsoft Excel spread sheet for statistical analysis.

RESULTS

Increasing pH facilitates the *in vitro* methylation of select substrates.

We previously hypothesized that since histone lysine protein methyltransferases such as DIM5 and protein arginine methyltransferases PRMT1 and PRMT3 are especially active at alkaline pH (Cheng et al., 2005; Zhang and Cheng, 2003) increasing the pH of *in vitro* methylation reactions should increase PRMT-dependent incorporation of ^{3}H-SAM into its substrates (Denman, 2006). Figure 1A illustrates the results of the *in vitro* methylation of two truncated FMRP splice variants, FMRP$_{Ex15a}$ and FMRP$_{Ex15b}$, by the protein methyltransferases in rabbit reticulocyte lysate (RRL)

assayed in PBS buffer (reaction pH 7.0) and HMTase buffer (reaction pH 8.3). It is quite evident in this case that the higher pH favors incorporation of ^{3}H-SAM into these proteins. In fact, by examining a range pH values it can be shown that maximal ^{3}H-SAM incorporation into FMRP$_{Ex15a}$ occurs at a pH of 8.3, Figure 1B. Similar results were observed when we substituted an HMTase buffer (reaction pH 7.0) for PBS indicating that the effects were pH- rather than buffer-dependent. Thus, all subsequent experiments used this buffer for *in vitro* methylation reactions at the lower pH.

These data agree with that reported for the pH dependence of myelin basic protein (MBP) *in vitro* methylation by CARM1 (Denman, 2008). To examine whether other substrates were similarly affected by altering the reaction pH we compared the methylation of the FXRP family of proteins (FMRP, FXR1P, FXR2P), along with PABP by RRL at pH 7.0 and pH 8.3. Interestingly, in each of these cases, ^{3}H-SAM incorporation into the full-length proteins preferentially occurred at pH 7.0 rather than at pH 8.3, Figure 1C; nevertheless, the range of differences were profound. Full-length FMRP was much more significantly affected by changing pH than were the other FXRPs, whereas PABP methylation was not greatly affected. Similar results were obtained when we assessed the methylation of histone H3, histone H4, MBP and NOLA by RRL at pH 7.0 and pH 8.3, Figure 1D. Finally, we also examined the effect of pH on the methylation of the cold-inducible RNA binding protein, CIRP, expressed in

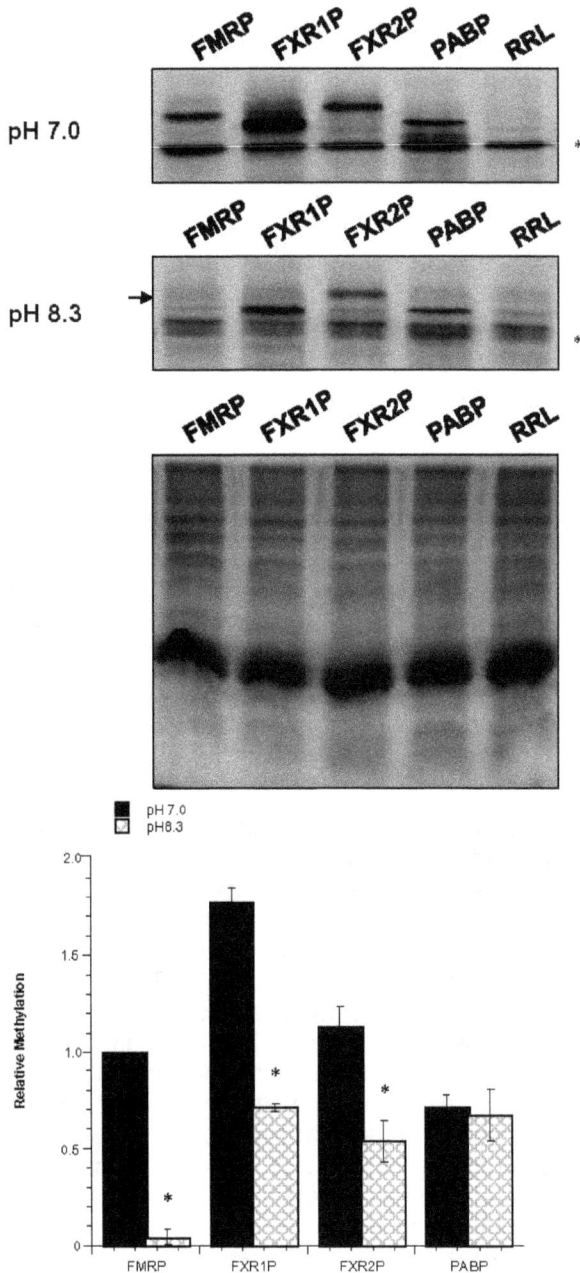

Figure 1C. Increasing pH facilitates FMRP *in vitro* methylation. Rabbit reticulocyte lysates (30 µg) programmed to express full-length FMRP, FXR1P, FXR2P, PABP or no protein were incubated in the presence of ³H-SAM at pH 7.0 and pH 8.3. The methylated proteins were resolved by SDS-PAGE and subject to fluorography. The asterisk marks a non-specific methylated RRL protein. The arrow in the second panel marks the location of FMRP. The relative protein loads are shown by the Coomassie gel in the third panel. The graph shows the relative methylation of each protein normalized to FMRP at pH 7.0. Values represents the means ± sem for three determinations/protein. For each member of the FXRP family methylation preferentially occurred at pH 7.0 (P=0.007 for FMRP, P=0.003 for FXR1P, P=0.05 for FXR2P by ANOVA).

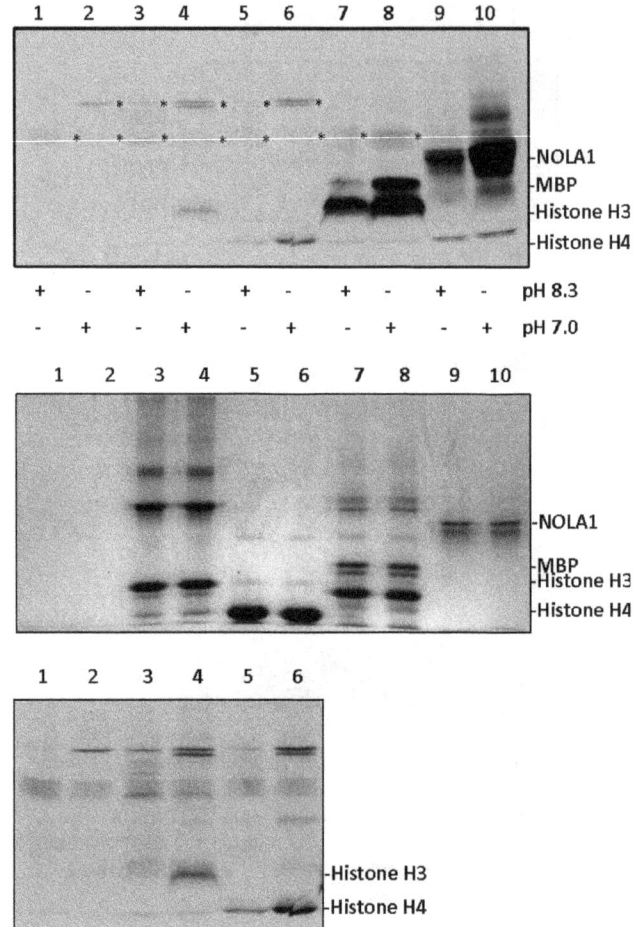

Figure 1D. Increasing pH facilitates FMRP *in vitro* methylation. Rabbit reticulocyte lysates (30 µg) were incubated in the presence of ³H-SAM and in the absence or presence of 1 µg of histone H3, histone H4, myelin basic protein (MBP) or NOLA1 at pH 7.0 and pH 8.3. The methylated proteins were resolved by SDS-PAGE and subject to fluorography. The asterisks mark non-specific methylated RRL proteins. The relative protein loads are shown by the Coomassie gel beneath the fluorgram. An over exposure of lanes 1-6 is shown below the protein loads.

HEK293 cells. In this case, the protein was more highly methylated at pH 8.3 than at pH 7.0, Figure 1E.

pH uniquely modulates the methylation of cellular proteins.

The aforementioned data suggested that *in vitro* methylationof different protein substrates is uniquely modulated by pH. To explore this question more generally we compared the ability of the endogenous protein methyltransferases in *Spodoptera fugiperda* cell lysates to methylate insect cell proteins at pH 7.0 and pH 8.3. We chose insect cell lysates because unlike most cell lysates there are a host of endogenous proteins that

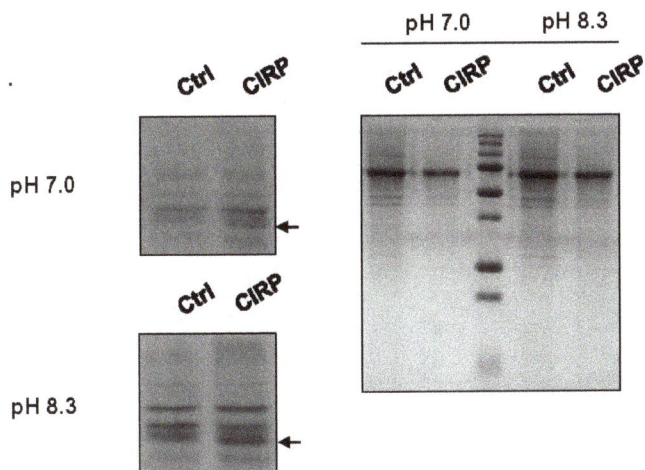

Figure 1E. Increasing pH facilitates FMRP *in vitro* methylation. HEK293 control and CIRP overexpressing cell lysates (30 µg) were incubated in the presence of ^{3}H-SAM at pH 7.0 and pH 8.3. The methylated proteins were resolved by SDS-PAGE and subject to fluorography. The arrow marks CIRP. The relative protein loads are shown by the Coomassie gel to the right of the fluorgram.

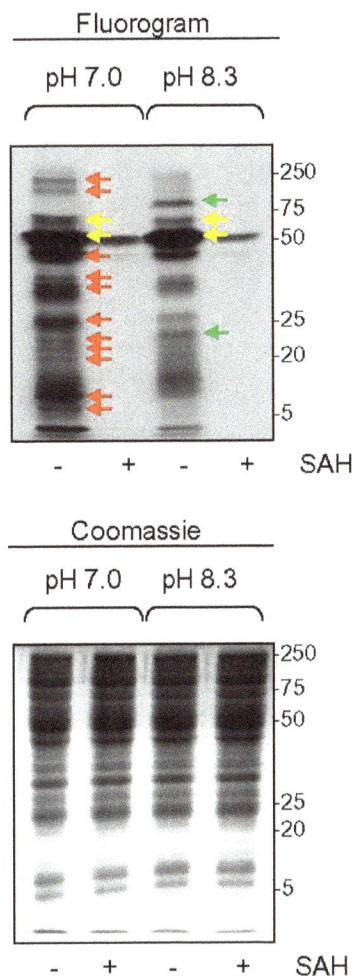

Figure 2. Alterations in pH affect the *in vitro* methylation of native insect cell proteins. Native insect cell lysates (30 µg) were incubated in the presence of ^{3}H-SAM and in the absence and presence of 1000 µM SAH at pH 7.0 and pH 8.3. The methylated proteins were resolved by SDS-PAGE and subject to fluorography. Red arrows indicate substrates that are preferentially methylated at pH 7.0; green arrows indicate substrates that are preferentially methylated at pH 8.3; yellow arrows indicated substrates that are equally methylated at both pHs. The relative protein loads are shown by the Coomassie gel beneath the fluorgram.

are hypomethylated, yet the lysates have abundant protein methyltransferase activity (Denman, 2008). In addition, to insure that the incorporation of ^{3}H-SAM into protein represented bona fide methylation, the reactions were carried out in the absence or presence of the methylation inhibitor S-adenosyl-homocysteine (SAH). Figure 2 shows the results of this experiment. In this case, the *in vitro* methylation of most *Spodoptera fugiperda* proteins was significantly better at pH 7.0; however, there were several notable exceptions to this rule and again, in some cases, the methylation of particular substrates was not substantially affected by the reaction pH.

These data indicated that the various protein MTs and/or substrates in the insect cell lysate were differentially affected by pH. To continue to evaluate the effect of pH on protein methylation we compared the ability of the MTs in RRL to methylate insect cell proteins. To insure that the methylation was a direct result of the input MTs from RRL we first denatured the insect cell proteins. Figure 3A shows that following denaturation at 75°C insect cell proteins are no longer able to be methylated by insect cell MTs. Two interesting results occurred when the MTs in RRL were used to methylate the insect cell proteins, Figure 3B. First, it was obvious that many of the insect cell substrates that were methylated by the endogenous insect cell MTs were not methylated by the MTs in RRL (compare Figure 2 with Figure 3B). Secondly, of the substrates that were methylated by RRL MTs, most were preferentially methylated at pH 8.3.

The difference between insect cell methylation patterns observed when the extracts were methylated by endogenous insect cell MTs as opposed to the RRL MTs might be due to changes in the type, amount or specificity

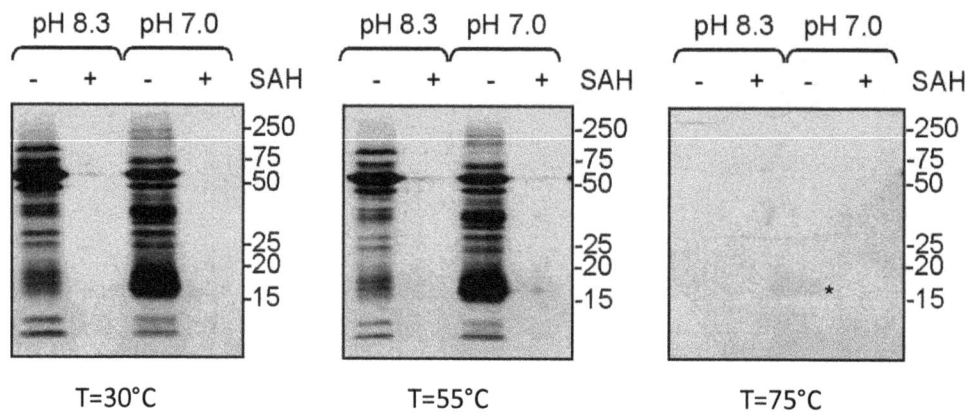

Figure 3A. Rabbit reticulocyte lysate differentially methylates denatured insect cell proteins depending on the pH of the reaction. Insect cell lysates (30 μg) were incubated in the presence of ³H-SAM, and in the absence and presence of 1000 μM SAH at pH 7.0 and pH 8.3 at the indicated temperatures. The methylated proteins were resolved by SDS-PAGE and subject to fluorography. The asterisk marks residual methylation at 75°C.

of the two populations of MTs, or it might be due to the fact that for the RRL methylation experiment the insect cell proteins were denatured by the heat inactivation of the insect cell MTs. To examine this question more closely we next assessed the effect recombinant human PRMTs had on denatured insect cell proteins at pH 7.0 and pH 8.3 to determine whether the methylation patterns produced would mimic those with the endogenous insect cell MTs or RRL MTs. Specifically, we examined two type I (hPRMT1 and hPRMT4) and two type II (hPRMT5 and hPRMT7) protein arginine methyltransferases.In every case, many more proteins were methylated by the recombinant PRMTs than were methylated by the MTs in RRL, Figure 4. Also, with the exception of hPRMT5, most of the methylation that occurred preferentially occurred at pH 7.0; nevertheless, for each hPRMT there were multiple substrates that were more highly methylated at pH 8.3.Again, these data demonstrate that pH uniquely modulates *in vitro* substrate methylation.

Since none of the methylated insect cell lysate proteins are known we decided to examine the effect pH had on the methylation of a well-characterized substrate by the human recombinant PRMTs. For this we chose a peptide corresponding to the RG-rich region of the SmD1 protein (Friesen et al., 2001). This protein is methylated by both type I and type II PRMTs *in vitro* and is differentially methylated by type II PRMTs in the cytoplasm and type I PRMTs in the nucleus (Miranda et al., 2004). Using an affinity-capture assay we demonstrated that the incorporation of ³H-SAM into the biotinylated-peptide was due to methylation as it could be competed out with the methylation inhibitor SAH or with free biotin, Figure 5A. Next, we showed that the methylation of this particular substrate by the endogenous MTs in insect cell lysates

was preferentially active at pH 8.3, Figure 5B. Finally, we assessed the effect of pH on SmD1 methylation by human recombinant PRMTs, Figure 5C. Here we found that the pH that the peptide was preferentially methylated depended entirely on the human PRMT that was being assayed.

pH alters the distribution of monomethylarginine and asymmetric dimethylarginine incorporated into cellular proteins

Although we have demonstrated that pH modulates the activity of *in vitro* methylation in both substrate choice and the extent of methylation we had no information regarding whether and if pH modulates the distribution of monomethylarginine (MMA), asymmetric dimethylarginine (aDMA) or symmetric dimethylarginine (sDMA) incorporated into cellular proteins. To begin to address this question we methylated NOLA1 using the endogenous protein methyltransferases in insect cell lysates at pH 7.0 and pH 8.3. Figure 6A shows that NOLA1, which migrates as a two bands ca. 22 and 18 kDa, is preferentially methylated at pH 7.0.A Western blot of these extracts, which was probed with an antibody that detects a subset of asymmetrically dimethylated arginine residues (ASYM24), demonstrated that (1) NOLA1 is specifically detected by anti-ASYM24 and (2) there is no corresponding increase in the level of aDMA in the pH 7.0 NOLA1-treated sample over the pH 8.3 NOLA1-treated sample, Figure 6B. These data imply that at pH 7.0a DMA is likely not incorporated into NOLA1.

To further examine this question we next hydrolyzed insect cell lysates that had been methylated at either pH

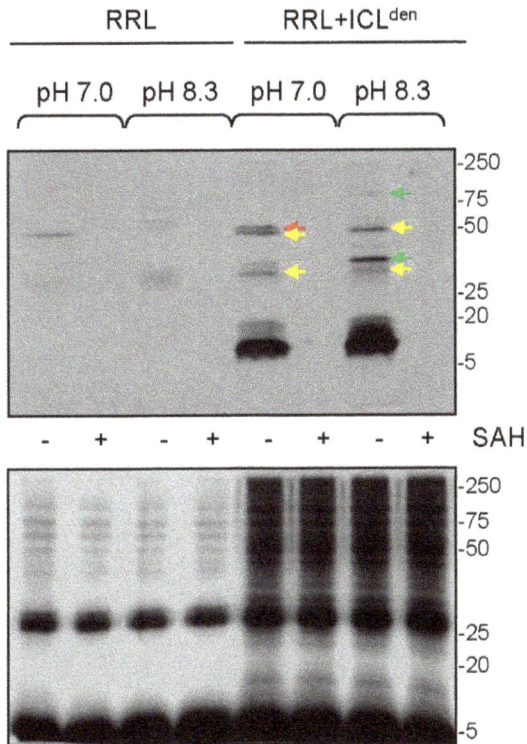

Figure 3B. Rabbit reticulocyte lysate differentially methylates denatured insect cell proteins depending on the pH of the reaction. Denatured insect cell lysates (30 μg) were incubated in the presence of ^{3}H-SAM, RRL and in the absence and presence of 1000 μM SAH at pH 7.0 and pH 8.3 as indicated. The methylated proteins were resolved by SDS-PAGE and subject to fluorography. Red arrows indicate substrates that are preferentially methylated at pH 7.0; green arrows indicate substrates that are preferentially methylated at pH 8.3; yellow arrows indicated substrates that are equally methylated at both pHs. Control lanes 1-4 show endogenous methylation of RRL proteins at both pHs. The relative protein loads are shown by the Coomassie gel beneath the fluorgram.

7.0 or pH 8.3 in the presence of ^{3}H-SAM into their amino acids and separated the methylarginine residues (MMA and aDMA) by TLC. The results, shown in Figure 6C, demonstrate that there is an approximate two-fold increase in the amount of MMA incorporated into protein at pH 7.0 relative to that at pH 8.3, confirming the results obtained with NOLA1 methylation.

Transient increases in intracellular pH produce sustained changes in cellular protein methylation

Can changes in intracellular pH drive alterations in protein arginine methylation? To begin to address this

question we used NH$_4$Cl to produce a transient intracellular alkalinization (Raley-Susman et al., 1991) in retinoic acid-differentiated mouse P19 cells. The alkalinization was monitored using SNARF-1AM, one of a family of fluorescent indicator dyes the ratio of whose emission wavelengths change as a function of pH (Roberts, 1999). Quite simply cells were loaded with SNARF-1AM and after a specified time the unabsorbed dye was washed out. The dye-loaded cells were then stimulated at a wavelength of 480 nm and the fluorescence intensities at 640 nm and 580 nm simultaneously recorded over time. Figure 7A shows that for untreated P19 cells the ratio the of fluorescence intensities at 640 nm and 580 nm (F_{640}/F_{580}) does not markedly change over the time we observed indicating that the intracellular pH was relatively stable. Nevertheless, when P19 cells were exposed to 100 mM NH$_4$Cl they exhibited a robust increase in F_{640}/F_{580} which lasted for 10 min, consistent with the hypothesis that NH$_4$Cl induces a sustained intracellular alkalinization, Figures 7A and 7B.

If NH$_4$Cl treatment alters either the extent or the specificity of substrate methylation one would expect to see a change in either the pattern or extent of methylation of the endogenous substrates in these extracts. To examine this question we first performed in vitro methylation reactions on P19 cell control and NH$_4$Cl-treated protein extracts in the presence of ^{3}H-SAM. The methylated proteins were then resolved by SDS-PAGE and subjected to fluorography. We discovered a general diminution of most of the methylated proteins, indicating that more of the P19 MT substrates had been methylated upon NH$_4$Cl treatment and were therefore unavailable for in vitro methylations than the control extracts, Figure 7C. To confirm that the implied changed occurred, we next assessed the PRMT activity in these extracts by supplementing them with ^{3}H-SAM and various exogenous protein substrates and comparing the extent of methylation. Strikingly, we again observed substrate-specific differences in the ability of these extracts to methylate particular proteins, Figure 7D. Specifically, NH$_4$Cl treatment greatly increased the ability of histone H4 and the SmD1 peptide to be methylated; whereas methylation of MBP was only marginally increased and NOLA1 methylation actually decreased slightly.

The differences in the methylation potential of the alkalinized and control P19 cell extracts could be explained by differences in either the activity or the expression of P19 PRMTs. To determine which PRMTs were expressed in P19 cells we used RT-PCR to assess PRMT mRNA expression. We found that both undifferentiated and retinoic acid-differentiated P19 cells express nearly identical amounts of PRMT1, PRMT3, PRMT4, PRMT5, PRMT7 and PRMT8 and that the splicing pattern between the two cell types also did not

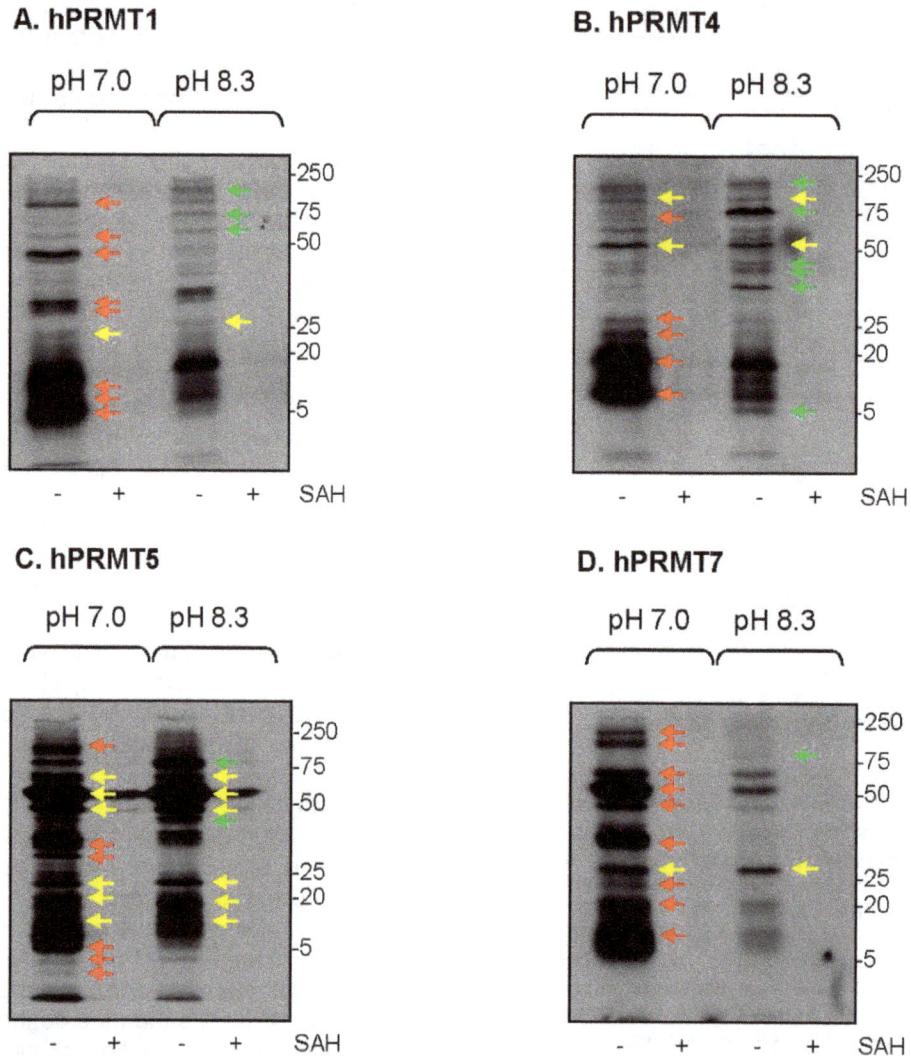

Figure 4. Recombinant human PRMT methylation of denatured insect cell proteins is affected by changes in pH. Denatured insect cell lysates (30 μg) were incubated in the presence of ^{3}H-SAM, a particular hPRMT and in the absence and presence of 1000 μM SAH at pH 7.0 and pH 8.3 as indicated. The methylated proteins were resolved by SDS-PAGE and subject to fluorography. Red arrows indicate substrates that are preferentially methylated at pH 7.0; green arrows indicate substrates that are preferentially methylated at pH 8.3; yellow arrows indicated substrates that are equally methylated at both pHs.

change significantly, Figure 8A. We then assessed the protein expression of PRMT1, PRMT3, PRMT4, PRMT5 and PRMT7 in the absence or presence of NH$_4$Cl, Figure 8B. PRMT1 and PRMT5 expression was robust and did not change as a function of treatment. On the other hand, we were unable to detect PRMT3 and PRMT7, indicating that these proteins are expressed at relatively low levels in P19 cells. Finally, PRMT4 protein expression, which was also weak, appeared to decrease slightly following NH$_4$Cl treatment. While not a comprehensive assessment of all 11 PRMTs these data indicate that the alterations in

substrate methylation seen following NH$_4$Cl treatment are most likely due to changes in the activities of the PRMTs rather than in their expression.

DISCUSSION

The family of SAM-dependent protein arginine methyltransferases has several well-conserved domains including a catalytic core, which has two conserved motifs a Rossman fold and a β-barrel, three signature

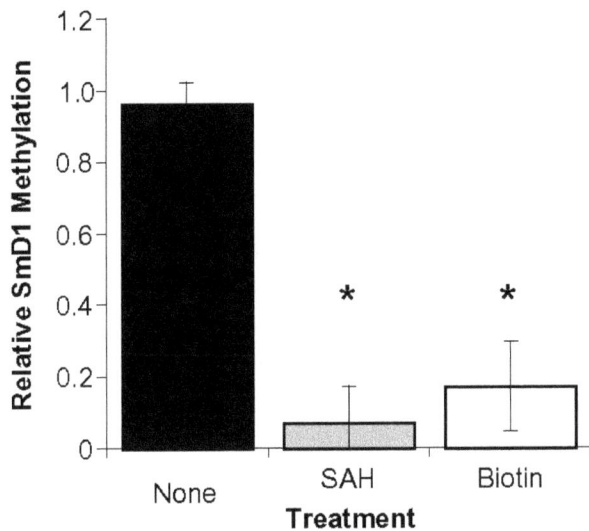

Figure 5A. Recombinant human PRMTs differentially methylate SmD1 at different pHs *in vitro*. SmD1 peptide was methylated by insect cell lysate(30 μg) in the presence of of 3H-SAM and in the absence or presence of 1000 μM SAH or 36 mM biotin as indicated. The incorporation of tritium into the peptide was measured by liquid scintillation counting.Values represent the means±sem for six determinations/group. (P=1.12x10^{-9} for biotin compared to control and P=3.1x10^{-6} for SAH compared to control by ANOVA.

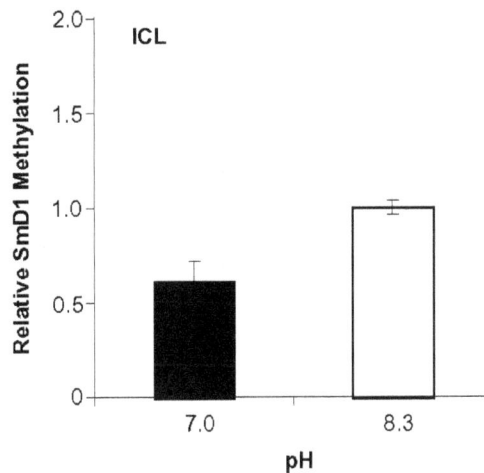

Figure 5B. Recombinant human PRMTs differentially methylate SmD1 at different pHs *in vitro*. SmD1 peptide was methylated by insect cell lysate (30 μg) in the presence of of ^{3}H-SAM at pH 7.0 and pH 8.3. Values represent the means±sem for six determinations/group. The extent of methylation at pH 7.0 was significantly different from that at 8.3 (P=1x10^{-3} by ANOVA). The incorporation of tritium into the peptide was measured by liquid scintillation counting.

SAM-dependent methylation motifs (I, II and III) and two PRMT-specific motifs (VRT and THWY) (Troffer-Charlier et al., 2007). The catalytic forms of these enzymes are dimers, although culture cell studies have shown that the PRMTs exist in a wide range of higher-order complexes and in multimers that are also active (Lim et al., 2005). Several PRMTs have been crystallized and the resulting structures have led to the hypothesis that deprotonation of the target ω-amino group, which is require for catalysis, occurs as a result of ionic interactions with the Ado-Met sulfonium ion, hydrophobic interactions and ionic interactions with a conserved Glu residue; all these serve to lower the pKa of the target ω-amino group (Zhang and Cheng, 2003).

The requirement that the target ω-amino group be deprotonated implies that these enzymes would be more active at alkaline pH and indeed, several studies have shown that particular PRMT substrates were more actively methylated when the pH of the reaction buffer was raised (Denman, 2008; Dolzhanskaya et al., 2006; Dolzhanskaya et al., 2008). Here, however, we have demonstrated that simply raising the reaction pH is not a general mechanism for accelerating protein arginine methylation. Rather, our data show that the extent of methylation depends on the substrate assayed. This was particularly evident when we examined the FXRP family of proteins. While full-length FMRP was much more

highly methylated at neutral pH, truncated forms of the protein were preferentially methylated at pH 8.3. This difference may be related to structural differences between the full-length and truncated proteins. X-ray crystallographic and NMR studies have shown that the C-terminal end of FMRP, which includes the methylation region, is unstructured (Adinolfi et al., 1999; Ramos et al., 2006). In contrast, molecular modeling studies of the truncated proteins have indicated that the sites of methylation lie within an extended β–sheet (Dolzhanskaya et al., 2008).

In addition to pH-induced substrate influences on methylation, we also demonstrated using a well-characterized single substrate, SmD1, that individual human recombinant PRMTs were differentially affected by changing pH, Figure 5. Thus, hPRMT1 was more active at pH 7.0 than pH 8.3, while hPRMT4, hPRMT5 and hPRMT7 were more active at pH 8.3 than at pH 7.0.

Cellular protein methylation occurs in a milieu containing multiple substrates and multiple methyltransferases. Using various cell lysates and cell extracts we next demonstrated that the pH at which *in vitro* methylation reactions were carried out greatly influence whether and to what extent endogenous proteins are methylated, Figures 2 to 4. Here, both the substrate variety and the extent of methylation varied significantly. At both high and low pH there were multiple substrates that were either preferentially or uniquely methylated. On the other hand, some substrates were

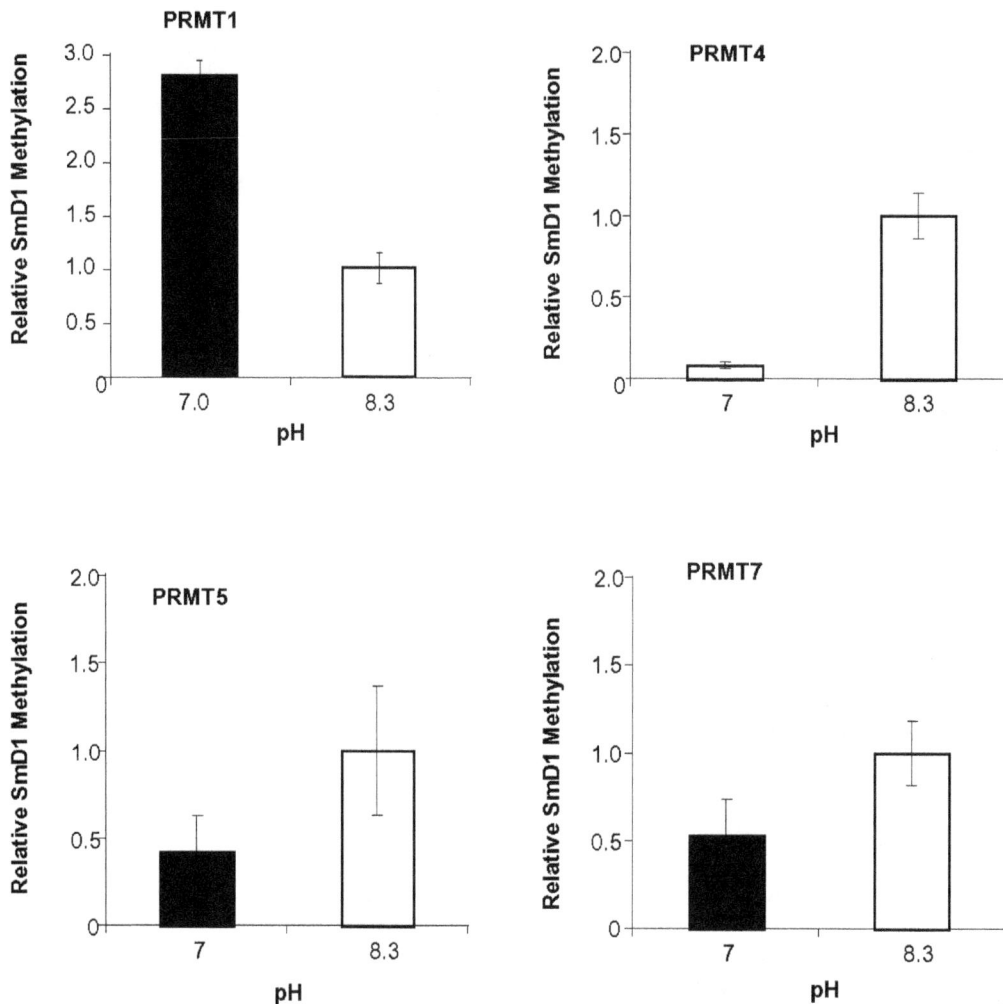

Figure 5C. Recombinant human PRMTs differentially methylate SmD1 at different pHs *in vitro*. SmD1 peptide was methylated by PRMT1, PRMT4, PRMT5 and PRMT7 in the presence of of ^{3}H-SAM at pH 7.0 and pH 8.3. The incorporation of tritium into the peptide was measured by liquid scintillation counting. Values represent the means±sem for four determinations/PRMT/group. The extent of methylation at pH 7.0 was significantly different from that at 8.3 ($P=2\times10^{-4}$ for PRMT1, $P=7\times10^{-5}$ for PRMT4, $P=9\times10^{-5}$ for PRMT5 and $P=1\times10^{-2}$ for PRMT7 by ANOVA).

not markedly affected by changing the pH. All of these data suggest that modulating a cell's intracellular pH should change both the extent and type of substrate proteins that are methylated.

What physiologically-relevant mechanisms do cells possess that would lead to pH-dependent methylation? As mentioned in the Introduction, intracellular and intra-organellar pH is tightly controlled and it is so because as little as a 0.5 unit change in pH can result in cell death (Garcia-Moreno, 2009). We investigated this question by artificially changing the extracellular pH that mouse embryonic P19 neurons were grown in using NH$_4$Cl. Previous studies had shown that bath application of

NH$_4$Cl to the medium that hippocampal neurons were grown in resulted in sustained alkalinization of the cytosol (Raley-Susman et al., 1991). Using various inhibitors these investigators determined the principal regulator of the alkalinization was the Na$^+$/H$^+$ antiporter, which is known to be, expressed in both retinoic acid differentiated and undifferentiated P19 cells (Bierman et al., 1987; Wang et al., 1997).

We first verified that NH$_4$Cl treatment of retinoic acid-induced P19 cells led to a sustained intracellular alkalinization, Figures 7A and 7B. More importantly however, we found that the alkalinization of the cytosol led to alterations in endogenous protein methylation and

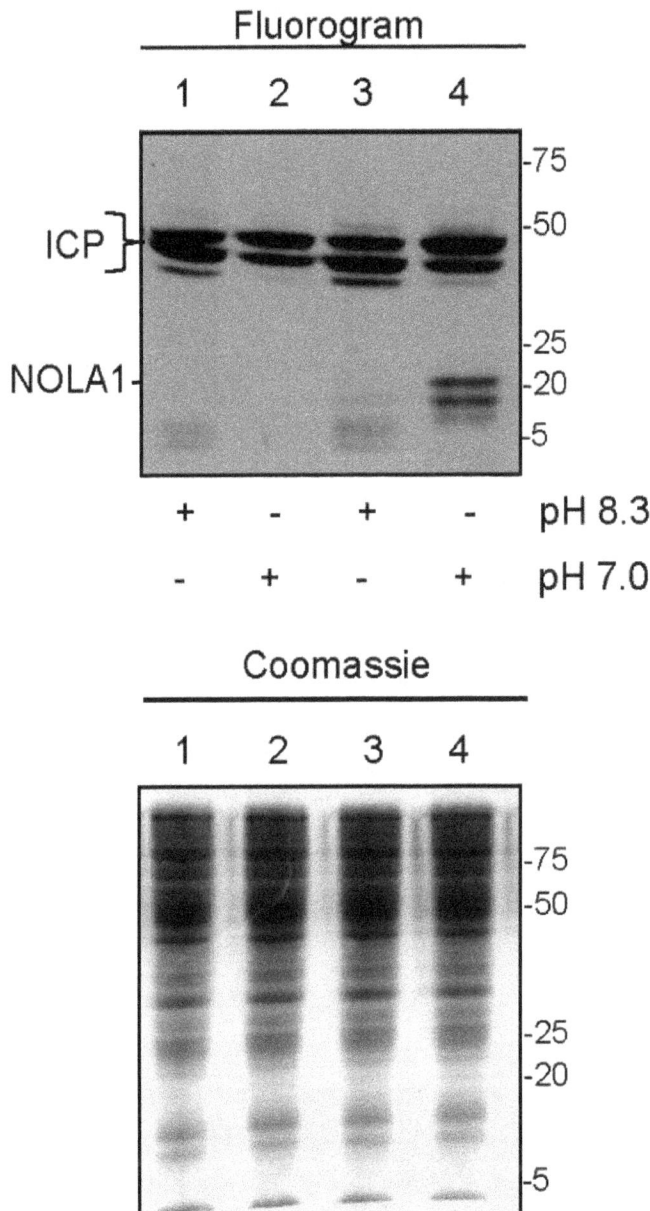

Figure 6A. pH alters the ratio of monmethyl to dimethylarginine. Insect cell lysates (30 μg) were incubated in the presence of ³H-SAM and in the absence and presence of NOLA1 (1 μg) at pH 7.0 and pH 8.3 as indicated. The methylated proteins were resolved by SDS-PAGE and subject to fluorography. ICP marks insect cell proteins that are prominently methylated during the reaction; recombinant NOLA1 migrates as a doublet centered at 20 KDa. The relative protein loads are shown by the Coomassie gel beneath the fluorgram. Coomassie gel to the right shows the NOLA1 protein before methylation.

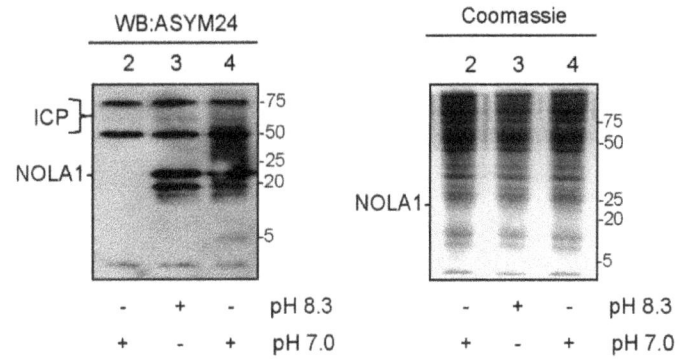

Figure 6B. pH alters the ratio of monmethyl to dimethylarginine. lanes 2-4 from Figure 6A were subject to Western blotting using anti-ASYM24.

alterations in PRMT activity. The data also suggest that other systems that induce transient or sustained changes in intracellular or intra-organellar pH will secondarily alter protein methylation and we are in the process of verifying this prediction.

How might the physiological pH changes accompanying the activation of the Na^+/H^+ antiporter or other pH sensors change protein arginine methylation without affecting PRMT expression? We speculate that the change in pH affects protein-protein interactions between various PRMTs and their regulators i.e. both activators and inhibitors. Precedence for this hypothesis can be found in papers by Jiang, Berthet and Guderian. Specifically, Jiang et al showed that adding the PRMT3/PRMT5 modulator Dal4.1B to in vitro methylation reactions of hypomethylated RAT-1 cells led to increases and decreases in the extent of methylation of particular substrates (Jiang et al., 2005). Likewise, BTG1, which binds to PRMT1, altered the extent of methylation of histone H2A (Berthet et al., 2002). Finally, RioK1, which binds to PRMT5, modulates the activity of the PRMT5 complex by enabling it to recruit and subsequently methylate the RNA binding protein nucleolin (Guderian et al., 2011). These changes in substrate methylation are reminiscent of the variations in substrate methylation we observe when the reaction pH is altered. Thus, it would be interesting to examine the influence pH has on the ability of each of these interactors to bind their respective PRMTs. Accordingly, one would expect that particular interactors would be either more or less associated with the particular PRMT at a given pH or that the expression levels of the interactors would fluctuate as a function of pH.

Besides changing substrate specificity and/or the extent of substrate methylation we also determined that pH influences the type of methylarginine product that is produced. Thus, at pH 7.0 there was an approximate

in the methylation of exogenously added substrates. Furthermore, the changes were not accompanied by changes in the expression of PRMT1, PRMT3, PRMT4, PRMT5 or PRMT7, implying the changes arise from

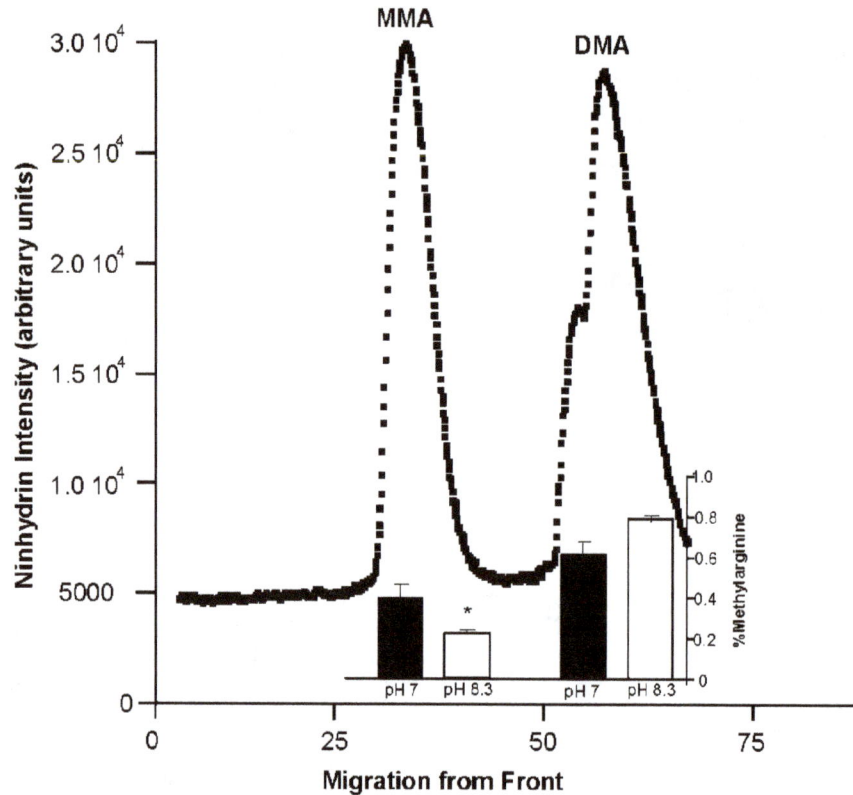

Figure 6C. pH alters the ratio of monmethyl to dimethylarginine. The distribution of MMA and DMA in *in vitro* methylated insect cell lysates at pH 7.0 and pH 8.3 was monitored byTLC. The graph shows the ninhydrin staining of the MMA and aDMA standards. The inset graph depicts the % distribution at the two pHs. Values represent the means±sem for four determinations/group.The distribution of MMA at pH 7.0 is significantly different from that at pH 8.3 (P=3×10^{-4} by ANOVA).

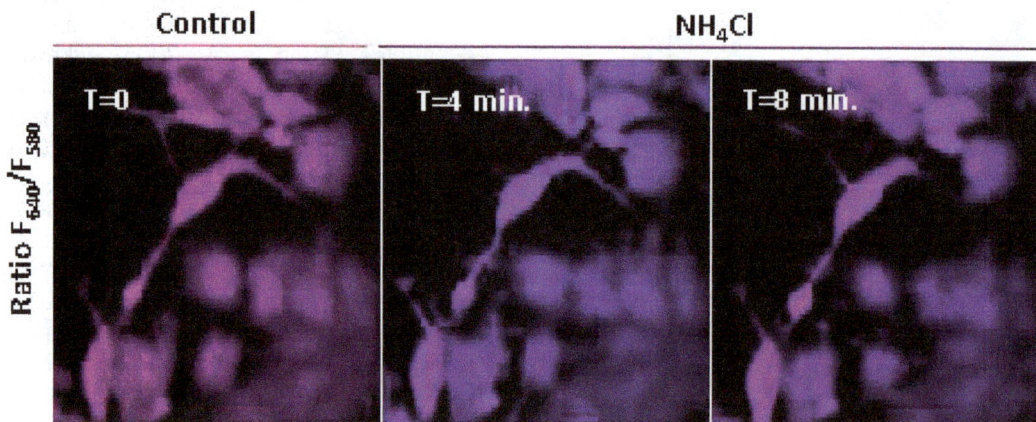

Figure 7A. Ammonium chloride induces a transient increase in intracellular pH and sustained increases in PRMT activity. Retinoic differentiated P19 cells (day 6) were loaded with the pH-sensing fluorescent dye SNARF-1AM. Subsequently, the cells were treated or not treated with NH_4Cl (100 mM) and subject to confocal fluorescence microscopy. The cells were excited at 488 nm and images recorded at 580 nm and 640 nm in 1 min intervals over the course of 12 min. (upper panel) Selected images of treated cells at 0, 4 and 8 min. are presented.

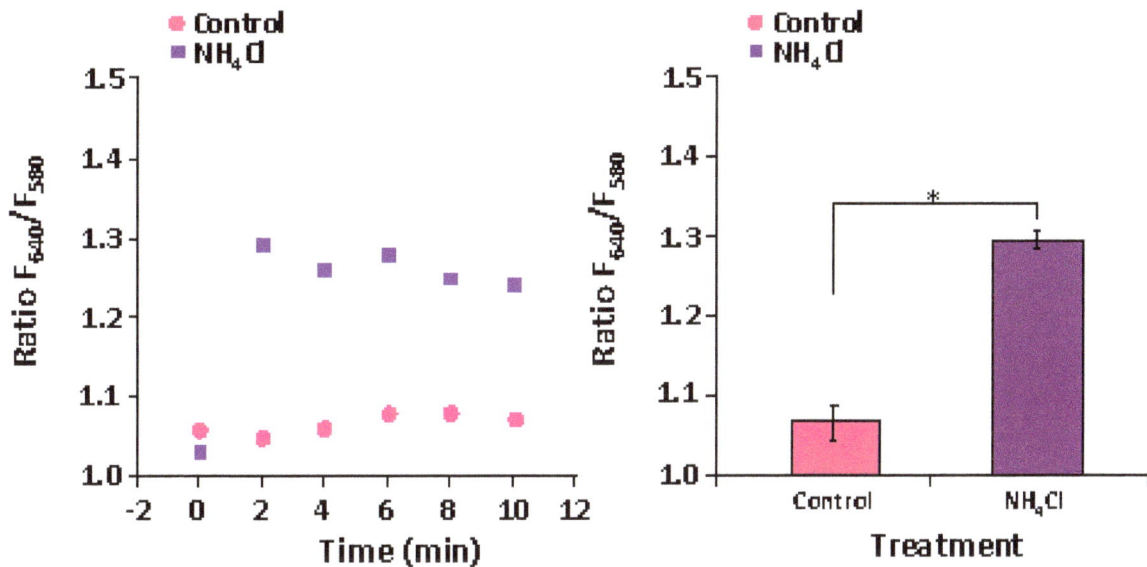

Figure 7B. Ammonium chloride induces a transient increase in intracellular pH and sustained increases in PRMT activity. The graph on the left shows representative values of the ratio of fluorescence intensities at 640 nm and 580 nm i.e. F_{640}/F_{580}, within the cells as a function of time for one experiment. The time-averaged data for three experimentsis shown in the graph on the right. The F_{640}/F_{580} ratio of NH4Cl-treated cells were significantly different from control cells(P=3×10^{-3} by ANOVA).

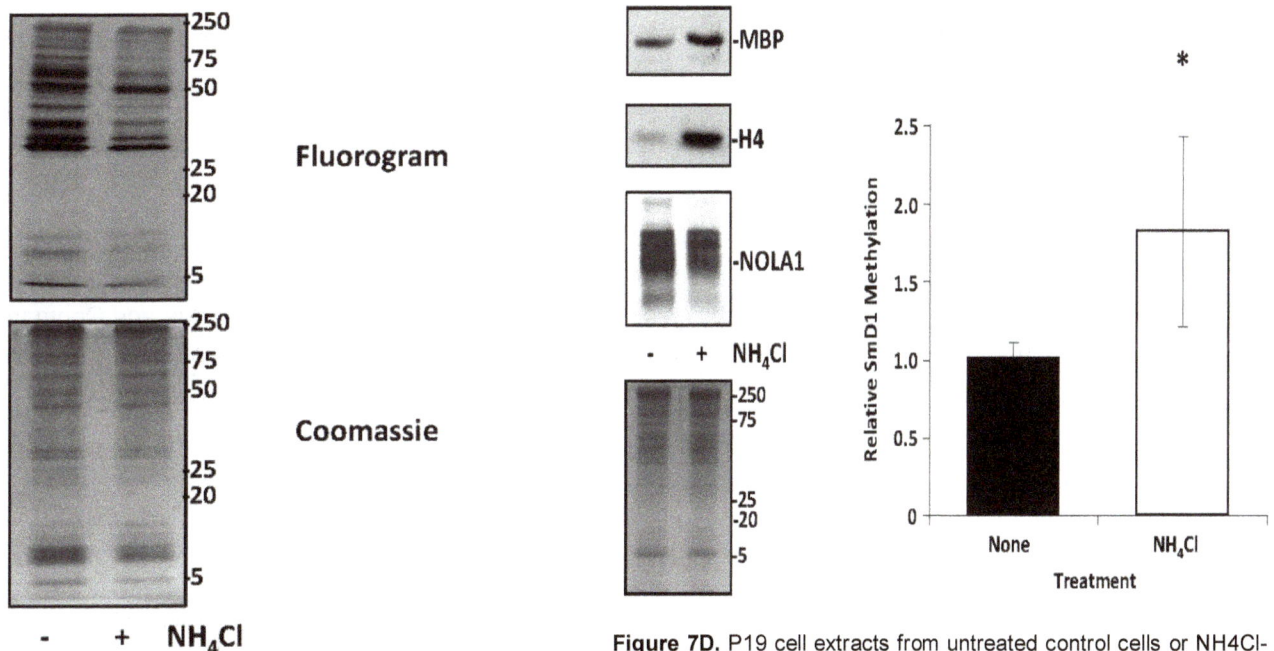

Figure 7C. P19 cell extracts from untreated control cells or NH4Cl-treated cells were supplemented with ^{3}H-SAM and the endogenous substrates in the extracts subject to *in vitro* methylation. The methylated proteins were resolved by SDS-PAGE and subject to fluorography.

Figure 7D. P19 cell extracts from untreated control cells or NH4Cl-treated cells were supplemented with ^{3}H-SAM and various substrate proteins (1 µg) or the SmD1 peptide and subject to *in vitro* substrate methylation as described above. The methylated proteins were resolved by SDS-PAGE and subject to fluorography or liquid scintillation counting. The relative protein loads are shown by the Coomassie gel beneath the fluorogram. The methylation of the SmD1 peptide in the NH4Cl-treated samples was significantly different from the control (P=3×10^{-3} by ANOVA).

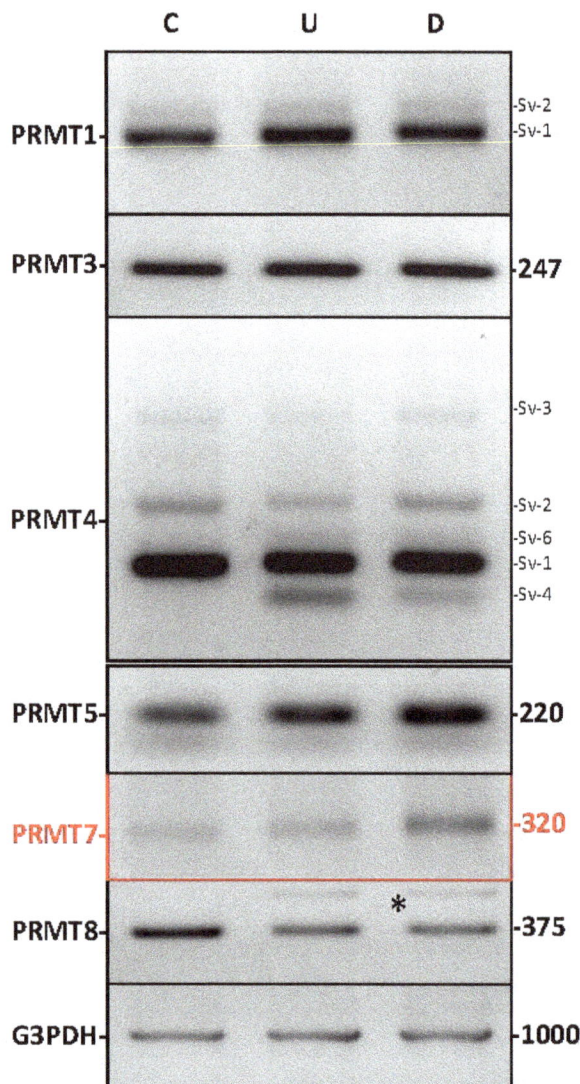

Figure 8A. PRMT mRNA expression in P19 cells. RNA isolated from mouse brain cortex (C), undifferentiated P19 cells (U) and retinoic acid differentiated P19 cells day 6 (D) was subject to RT-PCR using primers to PRMT1, PRMT3, PRMT4, PRMT5, PRMT7, PRMT8 and G3PDH. Primers amplifying PRMT1, PRMT4 and PRMT8 were designed to detect the various known alternatively spliced transcripts (sv) of each message. All amplifications were carried out with the same amount of cDNA for 35 cycles with the exception of PRMT7 (red box), which was carried out for 40 cycles in order to visualize the product.

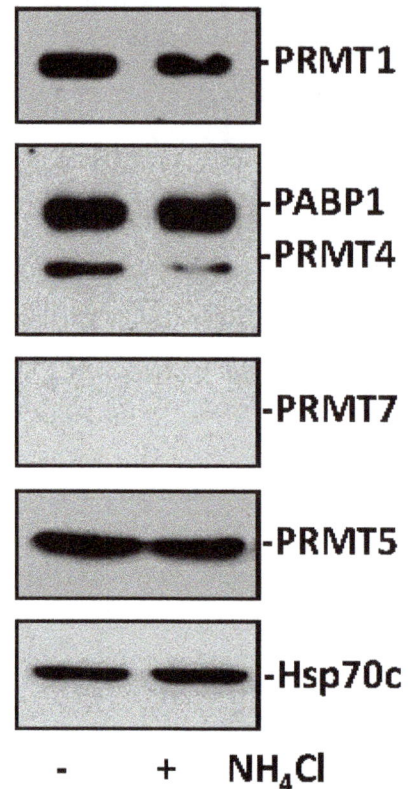

Figure 8B. PRMT mRNA expression in P19 cells. P19 cell extracts from untreated control cells or NH4Cl-treated cells were subject to Western blotting and probed for PRMT1, 3, 4, 5, and 7 expression; Hsp70cp and PABP1 are shown as a protein load controls.

methylproteome. As protein methylation can affect protein-protein, protein-RNA interactions and subcellular localization (for reviews see (Xie and Denman, 2011; Yu, 2011) altered substrate methylation resulting from pH-induced changes of PRMT activities coupled with low levels of demethylation may provide the long-term tag(s) necessary for the formation and maintenance of "molecular memory".

ACKNOWLEDGEMENTS

The authors would like to thank Dr. Noriko Murakami (New York State Institute for Basic Research in Developmental Disabilities) for invaluable assistance with the methylarginine amino acid analyses. This work was made possible through the support of the New York State Research Foundation for Mental Hygiene and a grant to R.D from the FRAXA Research Foundation.

two-fold increase in the amount of monomethylarginine (MMA) that is incorporated into insect cell proteins relative to the amount of dimethylarginine (DMA), Figure 6.

All together our data are consistent with a hypothesis in which transient changes in intracellular or intraorganellar pH produce sustained changes in a cell's

REFERENCES

Abad MFC, Di Benedetto G, Magalhaes PJ, Filippin L, Pozzan T (2004). Mitochondrial pH monitored by a new engineered green fluorescent protein mutant. J. Biol. Chem., 279(12):11521-11529.

Adinolfi S, Bagni C, Musco G, Gibson T, Mazzarella L, Pastore A (1999). Dissecting FMR1, the protein responsible for fragile X syndrome, in its structural and functional domains. RNA, 5(9):1248-1258.

Berthet C, Guehenneux F, Revol V, Samarut C, Lukaszewicz L, Dehay C, Dumontet C, Magaud J-P, Rouault J-P (2002). Interaction of PRMT1 with BTG/TOB proteins in cell signalling: molecular analysis and functional aspects. Genes Cells, 7(1):29-39.

Bierman AJ, Tertoolen LG, de Laat SW, Moolenaar WH (1987). The Na^+/H^+ exchanger is constitutively activated in P19 embryonal carcinoma cells, but not in a differentiated derivative. Responsiveness to growth factors and other stimuli. J. Biol. Chem., 262(20):9621-9628.

Chen YC, Milliman EJ, Goulet I, Cote J, Jackson CA, Vollbracht JA, Yu MC (2010). Protein arginine methylation facilitates cotranscriptional recruitment of pre-mRNA splicing factors. Mol. Cell. Biol., 30(21):5245-5256.

Chen YF, Zhang AY, Zou AP, Campbell WB, Li PL (2004). Protein methylation activates reconstituted ryanodine receptor-Ca^{2+} release channels from coronaryartery myocytes. J. Vasc. Res., 41(3):229-240.

Cheng X, Collins RE, Zhang X (2005). Structural and sequence motifs of protein (histone) methylation enzymes In: Annual Reviews of Biophysics Biomolecular Structure. vol. 34. Palo Alto, CA: Annual Reviews, pp. 267-294.

Chesler M (2003). Regulation and modulation of pH in the brain. Physiol. Rev., 83(4):1183-1221.

Chin HG, Estève PO, Pradhan M, Benner J, Patnaik D, Carey MF, Pradhan S (2007). Automethylation of G9a and its implication in wider substrate specificity and HP1 binding. Nucleic Acids Res., 35(21):7313-7323.

Deng X, Gu L, Liu C, Lu T, Lu F, Lu Z, Cui P, Pei Y, Wang B, Hu S (2010). Arginine methylation mediated by the Arabidopsis homolog of PRMT5 is essential for proper pre-mRNA splicing. Proc. Natl. Acad. Sci., 107(44):19114-19119.

Denman RB (2006). Improved PRMT substrate detection. In: Science STKE,eletter.http://stke.sciencemag.org/cgi/eletters/sigtrans;2001/2093/pl2001.

Denman RB (2008). Protein methyltransferase activities in commercial in vitro translation systems. J. Biochem., 144(2):223-233.

Dhaka A, Uzzell V, Dubin AE, Mathur J, Petrus M, Bandell M, Patapoutian A (2009). TRPV1 is activated by both acidic and basic pH. J. Neurosci., 29(1):153-158.

Dolzhanskaya N, Merz G, Denman RB (2006). Alternative splicing modulates PRMT-dependent methylation of FMRP. Biochemistry, 45(34):10385-10393.

Dolzhanskaya N, Bolton DC, Denman RB (2008). Chemical and structural probing of the N-terminal residues encoded by FMR1 exon 15 and their effect on downstream arginine methylation. Biochemistry, 47(33):8491-8503.

Dolzhanskaya N, Merz G, Aletta JM, Denman RB (2006). Methylation regulates FMRP's intracellular protein-protein and protein-RNA interactions. J. Cell Sci., 119(9):1933-1946.

Fersht A (1977). The pH dependence of enzyme catalysis. In: Enzyme structure and mechanism San Francisco: W. H. Freeman: pp.134-155.

Frankel A, Yadav N, Lee J, Branscombe TL, Clarke S, Bedford MT (2002). The novel human protein arginine N-methyltransferase PRMT6 is a nuclear enzyme displaying unique substrate specificity. J. Biol. Chem., 277(5):3537-3543.

Friesen WJ, Massenet S, Paushkin S, Wyce A, Dreyfuss G (2001). Smn, the product of the spinal muscular atrophy gene, binds preferentially to dimethylarginine-containing protein targets. Mol. Cell., 7(5):1111-1117.

Friesen WJ, Paushkin S, Wyce A, Massenet S, Pesiridis GS, Van Duyne G, Rappsilber J, Mann M, Dreyfuss G (2001). The methylosome, a 20S complex containing JBP1 and plCln, produces dimethylarginine-modified Sm proteins. Mol. Cell Biol., 21(24):8289-8300.

Garcia-Moreno B (2009). Adaptations of proteins to cellular and subcellular pH. J. Biol., 8(11):98.

Guderian G, Peter C, Wiesner J, Sickmann A, Schulze-Osthoff K, Fischer U, Grimmler M (2011). RioK1, a new interactor of protein arginine methyltransferase 5 (PRMT5), competes with plCln for binding and modulates PRMT5 complex composition and substrate specificity. J. Biol. Chem., 286(3):1976-1986.

Hung ML, Hautbergue GM, Snijders APL, Dickman MJ, Wilson SA (2010). Arginine methylation of REF/ALY promotes efficient handover of mRNA to TAP/NXF1. Nucleic Acids Res., 38(10):3351-3361.

Infantino S, Benz B, Waldmann T, Jung M, Schneider R, Reth M (2010). Arginine methylation of the B cell antigen receptor promotes differentiation. J. Exp. Med., 207(4):711-719.

Izzo A, Schneider R (2010). Chatting histone modifications in mammals. Brief. Funct. Genomics, 9(5-6):429-443.

Jankowski A, Kim JH, Collins RF, Daneman R, Walton P, Grinstein S (2001). In situ measurements of the pH of mammalian peroxisomes using the fluorescent protein pHluorin. J. Biol. Chem., 276(52):48748-48753.

Jiang W, Roemer ME, Newsham IF (2005). The tumor suppressor DAL-1/4.1B modulates protein arginine N-methyltransferase 5 activity in a substrate-specific manner. Biochem. Biophys. Res. Commun., 329(2):522-530.

Jones-Villeneuve EM, McBurney MW, Rogers KA, Kalnins VI (1982). Retinoic acid induces embryonal carcinoma cells to differentiate into neurons and glial cells. J. Cell Biol., 94(2):253-262.

Kowenz-Leutz E, Pless O, Dittmar G, Knoblich M, Leutz A (2010). Crosstalk between C/EBP[beta] phosphorylation, arginine methylation, and SWI/SNF/Mediator implies an indexing transcription factor code. EMBO J., 29(6):1105-1115.

Kuhn P, Chumanov R, Wang Y, Ge Y, Burgess RR, Xu W (2010). Automethylation of CARM1 allows coupling of transcription and mRNA splicing. Nucleic Acids Res., 39(7):2717-2726.

Lim Y, Kwon Y-H, Won NH, Min B-H, Park I-S, Paik WK, Kim S (2005). Multimerization of expressed protein-arginine methyltransferases during the growth and differentiation of rat liver. Biochim. Biophys. Acta., 1723(1-3):240-247.

Majumdar D, Bevensee MO (2010). Na-coupled bicarbonate transporters of the solute carrier 4 family in the nervous system: function, localization, and relevance to neurologic function. Neuroscience, 171(4):951-972.

Meister G, Eggert C, Bühler D, Brahms H, Kambach C, Fischer U (2001). Methylation of Sm proteins by a complex containing PRMT5 and the putative U snRNP assembly factor plCln. Curr. Biol., 11(24):1990-1994.

Miranda TB, Khusial P, Cook JR, Lee J-H, Gunderson SI, Pestka S, Zieve GW, Clarke S (2004). Spliceosome Sm proteins D1, D3, and B/B' are asymmetrically dimethylated at arginine residues in the nucleus. Biochem. Biophys. Res. Comm., 323(2):382-387.

Murer H, Hopfer U, Kinne R (1976). Sodium/proton antiport in brush-border-membrane vesicles isolated from rat small intestine and kidney. Biochem. J., 154:597-604.

Raley-Susman KM, Cragoe EJ, Sapolsky RM, Kopito RR (1991). Regulation of intracellular pH in cultured hippocampal neurons by an amiloride-insensitive Na^+/H^+ exchanger. J. Biol. Chem., 266(5):2739-2745.

Ramos A, Hollingsworth D, Adinolfi S, Castets M, Kelly G, Frenkiel TA, Bardoni B, Pastore A (2006). The structure of the N-terminal domain of the fragile X mental retardation protein: a platform for protein-protein interaction. Structure, 14(1):21-31.

Roberts EL (1999). Using hippocampal slices to study how aging alters ion regulation in brain tissue. Methods, 18(2):150-159.

Seksek O, Bolard J (1996). Nuclear pH gradient in mammalian cells revealed by laser microspectrofluorimetry. J. Cell Sci., 109(1):257-262.

Stetler A, Winograd C, Sayegh J, Cheever A, Patton E, Zhang Z, Clarke S, Ceman S(2006). Identification and characterization of the methyl arginines in the fragile X mental retardation protein Fmrp. Hum. Mol. Genet., 15(1):87-96.

Sung Y-J, Dolzhanskaya N, Nolin SL, Brown WT, Currie JR, Denman RB (2003). The fragile X mental retardation protein FMRP binds elongation factor 1A mRNA and negatively regulates its translation in vivo. J. Biol. Chem., 278(18):15669-15678.

Tabares L, Betz B (2010). Multiple functions of the vesicular proton pump in nerve terminals. Neuron, 68(6):1020-1022.

Tresguerres M, Buck J, Levin L (2010). Physiological carbon dioxide, bicarbonate, and pH sensing. Pflugers Arch, 460(6):953-964.

Troffer-Charlier N, Cura V, Hassenboehler P, Moras D, Cavarelli J (2007). Functional insights from structures of coactivator-associated arginine methyltransferase 1 domains. EMBO J., 26(20):4391-4401.

Wang H, Singh D, Fliegel L (1997). The Na^+/H^+ Antiporter Potentiates Growth and Retinoic Acid-induced Differentiation of P19 Embryonal Carcinoma Cells. J. Biol. Chem., 272(42):26545-26549.

Xie W, Denman RB (2011). Protein methylation and stress granules: Post-translational modifier or innocent bystander? Molecular Biology International, Article ID: 137459:1-14.

Xie W, Dolzhanskaya N, LaFauci G, Dobkin C, Denman RB(2009). Tissue and developmental regulation of fragile X mental retardation protein exon 15 isoforms. Neurobiol. Dis., 35(1):52-62.

Yu MC (2011). The Role of Protein Arginine Methylation in mRNP Dynamics. Molecular Biology International, Article ID: 163827.

Zhang X, Cheng X (2003). Structure of the predominant protein arginine methyltransferase PRMT1 and analysis of its binding to substrate peptides. Structure, 11(5):509-520.

Permissions

List of Contributors

M. C. Pagano
Electron Microscopy and Microanalysis Laboratory, Physics Department, Federal University of Minas Gerais, Brazil

A. I. C. Persiano
Electron Microscopy and Microanalysis Laboratory, Physics Department, Federal University of Minas Gerais, Brazil

M. N. Cabello
National University of La Plata, Argentina

M. R. Scotti
Institute of Biological Sciences, Federal University of Minas Gerais, Av. Antônio Carlos, 6627, Pampulha, CEP: 31270-901, Belo Horizonte, MG, Brazil

Natalia Dolzhanskaya
Department of Molecular Biology and New York State Institute for Basic Research in Developmental Disabilities, 1050 Forest Hill Road Staten Island, New York 10314, U.S.A

Wen Xie
Department of Molecular Biology and New York State Institute for Basic Research in Developmental Disabilities, 1050 Forest Hill Road Staten Island, New York 10314, U.S.A

George Merz
Department of Developmental Neurobiology, New York State Institute for Basic Research in Developmental Disabilities, 1050 Forest Hill Road, Staten Island, New York 10314, U.S.A

Robert B. Denman
Department of Molecular Biology and New York State Institute for Basic Research in Developmental Disabilities, 1050 Forest Hill Road Staten Island, New York 10314, U.S.A

Aintzane Cabo-Bilbao
Unidad de Biofísica (UBF, CSIC-UPV/EHU) and Departamento de Bioquímica y Biología Molecular, UPV/EHU.
Barrio Sarriena s/n, E-48940, Leioa, Spain

Ariel E. Mechaly
Unité de Biochimie Structurale, Institut Pasteur, 25 rue du Dr. Roux, F-75724 Paris, France

Jon Agirre
Unidad de Biofísica (UBF, CSIC-UPV/EHU) and Departamento de Bioquímica y Biología Molecular, UPV/EHU.
Barrio Sarriena s/n, E-48940, Leioa, Spain

Silvia Spinelli
AFMB-CNRS, UMR 6098, 163, Av. de Luminy, 13288 Marseille Cedex 09, France

Begoña Sot
Centro Nacional de Biotecnología, CSIC, Campus de la Universidad Autónoma de Madrid, Darwin, 3, 28049 Madrid, Spain

Arturo Muga
Unidad de Biofísica (UBF, CSIC-UPV/EHU) and Departamento de Bioquímica y Biología Molecular, UPV/EHU.
Barrio Sarriena s/n, E-48940, Leioa, Spain

Diego M.A. Guérin
Unidad de Biofísica (UBF, CSIC-UPV/EHU) and Departamento de Bioquímica y Biología Molecular, UPV/EHU.
Barrio Sarriena s/n, E-48940, Leioa, Spain

Zahra Shahba
Payame noor Najafabad University, Najafabad-Isfahan, Iran
International Center for Science, High Technology and Environmental Sciences, Kerman, Iran

Amin Baghizadeh
International Center for Science, High Technology and Environmental Sciences, Kerman, Iran

Vakili Seid Mohamad Ali
Islamic Azad University - Jiroft Branch, Iran

Yazdanpanah Ali
International Center for Science, High Technology and Environmental Sciences, Kerman, Iran

Yosefi Mehdi
Payame noor Najafabad University, Najafabad-Isfahan, Iran

S. B. Muley
Department of Biotechnology and Bioinformatics, Padmashree Dr. D. Y. Patil University, Navi Mumbai, Maharashtra, India. – 400614

V. Bastikar
Department of Biotechnology and Bioinformatics, Padmashree Dr. D. Y. Patil University, Navi Mumbai, Maharashtra, India. – 400614

S. Bothe
Department of Biotechnology and Bioinformatics, Padmashree Dr. D. Y. Patil University, Navi Mumbai, Maharashtra, India. – 400614

A. Meshram
Department of Biotechnology and Bioinformatics, Padmashree Dr. D. Y. Patil University, Navi Mumbai, Maharashtra, India. – 400614

N. Roy
Department of Biotechnology and Bioinformatics, Padmashree Dr. D. Y. Patil University, Navi Mumbai, Maharashtra, India. – 400614

Y. Dhanusha Yesudhas
Department of Bioinformatics, Karunya University, Coimbatore India

Mamdouh M. Shawki
Bio-Medical Physics Department, Medical Research Institute, Alexandria University, Egypt

Abdel-Rahman M. Hereba
Bio-Medical Physics Department, Medical Research Institute, Alexandria University, Egypt

L. S. Taura
Department of Physics, Bayero University, Kano, Nigeria

I. B. Ishiyaku
Department of Physics, Gombe State University, Gombe, Nigeria

A. H. Kawo
Department of Biological Sciences, Bayero University, Kano, Nigeria

P. Anil Kumar
National Institute of Nutrition, Biochemistry, Jamai Osmania, Tarnaka, Hyderabad, AP 500007 India

P. Yadagiri Reddy
National Institute of Nutrition, Biochemistry, Jamai Osmania, Tarnaka, Hyderabad, AP 500007 India

P. Suryanarayana
National Institute of Nutrition, Biochemistry, Jamai Osmania, Tarnaka, Hyderabad, AP 500007 India

G. Bhanuprakash Reddy
National Institute of Nutrition, Biochemistry, Jamai Osmania, Tarnaka, Hyderabad, AP 500007 India

Almaasfeh Sultan
Physics Department College of Science, Al Hussein Bin Talal University, Jordan

Sampath Natarajan
Department of Advanced Technology Fusion, Konkuk University, 1 Hwayang-dong, Gwangjin-gu, Seoul, 143-701, Korea

Rita Mathews
Department of Advanced Technology Fusion, Konkuk University, 1 Hwayang-dong, Gwangjin-gu, Seoul, 143-701, Korea

Rita Ghosh
Department of Biochemistry and Biophysics, University of Kalyani, Kalyani –741235, W. B., India

Dipanjan Guha
Department of Biochemistry and Biophysics, University of Kalyani, Kalyani –741235, W. B., India

Sudipta Bhowmik
Department of Biochemistry and Biophysics, University of Kalyani, Kalyani –741235, W. B., India

Angshuman Bagchi
Department of Biochemistry and Biophysics, University of Kalyani, Kalyani –741235, W. B., India

Sofiene Mansouri
Biophysics Research Unit, Faculty of Medicine, Sousse, Tunisia

Halima Mahjoubi
Biophysics Research Unit, Faculty of Medicine, Sousse, Tunisia

Ridha Ben Salah
Biophysics Research Unit, Faculty of Medicine, Sousse, Tunisia

I. V. Zhigacheva
Russian Academy of Sciences, N. M. Emanuel Institute of Biochemical Physics, ul. Kosygina 4, 119334 Moscow, Russia

E. B. Burlakova
Russian Academy of Sciences, N. M. Emanuel Institute of Biochemical Physics, ul. Kosygina 4, 119334 Moscow, Russia

I. P. Generozova
Russian Academy of Sciences, K. A. Timiryazev Institute of Plant Physiology ul. Botanicheskaya 35, 127276, Moscow,Russia

A. G. Shugaev
Russian Academy of Sciences, K. A. Timiryazev Institute of Plant Physiology ul. Botanicheskaya 35, 127276, Moscow, Russia

S. G. Fattahov
Russian Academy of Sciences, A. E. Arbuzov Institute of Organic and Physical Chemistry, ul. Akademika Arbuzova 8, 420083 Kazan' Research Center, Kazan', Tatarstan, Russia

A. Misra
Central Institute of Medicinal and Aromatic Plants, P. O. CIMAP, Lucknow-226015, India

N. K. Srivastava
Central Institute of Medicinal and Aromatic Plants, P. O. CIMAP, Lucknow-226015, India

A. K. Srivastava
Central Institute of Medicinal and Aromatic Plants, P. O. CIMAP, Lucknow-226015, India

S. K. Chattopadhyay
Central Institute of Medicinal and Aromatic Plants, P. O. CIMAP, Lucknow-226015, India

M. Dada
Department of Physics, Federal University of Technology, Minna, Niger-State, Nigeria

O. B. Awojoyogbe
Department of Physics, Federal University of Technology, Minna, Niger-State, Nigeria

K. Boubaker
Unité de Physique de Dispositifs à Semiconducteurs -UPDS- Faculté des Sciences de Tunis, Campus Universitaire 2092 Tunis, Tunisia

O. S. Ojambati
Department of Physics, Federal University of Technology, Minna, Niger-State, Nigeria

D. Kshatresh Dubey
Biophysics Unit, Department of Physics, DDU Gorakhpur University, Gorakhpur 273009, India

K. Amit Chaubey
Biophysics Unit, Department of Physics, DDU Gorakhpur University, Gorakhpur 273009, India

Azra Parveen
Biophysics Unit, Department of Physics, DDU Gorakhpur University, Gorakhpur 273009, India

P. Rajendra Ojha
Biophysics Unit, Department of Physics, DDU Gorakhpur University, Gorakhpur 273009, India

A. Y. Borisov
A. N. Belozersky Institute of Physico-Chemical Biology in M. V. Lomonosov Moscow State University. Vorob'ev hills, 119992 Moscow

Abu Teir M. M.
Department of Physics, Faculty of Science, Al-Quds University, Jerusalem, Palestine

J. Ghithan
Department of Physics, Faculty of Science, Al-Quds University, Jerusalem, Palestine

M. I. Abu-Taha
Department of Physics, Faculty of Science, Al-Quds University, Jerusalem, Palestine

S. M. Darwish
Department of Physics, Faculty of Science, Al-Quds University, Jerusalem, Palestine

M. M Abu-hadid
Department of Immunology, Faculty of Medicine, Al-Quds University, Jerusalem, Palestine

G. Rezaei Behbehani
Chemistry Department, Imam Khomeini International University, Qazvin Iran

A. Divsalar
Institute of Biochemistry and Biophysics, University of Tehran, Tehran, Iran
Department of Biological Sciences, Tarbiat Moallem University, Tehran, Iran

A. A. Saboury
Institute of Biochemistry and Biophysics, University of Tehran, Tehran, Iran

Z. Rezaei
Chemistry Department, Islamic Azad University, Gachsaran, Iran

Wen Xie
Division of Hematology and Medical Oncology, Department of Medicine, Weill Medical College of Cornell University, New York, NY 10065, USA

George Merz
Department of Developmental Neurobiology, New York State Institute for Basic Research in Developmental Disabilities, 1050 Forest Hill Road, Staten Island, NY 10314, USA

Robert B. Denman
Department of Developmental Neurobiology, New York State Institute for Basic Research in Developmental Disabilities, 1050 Forest Hill Road, Staten Island, NY 10314, USA

www.ingramcontent.com/pod-product-compliance
Lightning Source LLC
Chambersburg PA
CBHW050457200326

41458CB00014B/5215